The Glamour Factory

Inside Hollywood's Big Studio System

Ronald L. Davis

Southern

Methodist

University

Press

DALLAS

The Glamour Factory

Library of Congress
Cataloging-in-Publication Data

Davis, Ronald L.
  The glamour factory : inside Hollywood's
big studio system / Ronald L. Davis — 1st ed.
    p.   cm.
  Includes bibliographical references and
index.
  ISBN 0-87074-357-0 — ISBN 0-87074-358-9
(pbk.)
  1. Motion pictures—California—Los
Angeles—History. 2. Motion picture industry—
California—Los Angeles—History. 3. Motion
picture studios—California—Los Angeles—
History. I. Title.
PN1993.5.U65D34   1993
384'.8'0979494—dc20   93-8861

In memory of my parents

Ruth L. and E. Leroy Davis

who shared the fantasy,

even when the reality made

little sense

Contents

Preface

Hollywood's big studio system is unique in Western civilization, its size and impact unparalleled, its product the glamorous films that are now viewed as the industry's Golden Era. If film was the first art to utilize technology for humanistic purposes, that same technology made the Hollywood output available around the world, satisfying the global thirst for American movies. Yet the Hollywood phenomenon, with its hype and hyperbole, fits into the mainstream of American culture, its mores less exceptional than magnified. The nation's materialism, devotion to profit and social advancement, belief in dynamic leadership, technological know-how, practical invention, and expanding markets all emerged as goals and values within the old studios. At its height the movie industry mirrored the corporate ideology that catapulted the United States into economic prominence. Business practices within the big studio system were not unlike those of other American factories, and the moguls in motion pictures resembled those in railroading, banking, real estate, and oil during the late nineteenth and early twentieth centuries.

The Old West was disappearing as Hollywood was establishing itself as a movie colony, but frontier attitudes prevailed, creating an atmosphere much like that of the California gold fields. "The town was full of irresponsible agents, who reminded me of gamblers and confidence men in a gold rush camp," film pioneer Adolph Zukor observed in his memoirs, *The Public Is Never Wrong*.

Like the mining camps, Hollywood attracted a cosmopolitan population: movie people came from England, all over Europe, Latin America, Australia, and the Orient, as well as the United States and Canada. Claim jumping in early Hollywood, as in the gold fields, often resulted in fights and occasionally murders, some never officially solved. The most famous was the William Desmond Taylor killing in 1922, in which stars Mabel Normand and Mary Miles Minter were implicated. In his book *A Cast of Killers*, Sidney D. Kirkpatrick convincingly shows that director King Vidor had privately solved the crime, revealing that the killer was Mary Miles Minter's mother. "Everybody knew who shot Taylor," columnist Adela Rogers St. Johns said in an interview in 1979. "There was never any doubt in anybody's mind. We were in the wild and woolly West, and California had some unwritten laws. One was that a mother had a perfect right to shoot a man who had debauched her little girl."

Within a decade after 1913, Hollywood was transformed from a mining camp into a factory town. By the mid-1920s power was consolidated into four major studios: Metro-Goldwyn-Mayer, Paramount, Fox, and Warner Bros., all appropriating the assembly line approach of the Detroit automotive companies. Like Americans generally, most Hollywood employees were hardworking, well aware that success demanded long hours as well as talent and luck. But the glamour Hollywood exuded led a dream-starved public to accept the myth that life in the movie capital was a glorious holiday, where riches were abundant and lovemaking a constant preoccupation. For a nation growing up on the American dream of fame and fortune, Hollywood loomed as a twentieth-century Eldorado, where romantic individualism — perhaps the most compelling concept in American history — reached its epitome.

Americans wanted to see enticing images of themselves on the screen, projected in baroque movie palaces with lights twinkling from the ceiling, thick carpets, exotic alcoves, grand staircases, ivory elephants stationed at entrances, and ushers in splendiferous uniforms waiting to escort them to velvet seats. Those were images of America the world saw on film and tried to emulate, having absorbed the nation's fantasy in its most colossal statement. "Everybody wanted to come to America and see the streets of gold," actor Robert Cummings said, and many did come, expecting life in the United States to be what it was in the movies. Hollywood stars became idols for youngsters around the globe, emerging as mythic heroes. "When a person becomes a model for other people's lives, he has moved into the sphere of being mythologized," Joseph Campbell declared in *The Power of Myth*. "There is something magical about films. The person you are looking at is also somewhere else at the same time. That is a condition of the god. If a movie actor comes into the theater, everybody turns and looks at the movie actor. He is the real hero of the occasion. He is on another plane. He is a multiple presence."

Yet most stars, at least most of those who survived in Hollywood, recognized stardom as a business — an interesting business that paid well and provided luxuries as well as avenues for creative expression. "It's better than working in a bank," actor Robert Mitchum is supposed to have commented. Movie tough-guy Edward G. Robinson summed up the veteran's attitude for young Florence Henderson on the set of *Song of Norway* (1970): "My dear, stars are in heaven. Just do your job and be professional. That's what it's all about." Mythic figures though Hollywood's celebrities may have been to the outside world, studio employees knew them as workers, part of a team hired by factories whose product was motion pictures and glamour.

The old studio heads were showmen but they were also businessmen, whose business happened to be entertainment. "When I started, I realized early that moviemaking, no matter how they tried to dress it up, was not an art form," screenwriter William Ludwig said. "It was a business, and the business was manufacturing entertainment. It's the only manufacturing business where the capital assets go home at five-thirty, where they can get drunk, where they can get killed. That can't happen in an automobile plant; the presses don't go home, the steel doesn't get drunk. And it's the only manufacturing business where you don't even know if you have a product until it's too late. There are no samples. When people see it, that's it—for better or worse. Movie making is peculiar, but it's a business."

The Glamour Factory is the story of that peculiar business, told with the indispensable help of hundreds of insiders who watched and contributed to the industry while magic was being made. Most of this story is drawn from the Southern Methodist University Oral History Collection on the Performing Arts, which the author founded and continues to direct. Unless otherwise indicated, direct quotes in the text come from these interviews, taped between 1973 and 1992. The intent has been to utilize oral history to humanize the Golden Age of Hollywood by conveying the experiences of the gamut of film workers—from stars, directors, and producers to stuntmen, hairstylists, makeup artists, and publicists.

The SMU Oral History Program was launched in 1972 to gather data for future writers and cultural historians. Through the years the emphasis increasingly came to focus on motion pictures. The chronicle of Hollywood during its heyday grew sufficiently rich that a book seemed inevitable. While *The Glamour Factory* suggests the scope of the interviews in the university's collection, in each case only the surface has been tapped. Most of the five hundred transcripts presently comprising the SMU archives remain an unspent source for continued research.

Additional interviews and memoirs have been used to round out the picture of life in the big studios, and relevant secondary information has been incorporated. Oral histories are listed alphabetically in two categories in the section on sources, serving as a guide to the reader who wants to know where and when a quoted statement was made. Listed first are the interviews in the SMU collection, followed by those used from other oral history archives, particularly the American Film Institute, as well as published interviews. A third section lists books and articles consulted in preparation for SMU oral history interviews, most of which were consulted again during the writing of this book. The aim in *The Glamour Factory* has been to present as comprehensive a portrait of the big studio system as possible, yet to keep the intimate texture established by candid personal accounts.

What may seem to be sexist language in parts of the book is actually a reflection of the times, since sexism ran rampant during Hollywood's Golden Age, on screen and off. The major directors were men, with former film editor Dorothy Arzner an exception, and all of the top cinematographers were male. The masculine pronoun used in reference to those groups is therefore historically accurate; to use both genders would be anachronistic.

The number of people who contributed to this volume is too large to mention each one individually, as a glance at the list of SMU oral histories will reveal. Collectively the persons interviewed have provided the author and his staff with a magical tour. I take pleasure in thanking all the people interviewed for their frankness, kindness, and insight.

Tom Culpepper and Eleanor Solon, who transcribed most of the interviews used for this book, have been dedicated assistants. Both share an emotional investment in the work at hand, and without them this volume never would have reached print. Suzanne Comer, senior editor of the SMU Press, believed in the project from the outset and remained determined to make the book "a good read." Her death in the summer of 1990 was a blow to all who knew her, certainly to her authors. Fortunately Kathryn Lang proved a worthy successor. Kathie made revisions a pleasure, and I am indebted to her judgment and sharp eye. Paul Boller read the manuscript and, as always, proved generous with constructive criticism, while my colleagues in the history department at SMU have offered encouragement through two decades of interviewing. To them I extend my sincere gratitude. Special acknowledgment must be made to Jane Elder and Judy Bland, who read the revised manuscript with critical eyes and offered intelligent suggestions. The administration of Southern Methodist University and David Farmer, head of the university's DeGolyer Library, have been generous in their support over many years. My appreciation goes to them, and to the efficient personnel of the Mayer Library at the American Film Institute, the Margaret Herrick Library of the Academy of Motion Picture Arts and Sciences, and the Doheny Library at the University of Southern California for professionalism beyond the norm. Finally, I would like to thank my students, particularly those undergraduates who have taken my "American Society through Film" course. In most instances they were delving into Hollywood's Golden Era for the first time, and their enthusiasm, perhaps most of all, gave me the incentive to write the story that follows.

The Glamour Factory

1

Majors and Minors

Sunset Boulevard exterior of Famous Players-Lasky, the forerunner of Paramount, 1918. Courtesy of the Academy of Motion Picture Arts and Sciences.

By 1919, 80 percent of the motion pictures produced in the world were being made in southern California. The movie business had learned to mass-produce entertainment as other industries manufactured vacuum cleaners, automobiles, breakfast cereal, or toothpaste. By the early 1920s Hollywood studios were listed on the stock exchange, and filmmaking had become big business. Although D. W. Griffith had elevated an inexpensive amusement into a proletarian art with his epic *The Birth of a Nation* (1915), most film pioneers viewed their craft in commercial terms. "If you want art," director Rex Ingram told future montage expert Slavko Vorkapich in 1921, "stick to your painting. Film is an industry."

Around 1920, with Thomas Edison's monopoly broken, new motion picture companies emerged in California that consolidated the production, distribution, and exhibition arms of the business under aggressive leadership, ushering in the big studio era. By 1929, after the advent of sound, Hollywood stood on the threshold of its Golden Age, as major studios absorbed faltering rivals, amalgamated into forceful cartels, established autocratic leadership, developed contract systems, and acquired theaters as outlets for their product. In the process they brought the motion picture business to its maturity, making Hollywood the glamour capital of the world and creating stars that would outlast the studios.

No independent producer had defied Edison's monopoly more openly than Carl Laemmle, the founder of Universal Pictures. Calling his operation the Independent Motion Picture Company of America (IMP), Laemmle in 1910 acquired the services of director Thomas Ince and lured Florence Lawrence, the heretofore unbilled "Biograph Girl," away from her home studio, promising to make her a star. After fighting Edison's trust for years, Laemmle emerged victorious in April 1912, when he won a court battle that hastened the demise of the Motion Picture Patents Company, the Edison-controlled combine. Almost immediately Laemmle amalgamated his operation with three lesser outfits, eventually calling the company Universal, with himself as president.

Early in 1914 Laemmle purchased the 230-acre Taylor Ranch on the north side of the Hollywood Hills. The cost of the San Fernando Valley property was $165,000. Construction began that June on what would emerge in 1915 as Universal City, the studio's permanent home. Under Laemmle's leadership, Universal City thrived. In 1915 the company produced over 250 films, mostly two-reelers and serials, but successful features as well. Universal, however, owned no theaters, and the major Hollywood studios all would. Consequently Laemmle's company, along with Columbia and United Artists, became a respected member of what the industry called the "Little Three."

Paramount emerged as the leading production company of the silent era, at its height turning out as many as 156 pictures a year. Paramount's theater chain grew to over two thousand houses. Its flagship, the Paramount Theater in New York, was built in 1926, a deluxe showplace at the base of a Times Square skyscraper. In the theater's lobby relics from around the world were displayed, authenticated by various heads of state, making the Paramount something of a cultural shrine. "It was my father's concept," said Eugene Zukor, son of Paramount's mastermind, "that a theater should not only provide a place for amusement, but it should also have art and artifacts. It should have displays that children and educators would find interesting. He thought a theater should have a place in the community like a meeting house."

Not all of Paramount's theaters were as grandiose as the company's flagship. Regional differences were taken into account so that patrons would feel at home. "You couldn't standardize," Eugene Zukor explained. "The North consisted of cities where the population was concentrated; in the South more people had to travel to the theater by foot or in wagons or cars. You had to make provisions for those differences." Sometimes the designers miscalculated. When Paramount built a theater in North Platte, Nebraska, a railroad center, workers from the yards arrived in blue jeans and overalls. "I remember the night they opened this theater," the younger Zukor recalled. "Many of the men came directly from the roundhouse where they'd been oiling and greasing the locomotives, and they didn't want to step on the carpet. They just stood there and said, 'We can't go in there. We'll have to wait till Sunday after church, when we'll be out of uniform.' The kids just stared at the ushers in tuxedos; they'd never seen these creatures outside of pictures in magazines. Finally we had to bring it down to a level that was fitting to that community."

The Paramount amalgamation initiated the vertical integration of production, distribution, and exhibition that created a new type of trust in the film industry. Through block booking, the company forced theaters to take its mediocre pictures along with the good, booked even before a script was shot, sold on the drawing power of the studio's stars.

Metro-Goldwyn-Mayer, the production wing of Loew's Inc., dedicated in 1924, became within three years one of Paramount's major rivals. Marcus Loew, head of the parent company, controlled some four hundred movie houses by 1912, although his chain would remain the smallest of the five majors. But Louis B. Mayer, who ran the MGM studio, demonstrated a flair for organization and possessed an astute business sense. After 1923 Mayer's management abilities were complemented by the creative instincts of his principal aide, Irving Thalberg, who became Metro's production head. From the first, employees at the Culver City studio sensed a familial atmosphere. "It was a small company then," recalled Joseph Newman, who became a director in the sound era. "There were very few executives. In those days they didn't have producers. The motion picture business was a relatively new industry at the time, and practically everyone was learning. It was essentially a young group of people, and they were very outgoing and willing to let other people learn with them. The jealousies that came later hadn't developed."

Mayer's policy was to hire the best. "Good people make me look good," he said. "I'll go down on my knees to talent." The company grew and streamlined its operation, until by 1934 Metro-Goldwyn-Mayer was recognized as the most sophisticated of Hollywood's dream factories, containing twenty-three sound stages and a lot of 117 acres, which included office buildings, standing exterior sets, a lake, a park, a miniature jungle, and the world's largest film laboratory.

William Fox operated Fox Film Corporation on a more modest scale, except that by 1927 his network of theaters numbered over a thousand, with the Roxy in New York as his flagship. Fox first made his pictures at a studio on Western Avenue, then added a second section to his operation off Pico Boulevard, where a huge backlot, known originally as the Tom Mix Ranch, stretched over 260 acres. Fox was a pioneer in the addition of sound, utilizing a sound-on-film process that proved superior to the discs used earlier.

But it was Warner Bros., a near-bankrupt company, that made the big stir in 1927 with its introduction of the Vitaphone process, a cumbersome sound technique developed by Western Electric that synchronized acoustical discs with filmed images. Warners first released their films through First National Pictures but later assumed control of First National and increased the company's theater holdings. The Piccadilly Theater on Broadway, renamed the Warner, became the company's principal exhibition house. In 1927 the four brothers equipped the Warner Theater for sound, but their company was unable to make similar installations in movie houses across the country, since the cost per theater ranged between $16,000 and $25,000. Harry Warner, the company's financial wizard, attempted to talk theater managers into investing in the Vitaphone process. By the close of the year, only about two hundred theaters in the United States were equipped with sound, but Warner Bros. risked its corporate future on the novelty. The other major studios followed suit, installing sound systems in their theaters, although Vitaphone soon proved awkward and unsatisfactory.

For filmmakers who survived the transition to sound, the Golden Age of Hollywood was just ahead. By 1926 the picture business had become one of the country's six largest industries. Going to the movies had become a national habit, and once the Depression subsided, weekly attendance increased still more. Yet sound had to be improved and the camera made mobile again. "It was an exciting time," Joseph Newman remembered, "because the whole industry was thrown into chaos. Everything was new and nobody knew all the basic technology."

With sound, the last of the major studios, RKO, joined Paramount, MGM, Fox, and Warner Bros., completing Hollywood's "Big Five." Always a marginal major, partly because of turnover in administration, Radio-Keith-Orpheum was formed in 1928, basing production at a thirteen-acre studio on Gower Street. Because of its corporate instability, RKO never developed a unique style, though it continued to expand, buying theater chains on both coasts and moving its eastern headquarters to Rockefeller Center in Manhattan. In 1932 the company leased Radio City Music Hall, a 6,200-seat art deco palace, giving RKO control of New York's most famous showcase for first-run productions.

Although crisis seemed a way of life at RKO, studio employees found working there pleasant. The RKO lot was small, the atmosphere friendly, even intimate. In the middle of the studio was a small park, although interior furnishings were generally not as luxurious as at the other majors. Eventually RKO acquired a five-hundred-acre ranch for location work in the San Fernando Valley near Encino, where large standing sets were constructed for exterior scenes. The studio had its own stars, but it didn't have the extensive contract roster that either MGM or Paramount did, forcing RKO to borrow big names from rival studios. "The fact that they didn't have the tremendous star system like MGM made it a warm family," observed actress Laraine Day, who went to RKO from Metro. "You knew everybody; you knew all the crews from other pictures, even from several stages down the way. It was more fun."

Like all the major studios, RKO did not expect to make money on every picture. Prestige films like *Of Human Bondage* (1934), *Alice Adams* (1935), and *Winterset* (1936) were balanced by more commercial pictures. In addition

RKO maintained an active short-subjects unit, which ground out two-reel comedies, sports films, and documentaries to supplement the company's feature productions.

By 1933 every Hollywood studio had suffered grave financial problems. William Fox faced bankruptcy in 1930 and was forced to sell his shares in the Fox Film Corporation to bankers. In 1941 he was sentenced to a year's imprisonment for allegedly bribing a judge during his bankruptcy hearing. Darryl F. Zanuck had become production head at Warner Bros., but left Warners in 1933 to organize 20th Century Pictures with Joseph Schenck. Zanuck's success was immediate, despite the fact that his company had neither a studio of its own nor a distribution arm. Fox needed dynamic leadership, which Zanuck offered. On May 29, 1935, the two operations merged. 20th Century-Fox was formed, with Schenck as president and Zanuck as vice president in charge of production.

For twenty years Darryl Zanuck oversaw every aspect of the company's major releases. From Fox he acquired a chain of theaters, the recently completed studio on Pico Boulevard, and a roster of stars headed by Will Rogers and Shirley Temple. Unfortunately Will Rogers died in an airplane crash in 1935, and Shirley Temple grew up. Zanuck struggled to bolster his stable, adding Tyrone Power, Alice Faye, Don Ameche, Sonja Henie, and later Betty Grable. Since his performers could not compete with those of MGM or Warner Bros., Zanuck made screenplays his main attraction, hiring such top writers as Nunnally Johnson, Lamar Trotti, and Philip Dunne.

Twentieth Century-Fox was one of the newer lots and became the most beautiful, with manicured gardens, ornate hedges, and a feeling of spaciousness. "It was like a jewel," actor George Montgomery said. Writers and musicians worked in comfortable bungalows; a tour of the backlot resembled a journey into fantasyland. "It was heaven," recalled actress Vanessa Brown, under contract to the studio in the late 1940s. "There were little oases like *The Keys of the Kingdom* set and *The Song of Bernadette* set—all the magical pictures they had made throughout the years. They had a lake where they did mock maneuvers, and they had an oil well. You could bicycle, but more often you used your car."

Policemen at the gates knew all the contract players, and an esprit de corps developed, though newcomers were often terrified. Coleen Gray, a Minnesota farm girl, signed a contract with 20th Century-Fox in 1945. Gray found the studio intimidating. "I never had a day without anxiety," she said. "One of the wardrobe women especially used to scare me to death. She scared everybody."

Paramount, for thirty years separated from RKO only by a fence, incurred a deficit of over $15 million in 1932, going into receivership the following year. The company was reorganized in 1935, and the next year doubled the $3 million profit earned the year before. Its success depended on Bing Crosby musicals, Cecil B. DeMille's spectacles, and the comedies of Mae West. During World War II, when the public's appetite for entertainment seemed limitless, Paramount's profits rose to $15 million, despite the loss of European and Far Eastern markets. In 1946 that figure zoomed to $39 million, the largest annual profit ever registered by a film company.

Paramount has sometimes been called "the country club of studios." There was much joking in its permissive atmosphere, yet rapport was strong. Executives respected talent. "The whole idea," said writer Melville Shavelson, "was not to make things difficult, but to make it easy to make films." The twenty-six-acre Paramount lot had lawns and patios and a row of dressing rooms that reminded actress Patricia Morison of English mews. Mary Martin's dressing room was on one side of Morison's, Paulette Goddard's on the other. Bob Hope's was down at one end, and Bing Crosby's was at the other. To Carolyn Jones, who went to Paramount as a young contract player in 1952, the lot seemed like a little New England town, with "just enough of the show business phony about it to make the place interesting." Actress Lizabeth Scott compared it to a huge ball, full of color.

Despite the frivolity, Paramount developed a reputation for turning out exceptionally finished pictures. Whatever was needed was there on the lot— carpenters, painters, electricians. At its height the studio employed some 3,000 workers, including 131 actors, 13 directors, 52 cameramen, 103 musicians, 27 hairdressers, 283 electricians, and 3 gardeners.

Metro-Goldwyn-Mayer emerged as the glamour studio, the Tiffany's of Hollywood. While most of the other studios struggled during the Depression, Metro-Goldwyn-Mayer thrived, becoming the largest of the Hollywood lots. Metro had the top stars, the longest contract list, the largest sets, the biggest rehearsal halls, the lengthiest shooting schedules; its talent was treated royally. "It had the climate of Eden," claimed Richard Ney, who played Greer Garson's son in *Mrs. Miniver* (1942). "I was in paradise, eating apples everywhere."

MGM's Lot One, where the sound stages were, was a cement jungle. Lot Two contained the streets and houses and standing outdoor sets; Lot Three was a forest of trees and vegetation, and one section housed a zoo. Metro provided the best of everything. Each department—music, design, props, camera—was supreme. While long-term employees became part of the close-knit family, outsiders sometimes found the studio cold, operating on a rigid caste system. Producers, writers, and designers at MGM, in addition to the stars, had achieved great stature. "They all had a snobbish attitude," makeup artist William Tuttle said. "This seemed to follow through with the entire crew of that studio, right down to the janitor. All of a sudden you found yourself taking the same attitude with strangers. It was contagious." Insiders were proud of belonging to one of the richest companies in the world. "Metro was colossal," reflected actress Jan Clayton. "One has to use that word once in a lifetime. To me, as to all of us, it was the Emerald City."

MGM's pictures became the lushest in Hollywood. Louis B. Mayer insisted that his stars look beautiful at all times and be filmed amid elegant surroundings. "The thing that astounded me," dancer Ray Bolger declared, "was that they bought people. It was like you would go into a grocery store and say, 'Give me four comics and three toe dancers, and I want five girl and five male singers. I want nineteen character actors, and I want some unique personalities.' You buy them and then put them on the shelf. We would never have been able to make *The Wizard of Oz* if they hadn't had all those geniuses on the shelf."

Although Metro was a factory, life at the studio was pleasant, even fun. Comedienne Nancy Walker recalled, "If you wanted to take sixteen weeks to do a number like 'I've Got Rhythm,' nobody was going to yell at you too much. There was the money for it." Director Edward Buzzell, earlier a song-and-dance man on Broadway, claimed, "If you didn't do well at Metro, you shouldn't be in pictures, because they gave you every advantage."

In contrast to MGM, Warner Bros. was sometimes referred to as the "San Quentin of studios." Warners had a reputation for paying lower wages and getting the most possible from their talent. The studio was harassed by labor disputes and in 1945 became the center of riots that broke out during an eight-month strike.

In 1931 Warners consolidated its executive staff into quarters at the former First National Studio in Burbank, using its Sunset Boulevard facilities, the Warner Ranch, and the old Vitagraph plant as auxiliaries. The newly expanded Burbank studio became the focus of the company's activities. Rather than striving for MGM's gloss or Paramount's sophistication, Warner Bros. concentrated on realism, particularly during the Great Depression. "For a while we were known as Murder Incorporated," producer Hal Wallis claimed, "because the pictures we were making were hard-hitting and tough."

The studio made extensive use of black-and-white, and even its actors seemed tougher. "Everybody there was a hail-fellow-well-met and masculine type," declared Ray Bolger, who worked at the studio during the late 1940s. "There was not the Norma Shearer elegance that you found at MGM."

Although Warners produced memorable pictures and eventually made big Technicolor musicals, budgets were lower than Metro's and the rehearsal periods shorter. "You went about getting your job done quicker at Warners," Ray Bolger noted. As Gene Nelson, who danced in several of the Doris Day–Gordon MacRae songfests, put it, "The whole imaginative and creative thrust was missing, because Warners was making a product, whereas MGM was making something artistic."

Universal, the largest of the minors, was a neighbor of Warners in the San Fernando Valley. As late as the early 1950s flocks of sheep grazed on the hills of the Universal lot. Employees earlier could buy eggs from the studio at a discount.

Universal for years suffered under poor administration. When Carl Laemmle, Jr., was appointed general manager in charge of production in 1929, the industry regarded him as a joke, a case of nepotism. Ogden Nash quipped, "Uncle Carl Laemmle has a big faemmle," for it was well known that "Uncle Carl" had given jobs even to distant relatives, allowing some of them to live on the lot. "Junior Laemmle," as his son was called, made sure the studio turned out its quota of superior films each year, including the classic *All Quiet on the Western Front* (1930), but stockholders complained he spent too much money. By 1936 the Laemmle connection was severed.

With the release of *Three Smart Girls* in 1936, Universal pinned its future on fifteen-year-old Deanna Durbin, whose MGM option had expired. The young soprano saved the studio from bankruptcy, attracting huge box-office grosses with a series of musicals produced by Joseph Pasternak. During the late 1940s Universal's output consisted of a couple of Durbin pictures, a number of low-budget horror movies, two Abbott and Costello comedies, several exotic adventure films (what the industry called "tits and sand" pictures, usually starring Maria Montez or Yvonne DeCarlo), and a string of westerns and cheap musicals.

Even the studio's big pictures were made faster than those at MGM or Paramount. Although Universal "gradually became kind of an A studio," as actor Rock Hudson put it, "they had the mentality of not putting too much money in and not expecting too much profit." The studio continued to believe in little pictures, making an occasional expensive movie for prestige. "In the early 1950s Universal was very informal, very friendly, especially for us younger people," recalled contract player Julie Adams. "Compared with Metro or Fox, it was a wide spot in the road," Rock Hudson said. "You never locked doors, everybody was a friend."

Eventually Universal became a haven for independent producers, as did Columbia, another of Hollywood's "Little Three." Located at Sunset and

Gower streets, Columbia was squeezed into an area not much bigger than one of MGM's sound stages. A ramshackle jumble of cubbyhole offices, cutting rooms, film vaults, and projection rooms, Columbia operated in a single block on three cramped stages. Work was serious business. Columbia veterans referred to the studio as "the black hole of Calcutta" and a flesh shop. "It was a tacky, tacky place," maintained actress Nina Foch. "There were little bunches of people who really tried to do good work. But on the whole it was a dreadful place, very tawdry."

Harry Cohn, the studio's production head, tried to bridge the gap between the major companies and the Poverty Row outfits. Rather than invest in real estate or maintain a huge stable of stars, Cohn opted to put his money into production, borrowing talent from rival studios. During its heyday Columbia had only one major star, Rita Hayworth, although Glenn Ford, Evelyn Keyes, Larry Parks, and Janet Blair became substantial box office attractions.

Since Columbia owned no theaters, Cohn emphasized pictures that appealed to small towns across America and neighborhood theaters offering double features to attract kids to Saturday matinees. Westerns were important to Columbia, and cowboy heroes like Charles Starrett became vital to the studio's fiscal stability. In 1937 Columbia purchased a ranch in Burbank, near Warner Bros., where most of the company's outdoor shooting took place. Inexpensive comedies and B pictures, seven to ten days in production, became the backbone of the studio's output. Often the same stories were remade with different settings. "If they made a picture for no more than $95,000 or $100,000," director Edward Dmytryk pointed out, "they knew they would make a profit. If they made it for $110,000, they would not make a profit or maybe break even. If they made it for $120,000, they lost money. So I had two weeks to make a picture."

Films directed by Frank Capra and Rita Hayworth musicals were Columbia's prestige pictures, on which Cohn was willing to spend money. "In the seventeen years I was there," Charles Starrett said, "I felt like I was in the hardware business, down in the basement. Rita Hayworth was up on the fifth floor, in ladies' apparel, and of course Frank Capra was in the penthouse. He got the budgets to make the great pictures, while we were making money down in the basement doing westerns."

United Artists, the smallest of the "Little Three," was formed in 1919 by the four biggest names in the industry at that time: Charlie Chaplin, Mary Pickford, Douglas Fairbanks, and D. W. Griffith. Originally the company was envisioned as a distribution network to market the pictures produced by those four giants. For a time during the 1920s United Artists owned a small chain of theaters. A tiny studio at Santa Monica and Formosa (where Pickford, Fairbanks, and Chaplin had all worked) became known as United Artists, large enough for only three or four pictures to be made at any one time. After 1935, when its lot became the Samuel Goldwyn Studio, United Artists remained a releasing organization.

Once Goldwyn took over the studio, he rented space to other independent producers but carefully supervised his own productions, concentrating on a picture or two at a time. In 1935, David O. Selznick began his own studio, Selznick International, taking over the old Thomas Ince headquarters in Culver City, down the street from MGM. "The day of mass production has ended," Selznick told reporters in October of that year. "It has become, in the making of good pictures, so essential for a producer to collaborate on every inch of script, to be available for every conference and go over all details of production that it is physically impossible for him to give his best efforts to more than a limited number of pictures." (Haver, *Selznick's Hollywood*)

That was Selznick's credo, as it was Sam Goldwyn's. Each maintained his own roster of stars, and each strove to make pictures of exceptional quality. Goldwyn produced *Wuthering Heights* (1939), *The Best Years of Our Lives* (1946), and the early Danny Kaye comedies; Selznick made *Gone With the Wind* (1939), *Rebecca* (1940), *Spellbound* (1945), and *Duel in the Sun* (1946).

For a brief period after World War II, Eagle-Lion appeared as a minor studio of importance. Formed by Aubrey Schenck and Bryan Foy, heretofore known for budget movies, Eagle-Lion benefited from a British connection for releasing pictures in Europe. Its studio was located on Santa Monica Boulevard, just west of the Samuel Goldwyn lot. Susan Hayward, Bob Cummings, and Robert Preston all made pictures for Eagle-Lion, but the company lasted only a few years, constrained by limited finances.

More durable, although operating on even less money, was Monogram, which appeared during the Depression with the rise of the double feature. A Poverty Row outfit, Monogram functioned on two sound stages on Sunset Boulevard, turning out series like *Charlie Chan* and *Joe Palooka*, B westerns, and serials—entertainment popular in small neighborhood theaters and rural movie houses. The studio's contract list included Gale Storm, Johnny Mack Brown, the Bowery Boys, director William Castle, and Belita, a British ice skater. "You really had to be a fast study at Monogram," remembered actor Milburn Stone, who later played Doc Adams on television's "Gunsmoke." "They wouldn't hire anybody who couldn't almost grab a script and learn it by braille. You really had to be able to rip out the dialogue, because they shot fast."

Most Monogram pictures were made on an eight-day schedule and cost around $40,000, about twice what Producers Releasing Corporation across the street spent. In the late 1940s, as the market for second features waned, Monogram was renamed Allied Artists and attempted more ambitious pictures, what the business called "nervous A's." The first of these was a vehicle for Belita called *Suspense* (1946). The studio hired John Mescall, the cameraman who had worked on several of the Sonja Henie pictures, and geared up to make an ice spectacle. Arrangements had been made for the ice to cover a big stage. Pipes to freeze the water broke down, and the ice melted the first day. Belita swam instead of skating. It took weeks to remedy this situation, making Allied Artists' entry into the big league a limited success. (Mirisch int.)

The best of the B studios was Republic Pictures, founded in 1935 by Herbert J. Yates, formerly an executive with the American Tobacco Company. Yates took over the old Mack Sennett Studios in the San Fernando Valley and specialized in budget westerns and serials. Within the first year Republic's grosses were over $2.5 million, most of its money coming from the Midwest, the South, and the Southwest. The studio committed itself to inexpensive action pictures reinforcing traditional values. Republic combined superior sound, professional camera work, straightforward direction, and skillful editing to produce a marketable sequence of films.

Occasionally Republic ventured into A productions—pictures such as *I've Always Loved You* (1946), *Wake of the Red Witch* (1948), and *Sands of Iwo Jima* (1950)—but its producers primarily chose subjects that appealed to rural audiences. The company made several country musicals, which drew from radio shows for material certain to be popular in the small towns. Consisting of a dozen sound stages, a backlot, and a western street, Republic used less expensive Trucolor instead of Technicolor. Although the studio lacked star-glitter, it was a comfortable place with a western flavor. "It was like Texas," remembered Texan Dale Evans, eventually hailed as Queen of the Cowgirls, "very down-home. Of course there were lots of cowboys, since westerns absolutely put Republic on the map. It was like a big family."

Gene Autry, one of Republic's biggest stars, found the studio much like "a big country picnic." Autry worked closely with the crew and got to know them all. According to actress Catherine McLeod, who had been under contract to Metro earlier, "I knew every fireman, every gateman, every guard, every electrician, every stand-in, every wrangler. And they all knew me. I was a little fish in the MGM pond, but I was a whale at Republic."

Short shooting schedules at Republic meant hard work. Crews rushed to finish a picture in two weeks and be ready to start the next in three days. Actress Peggy Moran, under contract to the studio in the early 1940s, once commented, "These pictures don't make any sense!" and was told that if the theaters booked them, the company would make a profit. While personnel at major studios jeeringly called Republic "Repulsive," Yates's company persisted in turning out the inexpensive "oaters" that remained that studio's stock in trade.

During Hollywood's Golden Era, each studio was like a miniature city, with its own directors, its own stars, its own designers, its own cinematographers. Larger studios were self-contained units, with a physician, barbers, a dentist, a schoolhouse, even firefighters. "If there had been a war and we had been shut in there," publicist Esmé Chandlee said of her years at MGM, "we had everything to keep going for a long time."

"An air of mystery hovers over a film studio after dark," composer Aaron Copland observed while working for Sam Goldwyn in the early 1940s. "Its silence and empty streets give off something of the atmosphere of a walled medieval town; no one gets in or out without passing muster with the guards at the gates." (Copland, *New Music*) Copland found the quiet conducive to serious work, writing most of *Appalachian Spring* and a good portion of his Violin Sonata on the Goldwyn lot at night.

Despite all the glamour, the big studios represented a businesslike system. Talent stayed in one place, felt a sense of security, and developed, becoming part of the studio community. Crews were the best available and welcomed challenges. Nothing seemed impossible for them, since money was no object. Although the studios of Hollywood's Golden Era annually churned out the required product, they also provided a creative arrangement by which those with talent could blossom into popular artists.

2

The Moguls

Hollywood mogul Samuel Goldwyn. Courtesy of the Academy of Motion Picture Arts and Sciences.

The moguls who built the American movie industry were gamblers, the last of the great financial robber barons. They were also showmen with a sense of what the public wanted in entertainment, alternating gaucherie with taste, at their best producing mass culture on an enduring plane. "The Zanucks, the Cohns, the Mayers may have been vulgarians," commented film editor Marjorie Fowler, daughter of screenwriter Nunnally Johnson. "They may have been tough sons-of-bitches, but they had instincts." In the free-for-all that was the early film business, the moguls recognized that fortunes could be made from movies and realized that this vernal industry was theirs for the taking.

Most of the Hollywood moguls were Jewish, and almost all of them came from the ghettos of Central Europe. In America they tried to divest themselves of their Jewishness, assumed new names, and eagerly set out to claim riches in a land known to reward hard work. Samuel Goldwyn began his career as a glove salesman, William Fox as a cloth-sponger, Adolph Zukor as a furrier, Louis B. Mayer as a junk dealer, but eventually all four were drawn to the entertainment field, where opportunities seemed boundless. In motion pictures they found wealth and acceptance; they established studios and ran their empires like dictators. Each year they turned out sufficient "product" to satisfy theater operators, offsetting art with commerce. In a ruthlessly competitive industry the Hollywood moguls rose to the top brandishing fists of iron. Or as Sam Goldwyn put it, "It's a dog eat dog business, and nobody's going to eat me." (F. Goldwyn int., Wagner)

Yet while the motion picture industry could be bitterly cutthroat, during Hollywood's Golden Era it was also curiously fraternal and courtly. Although the moguls coveted profits and were capable of crushing opposition, they repeatedly demonstrated a sense of honor. "You could trust the studio with your life," actress Catherine McLeod said. "You could not, I learned, trust agents." The Hollywood tycoons prided themselves on honesty and meeting contractual obligations to the letter. Old-timers insisted that a handshake was binding. "When Louis B. Mayer gave his word on something," said publicist-producer Walter Seltzer, "you were pretty certain that it was going to happen. Those guys knew what they wanted. If they had said, 'Not interested in this picture, forget it,' that was it. But when they went forward with a project, it was full speed ahead."

Hollywood's moguls not only wanted to make money, they also wanted to produce fine films. They respected talent and opened doors for those they believed had it. "Nothing was impossible," the renowned director Rouben Mamoulian said. "Everything was made available to those in whom studio executives had confidence." By the 1930s all the major studios had been built. The moguls had the money, the structure was complete, and once a picture was finished, it went immediately into the theaters, which were consistently full. Studio heads needed talent to keep up a steady flow. Whatever their limitations, the Cohns, the Zanucks, the Warners all loved films and rewarded talent they respected.

Carl Laemmle at Universal was shrewd with an intrepid will, despite his easy-going exterior. Born in 1867 at Laupheim, in the South German kingdom of Württemberg, Laemmle was the tenth of thirteen children. His father became a businessman, yet the family remained poor. At seventeen Laemmle left Laupheim for America, where he swept floors, washed bottles, ran errands, worked as a farmer and as a bookkeeper, and eventually became a clothier. In 1906 he entered the film business, first as a nickelodeon operator in Chicago, later as a film distributor. By 1909 the Laemmle Film Service was the largest distribution agency in the country.

A tiny, nearsighted man with a round face, thick German accent, and a tendency toward hypochondria, Laemmle seemed more like a village burgomaster than a powerful executive. In reality he was a fierce competitor, lavish in his praise of good work, withering in his rebukes when duties were performed poorly. His generosity became legendary throughout the industry. Punctilious in business details, Laemmle insisted that bills be paid promptly and demanded exact daily accounts of his operations. He was known for his straight dealing—Uncle Carl's word was as "good as gold."

Proud of his success but unspoiled by it, Laemmle remained modest in his tastes. He dressed with care, played the horses for amusement, and loved to see his name in print. He claimed no credit as a technical or artistic pioneer; his view of art was always unpretentious. He cared little for the glamour of Hollywood, preferring the company of a small group of cronies, mostly relatives or friends from the old country. When not in New York or Europe, he liked nothing better than to walk about the Universal lot, chatting with anyone not too busy to talk, whatever their position. At times he seemed uncertain about the respect he should command. Once when studio employees gathered at the train station to greet him after a long trip,

Laemmle was furious to find them absent from work. But the next time he returned and no one met him, Uncle Carl was miffed.

After 1915 Laemmle's career was in decline, although he enjoyed the rewards of his earlier achievements. In 1929 he made his twenty-one-year-old son, Carl Laemmle, Jr., director-in-chief of production at Universal, but the industry never accepted Junior Laemmle as a significant force. Some of the studio's major talent left, and Universal suffered serious financial reverses. Eventually the studio was completely reorganized.

At Paramount soft-spoken Adolph Zukor, the power behind the greatest production-distribution-exhibition combine of the silent era, proved himself a financial genius and an able administrator, skilled in selecting and managing subordinates. A businessman rather than a skilled filmmaker, Zukor was wise enough to delegate artistic authority. Hungarian by birth, he came from a family of small tradesmen in the village of Ricse. At fourteen, full of ambition, Zukor immigrated to the United States, having read of the freedom and opportunities here. He arrived at New York's Battery Park in the autumn of 1888, with little more than an eighth grade education but with determination and an orderly mind.

Zukor soon became an upholsterer's apprentice, then went to work as an errand boy. He moved to Chicago, where he entered the fur business, but in 1903 he embarked on a career in entertainment with a penny arcade. Later, on Fourteenth Street in New York, he began projecting motion pictures on the second floor of the arcade, transporting patrons on a moving staircase. He dreamed about making feature-length films to replace the one- and two-reelers currently in vogue. He visited Edison in his laboratories and gained permission to proceed with longer pictures. Within a decade Zukor had laid the foundation for the interlocking structure of production, distribution, and exhibition that was to become the foundation of the American film industry during its Golden Age.

While Paramount's filmmaking became centered at its studio on Marathon Avenue in Hollywood, Adolph Zukor solidified his empire from the ninth floor of the company's office building in New York. He made a practice of seeing virtually every picture made, quietly going downstairs into the Paramount Theater to study an audience's reaction to a particular film. Confident that "the public is never wrong," Zukor kept in touch with changing tastes by discussing current trends with various magazine editors, even consulting his children at the dinner table about who they thought would be the coming stars.

Once or twice a year Zukor went to California to visit the studio and meet the new people. He took great pleasure in dropping by sets, getting acquainted with players and technicians. "Besides putting me closer to production," he said, "I hoped that such visits would make everybody feel that the business office was more than a place where we made contracts and counted the money." (Zukor, *Public Is Never Wrong*) Aware of the growing importance of stars, Zukor courted fan magazines, keeping Paramount performers before the public eye. But his methods were dignified and gracious. "What a marvelous, marvelous gentleman he was," contract player Walter Abel commented, a sentiment shared by most of the Paramount family.

A reserved man behind his mask of confidence, Zukor could be crafty and resourceful in battle. He could show anger, even explode, but he would also listen with enthusiasm to new ideas. He held few grudges and demonstrated a remarkable capacity for relaxing. "The minute he came in the door at home the world outside was completely locked out," his son, Eugene, said. "He would sit down to dinner, smoke his cigar afterwards, go to bed, and sleep like a boy, while I used to pace the floor at night, thinking about his meeting the next morning. I think that's why he lived to be 103."

The pressures were great, the weight of decision-making often staggering. "Let's say you had a Gary Cooper contract that calls for $250,000," Zukor's son illustrated. "We have to notify Cooper by three o'clock tomorrow afternoon. We have another contract that comes up for decision by noon tomorrow. As for today, we have several writers' contracts that have to be signed or else those people are going with Fox and MGM. We have to call their agents within the hour. There are probably awaiting in today's files decisions worth $600,000 or $700,000. It is now ten o'clock in the morning. By six o'clock this evening all of these have got to be acted on. You'd better know what you're doing!"

And the elder Zukor was prepared to make such decisions, arriving at his office early every morning ready for the day. He saw no conflict between business and art, insisting on an emotional involvement with the pictures produced. Above all he felt that the screen should be entertaining; otherwise audiences would not come. When properly made, Zukor believed, motion pictures serve as a window to the world, as well as to literature and the drama of life. They could also present images of America to foreign countries. "We could *show* freedom and not *talk* about freedom," Eugene Zukor explained. "By portraying on the screen the West, the South, the North, the way people lived, the growing of wheat, the growing of cotton and tobacco, the production of oil, we didn't have to write an editorial; it was part of our story-telling. Viewers abroad saw refrigerators, people going to work in automobiles, parking lots that held three thousand cars in front of a small factory. This was America that we were selling in stories. And the world took it and duplicated many of the things we did."

More directly involved in film production than either Zukor or Laemmle was Louis B. Mayer, the baronial head of Metro-Goldwyn-Mayer, whose personality permeated that studio. Probably Mayer could just as successfully have run General Motors or Chrysler. "The only difference was that he had a smell for talent," said Metro's drama coach, Lillian Burns. "He loved talent." Mayer was in charge of the studio's business arrangements—deciding how much should be spent, who should be hired and fired, formulating company policy. He could be kindly and considerate, down-to-earth and human, at times jovial and fun, but no one on the lot ever forgot that L. B. Mayer was a power to reckon with, capable of ruthlessness.

On first meeting Mayer often appeared humble, but that was part of his persona. He loved theatrics and at times would put the most melodramatic actor to shame. Although he viewed himself as a father figure to Metro's thousands of employees, he nevertheless insisted on control and granted favors to those who did favors for him. At his worst he was bad-tempered, vindictive, Machiavellian. "I would see L. B. Mayer and run like hell," confessed one studio employee. "My first meeting with him was like going in to see the Good Lord," another said. "Being a Catholic, I didn't know whether to genuflect or kiss his ring."

Yet Mayer loved to talk about his "children," the studio employees who made up the Metro "family." Every year on Mayer's birthday Ida Koverman, his executive secretary, would gather the "family" on one of the giant sound stages to "surprise" their boss with a party. Every year, practically sobbing, Mayer feigned surprise that his "children" remembered. Since he celebrated his birth on July 4, a date chosen in a burst of patriotism for his adopted country, it was unlikely anyone could forget.

Born in the village of Demre in Lithuanian Russia, Mayer escaped the pogroms in 1888, when he was three years old, settling with his parents in St. John, New Brunswick. At twelve he dropped out of school and went to work in his father's scrap iron business. When the business failed, he sold tickets in a Boston movie theater. He was cocky, ambitious, persistent. In 1907 he rented a theater in Haverhill, Massachusetts, named it the Orpheum, and became convinced his future lay in the motion picture industry. By 1937 Mayer was the highest-salaried person in the country, making more money than the president of the United States.

Recognizing that his own talent lay in the executive area, Mayer wisely selected Irving Thalberg to head the creative aspects of Metro-Goldwyn-Mayer's productions. Together they made MGM the Grand Hotel of studios. Eventually the collaboration of Thalberg and Mayer deteriorated, but not until MGM was established as the foremost studio in Hollywood. With Thalberg's death Mayer assumed greater authority over production, taking control with a renewed sense of purpose.

Mayer rarely read scripts or story treatments, relying on his studio's Scheherazade, Kate Corbaley, to tell him the plots. His taste leaned toward romance and the sentimental, and he loved stories in which children appeared. The *Andy Hardy* series reflected his ideal of home and family. Mayer was married and had two daughters; by age fifty, however, he was eager to frolic with party girls. Although he never deserved a reputation as a womanizer, he tried to assume the role, buying expensive gifts for more showgirls than he took to bed.

Mayer's need for respect curbed his lechery; he was more comfortable in the parent role. He liked to have actors and actresses visit him in his office, just to chat or introduce visiting family members. If one of his stars got into trouble or dated someone Mayer didn't approve of, a personal explanation from the offender was mandatory. Mayer manipulated his employees into doing his bidding. "When he really wanted something, he walked it out of you," director George Sidney claimed. "He walked me up and down that lot, back and forth."

Ida Koverman, whose office was next to Mayer's, protected her boss from the less important demands on his time. Periodically Mayer would hold meetings on one of the large stages for all of MGM's employees. "He was a great talker," publicist Emily Torchia said, "and a great persuader. I'm sure he didn't tell us the important things, but he made us feel that he was telling us the future of MGM and how all of us would profit from it. We went out of there with our eyes shining—the stars, everybody."

"People who do their jobs have one for life," Mayer regularly insisted (S. Marx, *Mayer and Thalberg*), and Metro was the first to institute a retirement program for studio personnel. "They were all monsters," MGM contract player Patricia Morison conceded of the Hollywood moguls. "They were all bandits. But, my God, look at the talent Mayer developed by keeping performers in his domain and making them toe the line." Mayer's eye for talent and his ability to select the right roles for stars proved major assets, although he relied heavily on the judgment of his producers. "I get people around me who know more than I do," he confessed. "That's how I'm smart." (Zierold, *Moguls*) He also believed in sheer size, frequently lecturing Thalberg on the wisdom of having fifty stars rather then ten, thirty directors instead of five, and twenty stages rather than three. "Think big," he told his production head. "That's the path of the future." (Carey, *All the Stars in Heaven*)

Mayer hired his share of relatives, however. His brother, Jerry, became a unit manager, basically doing what he was told. Jack Cummings, one of

Metro's more powerful producers, was Mayer's nephew, the son of his sister Ida Mae. Publicist Ann Straus was Cummings's sister-in-law, while director George Sidney was the son of Metro executive L. K. Sidney.

"I had a perfect formula for getting along with Mayer," producer Pandro Berman said of his work at Metro. "I just stayed away from him." By the 1940s avoidance was easier, since Mayer had established himself as one of the leading thoroughbred owners in the country and became less involved with studio matters. Neither dancer Gene Kelly nor director Charles Walters, both pivotal figures in the heyday of MGM musicals, had much contact with Mayer on either an aesthetic or a personal level. Their discussions were mainly with producer Arthur Freed.

While Mayer eventually lost control of his empire, Jack L. Warner remained at the helm of Warner Bros. until 1966, outlasting the rest of the Hollywood tycoons. Warner desperately wanted to be part of an artistic world he lacked much capacity to understand. Buffoonery was his defense mechanism—terrible jokes behind which he hid his near illiteracy. Jack Benny claimed Warner "would rather tell a bad joke than make a good picture." (Freedland, *Warner Brothers*) "*Arrivederci*," an Italian negotiator once said to him. The mogul looked at him and replied in typical Warner fashion, "A dirty river to you, too." (B. Thomas, *Clown Prince*) Only those too insecure to do otherwise laughed.

Warner regretted his lack of formal education, yet seldom read a book, insisting that novels the studio was considering be reduced to a one-page synopsis. Crude though his approach was, Jack Warner controlled the Burbank studio like a pasha. Underlings feared arguing with him, knowing he had the power to ruin careers. He told the venerable Jesse Lasky, who had become a producer at Warners, that his position with the studio was impregnable, only to notify him a few hours later that his services were no longer needed. If an executive opposed him, Warner invariably had the last word: "Whose name is on the front of the building?" he'd ask. (Freedland, *Warner Brothers*)

In many ways Jack Warner was more vulnerable than Mayer, for his ship was not run as tightly. Yet like Mayer and the rest, Warner had an instinctive sense of what made successful pictures. Underneath his Dutch comedian's exterior was an efficient production chief with a penchant for detail. If his tastes lacked sophistication, Warner never lost sight of the fact that he was making commercial entertainment for mass audiences. "Okay, boys," Jack would say when story conferences got too long and involved, "I'm from the Bon Ton Woollen Underwear Company and haven't got a clue to what you are saying." (Freedland, *Warner Brothers*)

The Warner brothers came from the Polish village of Krasnashiltz, near the German border. Like other peasant families they suffered from the Cossack pogroms, living in brutal poverty. The family immigrated to the United States in 1882, settling first in Baltimore, then in Youngstown, Ohio. None of the four brothers had much education, nor did they spend time in serious thought. Their father had been a cobbler in Poland, but in Youngstown he opened a butcher's shop, where Jack and his brother Sam helped out. Meanwhile Harry and Albert, the other two Warner boys, sold newspapers and shined shoes on neighborhood street corners, although Harry eventually learned the cobbler's trade. Years later, when he was president of Warner Bros., actress Ann Sheridan noticed Harry bending down and picking up nails as he walked across the lot, popping them into his mouth as he had learned years before.

At eighteen Sam Warner decided to become a nickelodeon operator, acquiring a scratchy print of *The Great Train Robbery* (1903). Jack, accompanied on the piano by their sister Rose, entertained between showings singing the popular ballads of the day. At twenty-three Jack was running the Warner Bros. studio in California. Harry became the company's financial wizard, and Albert took charge of its theater division. Sam, perhaps the most brilliant of the brothers, later became involved with the development of sound and was largely responsible for the success of Vitaphone. He died two nights before *The Jazz Singer* (1927) opened in New York.

Whereas Albert and Harry worked mainly in the East, taking care of distribution and financial matters, Jack occupied himself with production in Burbank. The relationship between the brothers was not easy; Harry and Jack quarreled almost constantly. Each year Harry would visit the studio, much as Adolph Zukor did Paramount. Actor Pat O'Brien recalled walking with him to the Warner Bros. commissary. Harry had a little gimp and noticed O'Brien watching him. "You know, Paddy," he volunteered, "for years I was a cobbler. I guess that has stayed with me." The actor was touched: "Here was this multimillionaire telling me about his back trouble."

But Jack was the Warner who made headlines, negotiated with stars, and came to believe he held almost a divine-right power. Screenwriter Julius Epstein, who with his twin brother Philip worked for Jack, called it "affable arrogance." Warner dressed in natty suits and blazers, customarily sported a boutonniere in his lapel, and fancied himself a playboy. At home he was a charming host, yet in his art-deco office he could be a tyrant. Over the years he fought with Bette Davis, James Cagney, and Olivia de Havilland, insisting that since he was paying his stars huge sums of money, they should be subservient to his demands.

Like L. B. Mayer, Jack Warner had an instinct for casting and a sense of what the public wanted. He was convinced that for Warner Bros. to make money he must have good stories. He remained a businessman, turning out a salable product. Old-timers claimed he planted spies on every set, while writers complained he ran the studio like Alcatraz. Punctuality became an obsession with him. "Read your contract," Jack told the Epstein twins. "Bank presidents come in at nine o'clock. Railroad presidents come in at nine o'clock. You are coming in at nine o'clock." Nothing escaped his notice. When he saw Joan Blondell's gown in some rushes of *Convention City* (1933), he insisted she wear a brassiere to cover her breasts. "Otherwise," he wrote in a memo to production head Hal Wallis, "we are going to have these pictures stopped in a lot of places. I believe in showing their forms but, for Lord's sake, don't let those bulbs stick out." (Behlmer, *Inside Warner Bros.*)

After Hal Wallis left the studio, Jack played a more active role in production. Where Warner had previously seen only a few of the "dailies," he now saw everything. Each morning around eight-thirty he would phone his secretary, Bill Schaeffer, then call the publicity department to find out what they had scheduled for the day. He loved stars, enjoyed basking in their glory, and was flattered when directors referred to him as "Colonel." Warner often helped out studio employees financially when they were in need. Many times he'd send a subordinate down to the backlot to check on a grip whose wife had died or on someone who had been hurt, not wanting that side of his personality to be publicized.

William Fox as a New York theater owner had been among the most vehement opponents of Edison's monopoly. Never a particularly lovable character, Fox was born in Hungary in 1879 and grew up in poverty. He spent his boyhood on New York's Lower East Side and was a garment worker before entering the nickelodeon business. By 1929 Fox owned more than a thousand theaters, second only to Adolph Zukor. He ruled over his kingdom from an office in the Roxy Theater in New York, but entered film production to ensure that his expanding theater chain had sufficient product. In the process he overextended his credit and by the early 1930s faced financial ruin. Fox tried to convert his assets into cash, but mounting debts overwhelmed him.

Meanwhile Darryl F. Zanuck, who had preceded Hal Wallis as production head at Warner Bros., had quarreled with Harry Warner and resigned. To cut costs during the worst of the Depression, Warner had agreed to a 50 percent reduction in salaries for studio employees, as had other studio heads. When Warners refused to accept a ruling from the Academy of Motion Picture Arts and Sciences to restore the cuts, Zanuck found himself in an intolerable position. In April 1933, by then the hottest young filmmaker in Hollywood, he walked out on a $4,000-a-week job.

During the autumn of 1933 Zanuck with Joseph M. Schenck formed 20th Century Pictures. In need of stars and permanent studio facilities, they merged 20th Century with Fox's operation in 1935, moving to the Fox lot. Zanuck built new sound stages, expanded the contract roster, and from the moment the merger took place single-handedly ran 20th Century-Fox.

Born in Wahoo, Nebraska, a tiny farming community west of Omaha, Darryl Zanuck possessed a great sense of what the average American wanted in entertainment. Short in stature, Zanuck was feisty, ambitious, opinionated, often testing people just to see if he could intimidate them. He could be cruel, yet workers respected his honesty. Above all he loved movies and demonstrated an instinct for making them. "Zanuck was electrifying," said contract player Dorris Bowdon. He knew filmmaking, proved a good administrator, established a solid liaison with banking interests, and possessed a keen sense of publicity. "Nobody was too big for him to challenge," Bowdon insisted.

Probably to compensate for his size, Zanuck became an athlete and health fanatic. He had ridden horses since boyhood, adopted polo as his favorite sport, learned to ski, and played croquet. Whatever he did, he did with a passion. Sexually he seemed insatiable, attempting to seduce nearly every attractive woman on the lot, and he appeared to enjoy shocking newcomers with his exhibitionism. Select starlets he escorted into a secret chamber behind his desk, filled with hunting trophies and a bed with a tiger-skin bedspread.

No one doubted Zanuck's abilities as a filmmaker. He supervised every aspect of the business, was a brilliant film analyst, understood construction and story value, and possessed a quick mind that seemed to retain everything. "We always claimed you could give Zanuck the front page of any newspaper and he could get three stories out of it," film cutter Gene Fowler, Jr., said. He wanted to win arguments, but would test directors to see how much they believed in their point of view. If they argued convincingly, he was fair enough usually to give in. "He had a terrific eye," Rouben Mamoulian averred. "You couldn't fool him. He had a hand in everything and knew what was good and what was bad." Those easily frightened lost Zanuck's respect. "For a little guy, he was a giant," studio worker Joe Silver declared.

In addition to a group of high-powered writers and intellectuals, Zanuck surrounded himself with trusted studio personnel who became his private entourage. Sam Silver, Joe's father and Zanuck's barber, was among the production head's closest friends. Sam rode horses with his boss, hunted with him, and frequently attended screenings with him. Zanuck brought Otto Lang, his ski instructor at Sun Valley, to Hollywood and turned him into a producer. Hector Dods, head of the studio's editing department, was a polo player from Australia. "He knew more about polo than he knew about film," Gene Fowler, Jr., commented, "but Darryl became a good polo player." When Zanuck wanted to learn French, he brought in Jacques Surmagne, who drove to the studio with his employer every morning so they could speak French on the way. Fidel La Barba, another member of Zanuck's coterie, had been a flyweight champion, Henry Lehrman a silent comedy director. Gregory Ratoff, a character actor Zanuck found amusing, was eventually made a director. To those who were part of his team, Zanuck remained generous and loyal, relying on them for advice, using them as a sounding board to keep in touch with popular tastes.

Within three years of forming 20th Century-Fox, Zanuck was earning $260,000 a year. Like the other Hollywood moguls, he lived in royal splendor, entertaining lavishly at his beach house in Santa Monica. Contract player Vanessa Brown recalled an extravaganza at the Zanucks' with "Nubian slaves bringing flaming ice cream under glass." Virginia Zanuck was a gracious hostess and seemed to tolerate her husband's philandering.

But Zanuck was also a dedicated worker. His mornings began around ten o'clock as he read a script while he ate breakfast. Then his barber would shave him, and he'd arrive at the studio an hour or so before lunch. The afternoon was taken up with story conferences, usually with the writer, producer, and director of the picture present. "He was a whiz with scripts," director Henry Hathaway said, "not from a literary point of view, but just pure instinct—what would work and what wouldn't." Sometimes Zanuck paced the room, swinging his polo mallet and puffing on a cigar. This scene was too long, he might argue, that sequence too short; something else sounded silly; the script fell apart in the middle; certain sections of dialogue were too talky. Sometimes he'd act out the parts, playing the female roles in falsetto, growling the men's parts, while Molly Mandaville, his script coordinator, transcribed.

Often these conferences dragged on for hours, while Zanuck drove his points home. He tended to talk in cliches, using phrases such as "her love turns to hate," and "fear clutched at his stomach." He saw scripts in terms of plot, images, and personalities, but above all he realized that motion pictures should *move*. Zanuck's talent lay in simplification, and he depended heavily on remakes and sequels on the theory that if something worked, try it again. "The strange thing was that he was a very bad writer himself," screenwriter Philip Dunne observed. "But his sense of what was essential and germane to a story was right. He also understood that the script was the picture. If you had a good script, you were going to get a good movie."

When the day's activities slowed on the lot around six o'clock, Zanuck rested briefly, usually took a swim or went to the steamroom, where he had a massage. Then he would eat dinner and return to the studio around nine to watch rushes and first cuts in his private projection room. He viewed all of the A pictures the company shot and seemed to remember every frame. Night after night he sat in a leather chair with his hand on the control panel, chain-smoking cigars, usually surrounded by a group of cronies. After the lights came up he would ask the assemblage for opinions. Sometimes he encouraged criticism to undermine the confidence of the producer or director and make it easier to get the changes he wanted. If comment wasn't forthcoming, Zanuck would begin to dissect the film himself, pacing the floor as he talked. Often discussion went on until one or two o'clock in the morning, sometimes until three or four. If the session broke up early, Zanuck might corner a friend and say, "I've got an Italian movie I hear is good. Would you like to see it with me?"

From his office, Zanuck dictated memos to everyone associated with a film in production, offering criticisms and suggestions, sending copies to everyone concerned. He wrote detailed comments on plot synopses, story treatments, and drafts of scripts, making suggestions and offering casting possibilities. Zanuck even interviewed young contract players. Sometimes he was curt; other times he grew overly familiar. "We're going to do great things for you," he'd promise, "great things." Sometimes he did, sometimes he didn't.

But Zanuck was a genius at administration as well as creative filmmaking, a rare combination. "I thought Zanuck was the greatest studio head I had ever seen," director Henry King said. "He was great in stimulating imagination and helping his directors in every way possible." (King int., AFI) If a project was in trouble, Zanuck gave the director his opinion of the problem in detailed notes, and his judgment usually made sense. Although Fox directors answered to producers, they also knew that the producer was directly answerable to Zanuck. Some of the studio's best pictures Zanuck produced himself, but he never sloughed off a picture because his name wasn't on it.

More than the other studio heads, Zanuck was willing to take chances. He produced *Gentleman's Agreement* (1947), a strong attack on anti-Semitism, at a time when the Jewish moguls were unwilling to deal with the issue on the screen, fearing they might create antagonism for the industry. The picture was controversial, but it was also well received, winning an Academy Award.

Although Zanuck had to answer to a board of directors in New York, he ran a one-man operation. He hired the people he wanted, filmed what he wanted, decided when a picture should be previewed, and brooked no interference. "Of all the motion picture executives I have known," studio manager Raymond Klune said, "Darryl Zanuck was far and away the best of the lot." (Klune int., AFI).

Like Zanuck, Harry Cohn of Columbia was involved with every major picture made on the lot; his power at Columbia was absolute. Although many studio employees confessed that they admired him, and most acknowledged his ability as a production head, Cohn was unpleasant to work with. His methods were crude, his reputation unsavory, his language foul. He tolerated few relationships he couldn't control and overpowered or fired those who opposed him. "I am the king here," Cohn said of his position at the studio. "Whoever eats my bread sings my song." Cohn seemed to thrive on unpopularity. "I don't have ulcers," he announced. "I *give* them!" (B. Thomas, *King Cohn*) Walter Seltzer called him "The ogre of Gower Street," while others called him names like "White Fang," "Harry the Horror," and "His Crudeness." "Harry did a lot of yelling," Seltzer explained, "and a lot of it was unreasonable and abusive and unnecessary. But he also had right on his side as often as not. He was a good picturemaker."

Cohn occupied a huge office, and his desk was set on a platform that made him look bigger. Performers complained he was a tyrant, that he thrived on combat. "The whole atmosphere around Columbia," veteran actor Arnold Moss recalled, "was one of fear. You never knew what was going to happen." Actress Nina Foch was kinder. "Cohn was a dreadful bore," Foch remembered, "but I had a lot of respect for that man. He knew what he was doing." Many even claimed they grew fond of Harry.

Cohn preferred people who dealt with him honestly and threw his profanity back at him. Character actress Ann Doran remembered walking down his long office toward his desk. Cohn had a phone up to one ear, placing bets on horses, and was talking on an intercom at the same time. During their conversation Doran argued with him, and on her way out she slammed down the top of a white baby grand piano near the door. Cohn respected her spirit. "He liked somebody who would be as abrupt as he was," the actress declared.

When Gene Autry arrived on the Columbia lot from Republic, Cohn warned him that he'd hear a lot of stories about how tough he was to work for. "But you'll find you're going to be treated fairly here," the mogul told Autry. "If you have any problems, the first place to come to settle them is Harry Cohn's office." Autry worked harmoniously with Cohn for six years. "Everything was dealt with right on top of the table," the singing cowboy insisted. Another cowboy star, Charles Starrett, earned Cohn's admiration both as a performer and as an individual; Starrett found his boss to be approachable and imbued with a strong sense of loyalty.

"I know all the stories," assured Lillian Burns, who became Cohn's executive assistant after she left Metro-Goldwyn-Mayer. "I've read or heard them all, but they were *not* enacted in front of me." The worst profanity she ever heard Cohn utter was "goddam." Like L. B. Mayer, Burns said, the Columbia head had an eye for talent and knew how to develop it. "Certainly he was much more involved in actual filmmaking than Mr. Mayer," she claimed.

Cohn had an obsession for movies, approaching the business as both showman and gambler. He saw all the rushes, was involved in raising money, negotiated with performers, and held day-to-day conferences with producers, directors, and writers. He knew what was happening in the studio at all times and tolerated neither carelessness nor misconduct. Despite his crudeness, Cohn often demonstrated a remarkable sense of style, and he was wise enough to surround himself with elegant people. Much of his time was spent making deals, trading people, borrowing talent, keeping himself informed. He had big ideas and demanded the best. He operated on instinct, but over the years his instincts proved amazingly solid.

For Cohn street language was natural. Born in 1891 on Manhattan's Lower East Side, he was the son of a German father and a Russian mother. Like most Jewish immigrants, his parents wanted him to gain an education, but Harry dropped out of school at an early age. He became a pool hustler and later a song plugger on Tin Pan Alley. Along the way he adopted the speech and mannerisms of a gangster; in later years he boasted of entertaining Frank Costello and other mobsters and hinted at underworld connections. One day during his youth he noticed Samuel Goldwyn and Nicholas Schenck on their way to Goldwyn's studio in a chauffeur-driven Pierce-Arrow. "Someday I'm going to live like that," he announced. (B. Thomas, *King Cohn*)

Cohn entered the movie business, moved to Hollywood, and started his career there on Poverty Row. He borrowed sets and found endless means of economizing, scrambling for every booking he got, since Columbia owned no theaters. "Columbia, the Germ of the Ocean," stars from the major studios used to sneer. But Cohn gradually began to produce some fine pictures. After Frank Capra won the Academy Award for *It Happened One Night* (1934), the studio was suddenly elevated to a major minor. Cohn continued to make scores of westerns, B comedies, inexpensive musicals, cheap mysteries, and exploitation films, but his studio also showed an artistic side and earned the acclaim it received.

Cohn used the profits from his B pictures to finance the A productions that later became classics. He thought big, talked big, yet paid small salaries. With success, Cohn began to dress in a dapper fashion and to drive the most expensive cars. He made frequent trips to Europe, demanding luxurious accommodations. Muscular and modestly handsome, he could be charming—especially to women. He indulged in dalliances, but pursued most of his affairs off the lot. Still, when contract player Nina Foch met Cohn, his first words were, "If you're going to fuck anybody in this studio, it'd better be me!" Foch was stunned and spent the seven years she was under contract to Columbia telling Cohn how much he reminded her of her father to keep him at a distance. "It was hard in those days to get through a studio and not sleep with the management," the actress commented.

Cohn's primary concern as head of Columbia was the A films that gave his studio prestige. He loved glamour, demanded the best from every department, and harassed his writers relentlessly. He hired the brightest screenwriters he could find, aware that good pictures depended on good scripts, then barked at them unmercifully. "You smart college bastards," he'd often begin a tirade, in an effort to force brilliant minds to his level. "He was a blunt instrument," screenwriter Edmund North noted. Cohn's office was directly across a courtyard from the writers' building. Periodically he would look out his window to make sure everyone was working; if a writer sat idle, he or she likely received a phone call. If they forgot to turn out the lights when they left the studio, the delinquent party usually received a sharp note in the morning.

What Cohn envisioned was a little Metro-Goldwyn-Mayer, and he delighted in stealing Metro's talent whenever he could. "There wasn't the class of MGM," said actress Audrey Totter, who worked at both studios. At Columbia there simply wasn't enough room for big dressing rooms, nor was there a commissary. Columbia was just four city blocks in the middle of Hollywood, but out of it came some remarkable work.

Like the other Jewish moguls, Harry Cohn played down his Semitic heritage. "Around this studio," he said, "the only Jews we put into pictures play Indians." (Zierold, *Moguls*) But Cohn was known as a man who could be trusted. "I would rather have had Harry Cohn's handshake than somebody else's signature," one worker said. "He never went back on his word." And he remained faithful. If a studio worker became sick, Cohn made sure he or she stayed on the payroll. But he expected loyalty in return.

Of the independent producers in old Hollywood, Samuel Goldwyn ranked at the top, eventually forming his own studio, which most regarded as a model. Called the "Celluloid Prince" because of his aristocratic ways, Goldwyn was a maverick, difficult to deal with unless he got his own way. Yet he possessed a sense of quality and good taste that came to be labeled the Goldwyn Touch. Goldwyn demanded excellence of every picture, emphasized writers as well as stars, and respected great authors more than the other moguls in the business. Lillian Hellman, Sinclair Lewis, Robert Sherwood, MacKinlay Kantor, Ben Hecht, and Charles MacArthur all worked for Goldwyn, and most of his outstanding films had their origin in literary sources. He demanded the best and paid for the best. "Goldwyn was amazing," declared Edmund North. "He was a wild man, absolutely impossible to deal with. But he had innate good taste about film."

Born Samuel Goldfisch in 1884 in a Warsaw ghetto, Goldwyn grew up in poverty, the firstborn son of Polish Jews. At eleven he went to work as an office boy in Poland, and four years later he was on the road in America selling gloves, full of nervous energy and routed toward success. In 1910 he married Blanche Lasky, sister of filmmaker Jesse Lasky, and, impressed by the profits to be made from movies, shortly became a partner with his brother-in-law and Adolph Zukor in Famous Players-Lasky, a forerunner of Paramount.

Soon there were problems. "Goldwyn was a very aggressive salesman type character in those days," recalled Eugene Zukor. "He hadn't been long out of the glove business, and he asserted his authority to an extent that was unpleasant." Immediately it was evident that Goldwyn wanted to be the decision-maker. "Before long," Adolph Zukor wrote, "I was convinced that Goldwyn disagreed many times only for the sake of argument." (Zukor, *Public Is Never Wrong*) So Zukor bought him out. More determined than ever, Goldwyn vowed to make only great pictures. Following a brief association with the Selwyn brothers, he changed his name from Goldfisch to Goldwyn and, after selling his interest in the Goldwyn Picture Corporation to Metro, went on his own, assuring his independence by financing his pictures personally.

From 1939 on Goldwyn released no more than two pictures a year, lavishing each production with money and close personal attention. He paid big salaries for stars he wanted, yet seemed incapable of remembering how to pronounce their names. "In all the time I've worked in films," composer David Raksin said, "Goldwyn is the only man I ever met who intimidated me even slightly." Without question he was a powerful presence. Goldwyn lived in a palatial, sixteen-room Georgian mansion in Laurel Canyon, where he and his second wife, Frances, entertained amid regal opulence. At one point

Goldwyn was reportedly earning $75,000 a week. He loved publicity and made sure he received it.

Goldwyn's mutilation of the English language became legendary. While many of the mangled remarks attributed to him were undoubtedly products of his publicity department, Mrs. Goldwyn confirmed that about half of the more outrageous sayings she had actually heard her husband utter. "I don't think anybody should write his autobiography until after he's dead," the mogul supposedly declared. When warned by an associate that a certain property the studio had under consideration was too caustic for films, Goldwyn replied, "To hell with the cost. If it's a good story, I'll make it." But he could be practical: "A verbal contract isn't worth the paper it's written on." On another occasion Goldwyn claimed, "I had a monumental idea this morning, but I didn't like it." In a generous mood he exclaimed, "I don't care if my pictures don't make a dime, so long as everyone comes to see them." Conferring with director King Vidor he suggested, "Let's bring it up to date with some snappy nineteenth-century dialogue." Later in his career the studio's head bookkeeper complained, "Mr. Goldwyn, our files are bulging with paperwork we no longer need. May I have your permission to destroy all our records before 1945?" Sam thought it over. "Certainly," he said. "Just be sure to keep a copy of everything." (F. Goldwyn int., Wagner; Zierold, *Moguls*) With such riches to draw from, the press always found plenty to write.

Despite his association with literary greats, Goldwyn read at the pace of a second grader, sounding out every word. During the filming of Lillian Hellman's hit play *The Little Foxes*, he persisted in referring to it as "The Three Little Foxes." Yet Goldwyn was genuinely committed to bringing smartly packaged, artistic films to large audiences. For him a film could not be considered worthwhile until it had been seen by the masses. In that regard he remained the glove salesman, determined to sell his product on a volume market.

Goldwyn not only maintained control over matters of policy, he also demonstrated an eye for detail. It was not unusual for him to watch a sequence in a day's rushes and complain, "I don't like the drapes—they're wrong. Reshoot it." As sole stockholder in his company, Goldwyn could demand whatever he wanted. "In my thirty years in the film business I've never read a script so bad," he told writer Melville Shavelson. "Mr. Goldwyn, what is it you want?" Shavelson pleaded in desperation. "I don't know," Goldwyn replied, "but when I see it, I'll know."

Sometimes his stubbornness led to disaster, as with his attempt to turn Polish actress Anna Sten into another Garbo, while his disagreements with directors of the stature of John Ford and William Wyler resulted in brutal arguments and resignations. "Congenial is not a word I'd use to describe Goldwyn," Wyler said.

David O. Selznick, another independent producer with his own studio, was no less difficult to work for and, like Goldwyn, kept a tight rein over everything concerned with his productions, as well as the Selznick roster of players. In 1935, after cementing a reputation as the successor to Irving Thalberg, Selznick left MGM and moved his operation a half-mile east to the old Thomas Ince studio, using its white-columned colonial mansion as his company's emblem. Backed financially by Jock Whitney, Selznick International's films—most notably *Gone With the Wind*—exhibit an opulent, stylish look. Like Goldwyn, David Selznick was a man of exquisite taste, extravagant both in his personal life and in his craft. He was a pioneer in the use of color and became an early enthusiast for dramatic uses of sound. A tireless worker, he chaired endless production conferences and dictated memos that grew longer as the years went by.

Selznick's father, Lewis J. Selznick, had been a movie mogul during the silent era. A Russian Jew, the elder Selznick immigrated to the United States as a youth and became a jeweler in Pittsburgh, where David was born in 1902. The family moved to New York just as movies were gaining respectability, and Lewis soon went to work for Universal. In 1914 he joined the World Film Corporation, then formed Selznick Picture Corporation, a strong competitor to Adolph Zukor's Famous Players. Selznick derided his competition and flaunted his own success. "Selznick Pictures Make Happy Hours" became the slogan he used to entice audiences, while Zukor vowed to crush him.

David O. Selznick spent his boyhood in his father's offices, absorbing the excitement of the movie business. At eighteen he was running the publicity department of Selznick Pictures. But in 1923 the Selznick company was bankrupted by Lewis J.'s inordinate spending and inability to secure adequate distribution for his pictures on the larger circuits. To David Selznick and his older brother Myron, it seemed the industry had ousted their father, perhaps even set out to maneuver him into bankruptcy. Convinced that their father had been the victim of a conspiracy, David and Myron Selznick swore vengeance. Myron would become one of Hollywood's most powerful agents, arrogantly forcing studio heads to pay exorbitant fees for the talent he represented and savoring their discomfort, while David pledged that the Selznick name would one day return to the world's movie screens.

David Selznick arrived in California in 1926, beginning as an assistant story editor at Metro-Goldwyn-Mayer, where he learned to tell stories in vivid images. In 1928 he was offered a position at Paramount as B. P. Schulberg's assistant and spent the next three years shepherding that studio through the sound revolution. Restless, Selznick left Paramount in 1931 to head production at RKO. There he kept costs low and attracted major talent to the studio, creating a system whereby independent producers received little interference from RKO's front office. But Selznick dreamed of heading his own production unit, and in 1933 Louis B. Mayer granted his wish at MGM.

Having been a distributor for Lewis J.'s Select Pictures in Boston, Mayer had his own reasons for disliking the Selznick name. Their association had ended in an angry name-calling session, yet over Mayer's objections his daughter Irene married David in 1928. With his father-in-law's support David Selznick produced sophisticated pictures at Metro, bringing to the screen many of the literary classics David had loved as a boy. His glossy, romantic style seemed what Depression audiences wanted, and his standing within the industry rose accordingly. David and Irene Selznick became Hollywood's royal couple, entertaining lavishly at their Santa Monica home, planning parties with the same style with which David created his films.

Like all the Hollywood moguls, Selznick thrived on activity. Benzedrine helped him stay alert and contributed to his superhuman energy, allowing him to work until one or two in the morning. He rarely came into the studio before noon, but expected his staff to report to their offices promptly at eight o'clock, even when they had worked with him late the night before. "David Selznick was a high-powered person indeed," recalled Margaret Tallichet Wyler, who was under contract to Selznick as an actress before she married William Wyler. "To me he was a godlike figure." Many were terrified of him. Director George Cukor, whom Selznick fired during the making of *Gone With the Wind*, held no grudges: "David was full of beans and full of enthusiasm. He was a marvelous, full-blooded, charming man — a tough businessman and talented."

Selznick developed an obsession that *Gone With the Wind* should not be his epitaph and determined to prove he could produce something even greater. Unfortunately he selected small stories that could not bear the weight of his mammoth productions. *Duel in the Sun* (1946), for instance, based on a western novel by Niven Busch, was certainly no epic. Selznick was determined to make it the biggest western ever filmed, with an all-star cast and gigantic sets. "When David got through, the cantina was almost like an Oriental bazaar," observed Gregory Peck, who starred in the film, "with Tilly Losch dancing in the middle of the bar. It was all very exotic." During the making of *The Paradine Case* (1948) the actors were handed pages to memorize on the morning scenes were to be shot, so that there was no time for them to plan performances. "David got to be a fanatic about doing everything, even writing the dialogue," Peck said. "He had an obsession that every Selznick picture had to be an epic, and eventually that ruined him."

Epics were out of Republic's league, but fortunately Herbert J. Yates, who ruled the studio, was comfortable with smaller pictures that consistently made money. A testy, callous businessman from New York, Yates knew little about filmmaking and was seldom directly involved with production. He would occasionally visit a set, but most of his time was spent on budgets and executive decision-making. Turning out a steady product, Yates relied on the staples—serials, westerns, adventure pictures, inexpensive musicals, and comedies; Republic rarely produced more than two or three A films a year.

Yates could be charming—jovial and generous. But once he sat behind his desk, he was a tough businessman, driving hard bargains. Yates barked and spit a lot, missing the spittoon more often than not. "He couldn't hear," contract actress Catherine McLeod remembered, "so you always had to sit on his left side. Mr. Yates admired sex and I wasn't sexy, so that bothered him where his pictures were concerned. I either had to ride a horse or be sexy."

After Yates fell in love with ice skater Vera Hruba Ralston, she was given preferential treatment at the expense of nearly everyone else on the Republic lot, with her films receiving the biggest budgets whether they were successful or not. Yet even Ralston's pictures Yates supervised from a distance. Unless problems arose, he limited his association with studio personnel to Christmas parties, banquets, and company functions that everyone was expected to attend. Among Republic's workers he was referred to as "Papa Yates" because of the paternal conversations he occasionally held with contract players about the direction of their careers.

Like the other studio heads, Yates was a dictator, capable of making decisions instantly and fearlessly. If he lacked the artistic instincts of some, he equaled them as a taskmaster, demanding and getting the weekly product that kept the nation's smaller movie theaters filled.

All the Hollywood moguls were men of daring, willing to take chances, brilliant in their way. If they were not themselves men of taste, they nonetheless sensed what should appear on film and knew what the public would pay to see. Although they courted monied interests, most of the movie moguls had a deep mistrust of financiers within the business itself, convinced that the making of films should be guided by seasoned professionals who knew the industry firsthand. Theirs was a crude kind of showmanship, but their approach worked and provided the foundation for the American film industry during its formative years. The moguls supplied the vision that guided Hollywood to the crest of the world market.

3

Executives and Producers

Production head Irving Thalberg, entertainer Maurice Chevalier, and Metro chief executive Louis B. Mayer (left to right). Courtesy of the Academy of Motion Picture Arts and Sciences.

Although the studio heads ruled their domains like feudal barons, they depended on underlings to implement their policies and supervise individual productions. Studio executives usually came from the entertainment industry—vaudeville, circuses, nickelodeons, amusement parks, and theatrical agencies—almost all accepting P. T. Barnum as their model for showmanship. Those who survived in the studio system were aggressive, hardworking, tough-minded, and blessed with sound instincts.

Reporting to each studio's head of production were supervisors, called producers, one of whom was in charge of every picture. The system of producers dominating the big studio era was invented at Metro-Goldwyn-Mayer shortly after the introduction of sound to insure sufficient product for the theaters. At its height Metro was known as a producer's studio. Although all reported to Louis B. Mayer, producers at MGM had a great deal of autonomy, which for the most powerful was virtually complete. They supplied the director with a finished script, assigned a budget and crew, helped select the cast, and enabled the director to shoot the film, consulting with him when necessary. As far as the director and company were concerned, the producer was the immediate boss on the project, and the best producers commanded respect, their involvement dependent in part on the strengths or weaknesses of the director.

"Producers had tremendous power at MGM," said Armand Deutsch, who himself was briefly a producer at the studio during the 1950s, "and they were structured very meticulously." The top group, producers like Pandro Berman, Sam Zimbalist, and Arthur Freed, received the best properties and major talent. Under them was another group, and under that, another. Deutsch was part of the third group, yet below him were the producers of B pictures.

The prince of production heads was Metro's Irving Thalberg, the model for F. Scott Fitzgerald's *The Last Tycoon*. Young, hypersensitive, full of ideas, Thalberg had entered the picture business as Carl Laemmle's secretary in New York. By age twenty he had become Universal's general manager, hailed as the boy wonder of Hollywood. Largely self-educated, Thalberg gave the impression of intellectual depth. A voracious reader, he persuaded Laemmle to film Victor Hugo's *The Hunchback of Notre Dame* (1923), which Thalberg envisioned as a spectacle rather than a horror picture, with legendary results. Soon he emerged as one of the greatest minds in the business.

In 1923, at age twenty-three, Thalberg left Universal, to Laemmle's dismay, to become vice president and production assistant of the Mayer Company. A year later, with a corporate merger, he became Metro-Goldwyn-Mayer's chief of production, content to let L. B. Mayer receive the major publicity, while he built the studio into the artistic jewel it became. Despite conflict from the beginning, for a decade Mayer and Thalberg remained the most successful partnership in Hollywood, developing a star system without equal. Impressed by his protege's taste and story sense, Mayer rarely interfered with script matters or casting decisions. Thalberg respected Mayer's powers as a negotiator and depended on him to execute policy decisions and deal with the New York office. As one observer put it, "Thalberg made the bullets and Mayer fired them." (B. Thomas, *Thalberg*)

Gentle, chivalrous, fun-loving, but shy, Thalberg gave the appearance either of aloofness or modesty. When Lillian Gish arrived on the lot, her mother handed him the keys to their trunk, thinking he was an office boy. "He looked all of eighteen," Gish recalled. "He was charming about Mother's giving him the keys; he said, 'Oh, that happens to me all the time.' I immediately liked him very much." Actress Joan Fontaine remembered that his eyes seemed to look right into you. "Whereas David Selznick came on very strong, rather like a social bull," Fontaine remarked, "Thalberg came on like a scientist."

He seemed to thrive on responsibility, quietly confident of his own abilities and judgment. His method was to shoot a picture quickly; then he'd examine the results and make the necessary adjustments. As publicist Walter Seltzer explained, "They hated the term 'retakes,' since that implied mistakes. So 'retakes' became 'added scenes.' But Thalberg was able to construct the films he supervised by a very sensible system of seeing what was lacking, considering what additional footage might accomplish, and then shooting it."

Whereas Mayer remained the manufacturer, Thalberg was a creative force. "Thalberg," said actor Jackie Cooper, "was the guy in the front room—as they say in the clothing business—and Mayer was the guy in the back room." Thalberg regularly visited sets and enjoyed watching rehearsals. He was unconventional and courageous, yet he knew how to handle people. Eventually jealousy, greed, and a thirst for power pulled the Mayer-Thalberg partnership apart. They had always been a paradoxical pair. Mayer was widely hated; Thalberg was genuinely loved. Mayer was older, robust, pugnacious; Thalberg was politically naive, frail, retiring. Mayer became convinced that his production head was out to lower his standing with the parent office in New York; Thalberg was certain that Mayer was thwarting his efforts artistically. Even after Mayer's protégé had been relieved of his duties as production head and assigned his own unit, tension existed between the two.

Thalberg and his wife, actress Norma Shearer, had been friends of writer Charles MacArthur long before MacArthur's wife, Helen Hayes, became an actress at MGM. The four of them went to Europe together and became close friends. "Thalberg was a man that I particularly loved," Hayes recalled years later, "and was great fun; we had a lot of laughs." But by then Thalberg and Mayer were locked in a struggle over the future of MGM. "The company's theory was to make cheaper pictures and charge less for them," Hayes explained. "Irving said, 'We'll make fewer pictures and better pictures, so that audiences cannot resist going to an MGM film.' Those were the two theories on how to handle the crisis presented by the Depression." The powers in charge opposed Thalberg's theory. "So they undercut him and they dumped him," Hayes continued. "We were abroad when it happened, and it was a terrible time. He was so ill anyway, and this was just the beginning of the end."

To outsiders the Metro collaboration appeared unchanged, but close associates knew better. Always delicate, Thalberg suffered a heart attack in 1932 and died four years later. During his tenure production at MGM was directly under his supervision. Lesser pictures Thalberg delegated to Harry Rapf, but during the silent era the two men oversaw everything. Later, with sound, a half-dozen producers, each responsible for three or four films a year, operated under Thalberg's guidance. All were attuned to his methods, and there seemed little outward resistance to his domination. He cast his associate producers with the same care he did Metro stars. Hunt Stromberg received the more erotic assignments, such as *Our Dancing Daughters* (1928) and *Red Dust* (1932). Lawrence Weingarten specialized in comedies, while Albert Lewin drew films of greater literary quality, and Harry Rapf supervised the tearjerkers and B pictures. For nearly twenty years, the producer was the dominant figure in the making of American films. Nowhere was the system more elaborately structured than at MGM, where it began under Irving Thalberg.

With Thalberg's exit, Louis B. Mayer decided to supervise the company's pictures himself, although he recognized his limitations and was realistic enough to know he could never accomplish what Thalberg had. Mayer knew he would have to depend heavily on his producers, establishing a unit system (each unit headed by an important producer) that lasted until his departure in 1948. Under Mayer were four assistants who constituted the core of MGM's decision makers. Edgar (Eddie) Mannix was the studio's general manager, Benjamin (Benny) Thau was in charge of labor relations, Sam Katz was a production specialist, and Al Lichtman was the distribution head.

Mannix joined Mayer in 1925 and stayed with the studio through the next thirty years. He had been a bouncer at the Palisades Amusements Park in Fort Lee, New Jersey, when it was owned by the Schenck brothers. An Irishman who smoked huge cigars, Mannix used foul language, had a battered face and raspy voice, and kept a shillelagh on his desk. Despite his gruff exterior, he had a generous disposition and earned a reputation for honest negotiations. Mannix was a watchdog for Nicholas Schenck, president of Loew's Incorporated, MGM's parent company. There was always tension between the New York office and Metro's production headquarters in Culver City, and few major decisions could be made in California without clearing them with Nick Schenck. Later Mannix became the company's chief negotiator with the various guilds. Anyone at MGM who got into trouble—a traffic violation or a sex offense—went to Mannix for help. Legally or illegally, he usually solved the problem.

Benny Thau was in charge of talent and casting. A lawyer by training, he negotiated the contracts with stars, directors, producers, and craft people. Thau's preference in actors was conservative, and his power was virtually beyond appeal. If an actor had production problems, he went to see Benny Thau rather than L. B. Mayer. "Everything went across Benny's desk," said director Edward Buzzell. Much of his time was spent settling differences. A gentleman, softspoken and dapper, Thau was universally respected. "When he shook his head, it was as good as a contract," Armand Deutsch said. "Everybody at Metro just accepted what Benny said as fact. He was an extremely uncommunicative fellow, and nodding his head was a major exercise for him." Yet Thau always seemed pleasant, seldom speaking above a whisper, occasionally bringing an actor a pound of butter from his farm. Later, when Dore Schary took over the studio, he found Thau's coolness bewildering. "I asked myself, 'When will Benny thaw?'" Schary wrote. "He never did." (Schary, *Heyday*)

Sam Katz, a former vaudeville manager, aided Thau with production and was directly in charge of making MGM musicals. "We are a rich company," Katz told screenwriter Felix Jackson. "If you give us one picture in two years, that's fine." Metro's musicals were the most lavish, finely crafted, and innovative in Hollywood history, parading the cream of the world's musical talent. During the 1940s and early 1950s it might take a producer eighteen months or more to complete a major assignment, but the results were usually outstanding.

Al Lichtman spent most of his time in the East, booking films into theaters, while Louis K. Sidney, another pivotal Metro executive, served as an aide to Mannix. A large, charming man, always impeccably dressed, Sidney officed between the suites of Eddie Mannix and Benny Thau in the Thalberg Building, MGM's administrative headquarters, known as the "Iron Lung." Metro's top administrative offices were on the third floor, connected by a network of intercoms, private telephones, and secret passages.

Under this team of executives were the producers who supervised Metro's pictures. The producers were powerful, drawing huge salaries, eating lunch in the executive dining room, and constituting a creative aristocracy. Most of them occupied offices on the second floor of the Thalberg Building, beneath Mayer, Mannix, and Thau. In this group were Edwin Knopf (brother of publisher Alfred A. Knopf), Sam Zimbalist, Larry Weingarten, Carey Wilson, Sidney Franklin, Pandro Berman, Arthur Freed, Joe Pasternak, and Jack Cummings, all key figures in the MGM hierarchy. Each selected his own material, developed it with writers of his choosing, chose his director and cast (the most important films commanding the choice directors and stars), supervised filming, and had the final say in editing and scoring the finished picture.

Metro boasted three major producers of musicals—Freed, Pasternak, and Cummings, each with his own style. Earlier at RKO, Pandro S. Berman had produced several of the Fred Astaire–Ginger Rogers movies, but he gave up musicals when he arrived at MGM in 1940, recognizing that the field was saturated there. "They needed me to do other things," the producer declared.

At Metro, Berman turned to dramas, comedies, and the classics—*The Three Musketeers* (1948), *Madame Bovary* (1949), and *Ivanhoe* (1952). He was involved with the construction of his films, and he collaborated closely with his writers. "We'd get writers who knew how to write plays or novels," Berman said, "[but] often they were lost when they got into the screen world unless somebody sat with them and constructed scenes for them. That was the producer's main job in my day." Louis Mayer used to say, "Give me a screenplay that I like, and you've done your job as producer." (Berman int., Steen) So long as the story was on paper, Mayer felt his assembly line of directors, actors, and technicians could accomplish the rest.

Occasionally Berman had to be tough with directors in order to meet schedules and stay within budgets. Yet he never considered himself a businessman or a promoter. Above all he was a working producer under the shelter of a major studio, which meant that his greatest contribution was to find the story, develop it into a screenplay, find a director, cast the picture, and make it, usually using studio personnel. At MGM the producer made or approved all decisions on pictures, both artistically and financially.

While making motion pictures was a collaborative effort, Pandro Berman felt that *somebody* had to put a personal stamp on every film. "It could be the writer," he maintained, "it could be the director, it could be the producer, it could even in some cases be the star if the star were of the stature of Doug Fairbanks, Sr. All pictures should contain some person's drive and some person's point of view." Berman thought his own stamp probably appeared on 50 percent of the 115 pictures he made, although there were many stories he wanted to film that studio policy or the New York office wouldn't permit, such as *From Here to Eternity* and *The Caine Mutiny*. "They wouldn't let me do *Picnic*," Berman lamented, "because Nick Schenck thought it was a dirty play. They wouldn't let me do *Blackboard Jungle* (1955); then I really fought. I went to New York and had meetings with Schenck and the head of the sales department. And I fought them. Finally we got permission to make the picture. We made it for only $900,000, and it grossed over $9 million."

Not all of MGM's pictures were big productions, although even its B movies had a glossier look than many of the A's turned out by other studios. Joe Cohn, who had been with Samuel Goldwyn earlier and was Metro's troubleshooter on production matters, supervised the *Andy Hardy* series and handled MGM's B movies. His office kept a chart noting the location of each Metro film in production—whether on one of the sound stages, on a nearby street in Culver City, or in the jungles of Africa. Cohn's department figured the costs; his office hummed with the chatter of estimators projecting expenses.

At more loosely structured Paramount, producers enjoyed greater independence but less power. Paramount lacked the dynamic one-man leadership provided by Darryl Zanuck or Harry Cohn, nor was there the well-defined hierarchy of Mayer's MGM. Paramount wasn't as political as Metro. Adolph Zukor, Paramount's founding force, spent most of his time in New York. In the early days Jesse Lasky, a kindly, decent man who was treated shabbily in Hollywood, was in charge of production. Lasky brought Maurice Chevalier, Pola Negri, and director Ernst Lubitsch from Europe, but in 1932, as Paramount felt the effects of the Depression, he was ousted from control and shortly became an independent producer. In 1928 B. P. Schulberg, writer Budd Schulberg's father, was named studio head—a dour, hardworking man, who tried to tighten Paramount's central administration. "I would come at night to do some work," Broadway director George Abbott recalled, "and Schulberg's light was always on. I learned then that his was a twenty-hour-a-day job." Schulberg left Paramount in 1932, along with Jesse Lasky.

Their replacement was Emanuel Cohen, who lasted only two years as the studio's production chief, but signed Bing Crosby and Mae West to long-term contracts. Cohen was followed in 1935 by the ill-suited Ernst Lubitsch, who stayed one year as production head, subsequently producing for Paramount the pictures he directed. Faced with bankruptcy, the studio reorganized in 1936; Barney Balaban was elected president, Y. Frank Freeman was placed in charge of production, and William LeBaron was chosen as Freeman's production assistant.

Balaban, formerly of the Balaban and Katz theater chain and Paramount's principal shareholder, was a quiet dynamo who until 1964 ran the Paramount empire from its New York headquarters. A staunch supporter of Adolph Zukor, Balaban insisted on his own man to head the West Coast operation, selecting for the job Y. Frank Freeman, a longtime exhibitor from Georgia who had previously been a partner in Coca-Cola. An affable Baptist, Freeman ran Paramount Studios effectively, but knew little about making movies and didn't pretend to. He stayed out of creative decisions, concentrating on contract negotiations and labor disputes. Freeman infrequently visited sets and seemed aloof to employees, yet he exerted a strong influence. In production matters he relied on his able assistant, William LeBaron. A literate, sophisticated man from the New York theater world, LeBaron had earlier headed production at RKO and at Paramount produced several films featuring Mae West and W. C. Fields. He proved an understanding executive, well liked by directors and stars. He worked closely with studio manager Henry Ginsberg, an ex-lawyer and former banker, who took care of Paramount's business affairs.

In 1941 LeBaron was replaced by B. G. (Buddy) DeSylva, an expansive man who knew talent and how to develop it. Formerly of the songwriting team of DeSylva, Henderson, and Brown ("The Birth of the Blues" and "The Best Things in Life Are Free"), he had produced such Broadway musicals as *Louisiana Purchase*, *Panama Hattie*, and *DuBarry Was a Lady*. A tough fighter, shrewd and fearless, DeSylva officed on the ground floor of Paramount's Tudor-style administration building, where he developed the profitable Crosby-Hope *Road* series and boosted the careers of Alan Ladd and Betty Hutton. In 1944 he suffered a heart attack and resigned, replaced by Ginsberg, Joseph Youngerman, and casting director William Meikeljohn.

So long as LeBaron or DeSylva turned out their quota of successful pictures, they had little interference from either the New York office or Y. Frank Freeman. An assortment of producers worked under them, ranging from the elegant Arthur Hornblow to pear-shaped Paul Jones, who specialized in comedy. None of them enjoyed the stature of the major MGM producers, although most were experienced craftsmen. As long as the budget seemed reasonable, and if the distribution arm of the company felt it could do something with the movie, it was easy at Paramount to get a property approved. "The producers' setup only worked if you were in harmony with the men who ran the studio," John Houseman observed. "If you weren't, it was bad business."

The Olympian figure at Paramount was Cecil B. DeMille, who controlled his own unit, independent of the Paramount structure. He used the studio's talent and facilities and distributed his films through Paramount channels, but otherwise DeMille's operation was autonomous. "It was like a little empire within an empire," actress Laraine Day remembered. DeMille had his own bungalow, chose his own projects, and was given top consideration by every department on the lot. He could be difficult to work with and played the role of movie potentate more fully than anyone else in Hollywood. While filming the first version of *The Ten Commandments* (1923), actors Theodore Roberts and James Neil stopped by to see DeMille in their biblical costumes. After sitting in his outer office for what seemed like an eternity, they sent word in that "Moses and Aaron are still waiting to see God." (Mayer, *Merely Colossal*) But DeMille was loyal to those who stayed with him, casting the same character actors time and again, even after they were beyond retirement age.

In 1944, after leaving Warner Bros., Hal Wallis enjoyed a similar independence on the lot, releasing his pictures under the Paramount emblem. Wallis built his own roster of contract players, had an uncanny eye for talent and story potential, and emerged as one of the foremost producers in the business. He preferred hard-hitting dramas like *The Strange Love of Martha Ivers* (1946) and *Sorry, Wrong Number* (1948), but also filmed such acclaimed Broadway plays as *Come Back, Little Sheba* (1952) and *The Rose Tattoo* (1955). Most of his films were stories with substance, and all were well crafted. Pragmatic in his approach, Wallis balanced a near-classic like *The Rainmaker* (1956) with a Martin and Lewis comedy or an Elvis Presley musical, preparing each carefully. "Wallis' productions were synchronized to perfection," said actress Lizabeth Scott. "He knew exactly what he wanted. He took cinema out of the trite world and made it more intellectual, but not ponderously so. He knew just where to stop to keep it entertaining."

Like the other studios, Paramount produced its quota of B pictures, first under Harold Hurley, then under Sol Siegel (who earlier had been at Republic), and after 1947 under William H. Pine and William C. Thomas.

Although Jack Warner ran the Warner Bros. studio with a firm, frugal hand, from 1928 until 1933 Darryl F. Zanuck was his executive producer and right-hand man, more involved with day-to-day production than his boss. Warner telephoned Zanuck every morning to get a rundown on the day's activities, but relied on his production chief to oversee scripts, supervise film editing, and make decisions on casting, wardrobe, and sets. During the Zanuck years Warner did not enter into the actual making of pictures. Under Zanuck a corps of "supervisors" handled details on specific productions and shepherded them to a final cut.

When Zanuck left the studio in 1933, Hal Wallis took over as Warner's executive producer. "My first decision as head of production," Wallis recalled, "was to place the emphasis in our pictures on contemporary subjects. . . . While I enjoyed the lavish escapism pictures made by Paramount and MGM, I wanted to tell the truth on the screen. Even a musical, I felt, could comment effectively on contemporary life." (Wallis and Higham, *Starmaker*) At that time the studio made fifty pictures a year. Wallis oversaw at least four pictures simultaneously, with four others in script preparation and four or more in editing.

His day normally began with story conferences, working with writers on scripts in progress and checking finished pages. He conferred with cameramen and scenic designers to establish the tone of a production, discussed costume sketches with designer Orry-Kelly, went through scores with composers Max Steiner or Erich Korngold, and reported any difficulties to Jack Warner. Eventually the studio installed projection equipment in Wallis's home so that he could watch rushes in the evening after dinner. If there were problem scenes, he ordered them to be reshot. Gradually Wallis mastered all phases of picturemaking, demonstrating an unusual ability to analyze other workers' jobs.

Not everyone was fond of Wallis, who was a strict disciplinarian. Few were close to him; many exchanged angry words with him. He wrote Michael Curtiz, by Wallis's own admission his favorite director, heated memos on every aspect of production. During the shooting of *Captain Blood* (1935) Wallis's patience with Curtiz neared exhaustion: "I have talked to you about four thousand times, until I am blue in the face, about the wardrobe in this picture. . . . I distinctly remember telling you, I don't know how many times, that I did not want you to use lace collars or cuffs on Errol Flynn. What in the hell is the matter with you, and why do you insist on crossing me on everything that I ask you not to do? . . . I want the man to look like a pirate, not a molly-coddle." (Behlmer, *Inside Warner Bros.*)

Even skilled directors, Wallis argued, benefited from a strong producer who gave a second opinion. If the director accepted that input, the film became a true collaboration. Practical and cost-conscious throughout, Wallis was brilliant in editing and post-production, which largely determined the pace and design of his pictures. Often in editing a film a director was too involved to be objective, "in love with scenes and bits of business that a strong producer might cut for the good of the picture," observed Wallis. (Wallis int., McBride)

His productions were superior. "To be worthy of the name," Wallis said, "a producer must be a creator. . . . Producers are men who, for better or for worse, merge the diverse talents of several hundred people in the common objective of making a film. He must be a diplomat, organizer, strategist, planner, businessman, psychiatrist, juggler, midwife, and manager of egos. Above all, he is the decision maker." (Wallis and Higham, *Starmaker*)

At Warner Bros., as later for his own company at Paramount, Wallis read constantly, searching for stories that would lend themselves to screen adaptation or could serve as vehicles for specific stars. He often received galleys directly from publishers, outlines of stories, and forecasts of what publishing houses were planning to bring out. He saw most Broadway shows, seeking both properties and talent. "I look for projects I feel will make good entertainment," he said. "Many of my films contain messages, but I try to see that the message is delivered entertainingly." (Wallis int., McBride)

Toward the end of his tenure at Warner Bros., Wallis headed his own unit, concentrating on one picture at a time, among them *Yankee Doodle Dandy*, *Now, Voyager* (both 1942), and *Casablanca* (1943). In 1944 he had an altercation with Jack Warner that made it impossible for Wallis to remain at the studio. At the Academy Awards ceremony that year, when *Casablanca* was announced as the best picture, both Wallis and Warner rose to accept the Oscar as the film's producer. Wallis was furious and handed in his resignation.

At that point Warner decided to assume greater control over production, selecting his former casting director, Steve Trilling, as his executive assistant. Trilling had neither the grasp nor the authority of Zanuck or Wallis, having served primarily as Jack Warner's troubleshooter. Fiercely loyal, Trilling until 1951 was Warner's buffer from top executives and producers. He was a quiet, polite man, not demonstrably talented, making the decisions Warner was too busy to make or preferred to avoid. "He was the hatchet man," dancer Gene Nelson said. Trilling performed many of Warner's unpleasant duties and took the criticism. "Steve was a bad influence for the company," director Irving Rapper claimed, "because he got the contract players in great parts and ruined stories. He was not for that job."

For years Bill Koenig was the studio manager at Warner Bros. He was noted for being economical, cutting costs wherever he could, yet operating efficiently. Later Tenny Wright, a gruff one-time professional prizefighter, headed the physical plant, assigning a unit manager to tally costs and keep up with props on each production.

If MGM became the Senate for producers, Warner Bros. was the House of Representatives, for the turnover was more rapid. An exception was Henry Blanke, who remained with Warners for nearly forty years, arriving in 1923 as Ernst Lubitsch's assistant. Ten years later he was named a supervisor and became associated with some of the top pictures on the lot—*Jezebel* (1938), *The Maltese Falcon* (1941), *The Treasure of Sierra Madre* (1948). Like other strong producers, Blanke's work began with the script, which he broke down sequence by sequence. He made a synopsis of every scene, then eliminated what he felt was unnecessary. He did no writing himself, but shaped the script with his writers by pruning the superfluous. Sometimes he initiated projects, sometimes they were assigned to him. Blanke accepted the fact that the big studio system depended on a certain number of mediocre pictures a year, and he was practical enough to supervise his share. Sometimes Blanke had seven to ten writers working on different assignments at one time, calling them together around four o'clock each afternoon to discuss problems.

Once a script was finished, it went to the budget department. If the cost estimate ran too high, Blanke and his writers either cut something, or the producer figured a way to scrimp on details during production. With the director he selected a cast, although in some cases actors were chosen before the director was. After shooting began the producer visited the set when called, smoothing over any friction. Blanke considered himself the dominant force on his films: "It's like a conductor," he explained. "A conductor of an orchestra must know every instrument, not that he has to play it, but he has to know what it represents. The conductor must know what the music should sound like. And it's the same when you come to shoot a picture; the producer must always have the entire thing in mind." (Blanke int., AFI)

Blanke started at Warners for $50 a week and vaulted his way to $5,500 a week, working for a salary, not a percentage. "I never was interested in money," he claimed, "only trying to make a good picture. If it's good and a box office hit, then it's automatic, you get more salary." As was customary within the big studio system, the director had the privilege of a first cut, but the producer reserved the right to make final editing decisions. Above all, a producer must have good taste, Blanke believed. Strongly influenced by Lubitsch, Blanke always tried to devise something different. "I did the first biographies, *Zola* and *Pasteur*," he said. "I tried like [Lytton] Strachey to debunk biographies, to show the man with some weaknesses, as you and I have. That was something new at the time." (Blanke int., AFI)

Jerry Wald, another top producer at Warner Bros. during the 1940s, became a prototype of his profession, a model for the driven, opportunistic producer in the Budd Schulberg novel *What Makes Sammy Run?* A skillful maneuverer, Wald appeared to be a gentle personality, but in reality he seldom let anyone stand in his way. "He was a supreme egotist," screenwriter Edmund North recalled, "and would steal anything that he had ever seen or heard about to put into a picture, bombarding his writers with ideas he had just seen the night before."

Like Hal Wallis, Wald read constantly, looking for suitable scripts. He subscribed to seventy-three publications, read synopses of novels and plays, scanned perhaps a dozen books a week, but spent most of his time reading magazines and newspapers. He came to Warner Bros. as a screenwriter when he was twenty years old and was elevated to producer in 1941 at age thirty, after writing more than two dozen successful films. By 1948 he was producing four to twelve pictures a year. Among the best-known of his films are *Key Largo, Johnny Belinda, The Adventures of Don Juan* (all 1948), and *Flamingo Road* (1949). Wald compared the studio to a newspaper. "Warner is the publisher," he said, "who determines studio policy and delegates money and authority. Steve Trilling is the managing editor, and I am in the position of city editor, supervising a staff of writers and artists, originating ideas, and handing out assignments." (Goodman, "Hollywood Producer") Those who wrote, directed, and acted in films were so close to their work that they often lost perspective. Wald felt his contribution was sympathetic, informed, and comparatively objective advice. His job was to ensure that the picture made money and at the same time entertained and perhaps even informed a potential audience of seventy million moviegoers.

As at MGM, there was much nepotism around the Warners lot during the Golden Era. Milton Sperling, another Warner Bros. producer—*Cloak and Dagger* (1946), *South of St. Louis* (1949), *Distant Drums* (1951)—was Harry Warner's son-in-law. Sperling had been secretary to both Darryl Zanuck and Hal Wallis. Mervyn LeRoy, who directed at Warners but produced *The Wizard of Oz* (1939) at Metro, was married to Harry's daughter Doris. William Orr, who later headed Warner Bros.' television production, after working at the studio for years, was Jack Warner's step son-in-law.

At its height Warner Bros.' B unit turned out twenty-five to thirty pictures a year, concocted to fill the bottom half of a double bill. Under the supervision of Bryan "Brynie" Foy, one of vaudeville's Seven Little Foys, the B division produced approximately half of the studio's yearly product. Costs were kept low, each picture being completed for $150,000 or less. B movies contained many inserts of newspaper headlines, instead of action, as well as medium shots and close-ups, allowing sets to be minimal. Brynie Foy, who became known as "Keeper of the B's," had a knack for spotting script ideas from newspaper stories. "A news event that happened today," publicist Bill Hendricks claimed, "Bryan Foy would have made into a picture three or four weeks later. Some of them were surprisingly successful."

The same stories were used repeatedly. Foy kept a stack of old scripts on his desk. Whenever he needed a new idea, he'd pull one off the bottom of the pile and hand it to his writers to rework, a practice known as "switches." A western of a few years before might be "switched" to a South Sea island picture, or a successful gangster movie turned into a B western. When Foy found himself in need of a story for the studio's Dead End Kids, he came up with a picture called *Crime School* (1938), consisting of half of *The Mayor of Hell* (1933), in which James Cagney had starred, and half of *San Quentin* (1937), which featured Humphrey Bogart, Ann Sheridan, and Pat O'Brien. *Crime School* was a box office hit, although it ran four or five days over production schedule. "In those days that was a criminal offense," writer Vincent Sherman said, "to be four or five days over budget on a B picture."

Unless a unit ran over budget, Jack Warner paid little attention to B movies. Often he didn't even watch the rushes. Many of the studio's writers and directors got their start in the B bracket, then worked their way up to nervous A's and eventually to top productions. It was a businesslike arrangement. When Bryan Foy left in the mid-1940s to become a producer at 20th Century-Fox, William Jacobs took over B pictures at Warner Bros.

At Fox, before the merger with 20th Century, Winfield Sheehan (husband of opera singer Maria Jeritza) served as production chief, running the studio like a general. The gate Sheehan entered was on the west side of the lot. As he passed, guards would stand at attention and salute. When Zanuck took charge, Sheehan was removed. He subsequently made a number of independent pictures.

From the time 20th Century-Fox was formed, Darryl Zanuck dominated the making of all the studio's A movies. There were important producers on the lot—William Perlberg, Lamar Trotti, Sam Engel, George Jessel, among others—but Zanuck assumed the major role in working with writers, casting, evaluating rushes, and supervising final editing, even deciding when a film was ready to preview. Joseph Schenck, as chairman of the board, was Zanuck's superior, but Schenck remained largely a father figure and didn't involve himself in production. Spyros Skouras, a one-time Greek shepherd boy, became the company's president in 1942, operating from New York, and although there was growing tension between the two, Zanuck ruled until his departure from the studio in 1956.

Under Zanuck was William Goetz, Louis B. Mayer's son-in-law, who as vice president of 20th Century-Fox became a vital force. When Zanuck went into the army during World War II to make training films, Goetz took over as the studio's production head. Although he did a good job, Zanuck felt he had seized more power than was necessary; he was furious that Goetz had moved into his private office and ordered it repainted. Shortly after Zanuck's return in 1943, Goetz left the studio to form International Pictures.

Lew Schreiber, another of Zanuck's executives, was a combination casting director, studio manager, and problem solver. Schreiber could be peppery or cold, yet had an endearing quality about him. He kowtowed to Zanuck, but knew his business. "I had relatively little contact with Schreiber," contract player Vanessa Brown remembered. "We only went in there [Schreiber's office] to get bawled out for something. Yet he really liked the players; I guess that's why we forgave him."

Sol Wurtzel, the one executive Zanuck retained from Fox's regime, supervised B pictures, at less than half the cost of Zanuck's A films. A boorish, sour person, Wurtzel had started as William Fox's secretary and general assistant and worked his way up to become executive producer during the 1920s. After the merger Wurtzel's operation was separate from Zanuck's. He had his own staff, his own projection rooms, his own cutting facilities, and even his own group of producers, among them Bob Lippert, who turned out seventy-minute pictures for $90,000 or less.

At RKO, studio heads toppled every year or two, an unsettling situation that made work difficult, since company policy also kept changing. Generally the new head fired most of the administrative personnel of his predecessor, preferring to hire his own, so that the environment at RKO bordered on chaos. William LeBaron, the studio's first production chief, had been a supervisor for Famous Players-Lasky before taking charge of that company's Long Island studios in Astoria. In 1927 he became production head of Film Booking Offices and two years later assumed the same post with FBO's successor, RKO. Although LeBaron had considerable silent film experience, too many of his early sound productions failed to please either the public or the critics. *Cimarron* (1931) won the studio enormous prestige, but lost $565,000. LeBaron found grinding out a picture or more a week difficult and came under increasing pressure from David Sarnoff, the corporation president. In 1931 Sarnoff replaced him with David O. Selznick, then an arrogant young genius, who immediately voiced his contempt for formula pictures and began to upgrade RKO's reputation.

Selznick strengthened the studio's directing staff by bringing in George Cukor from Paramount and employed such writing talent as Gene Fowler, Ben Hecht, and Dudley Nichols. A man of enormous energy, Selznick possessed imagination and a dedication to quality. Selznick was ambitious and eager to take part in every step of production. Sometimes he rewrote scripts himself, not always with beneficial results.

The economic storms that rocked the nation in the 1930s struck RKO during Selznick's regime, so that within a year his policies came under attack. The New York office questioned his judgment, the criticism becoming harsher as the company's financial situation worsened. In 1932 RKO suffered losses of over $10 million, with the result that Merlin Aylesworth, the corporation's current president, assumed final authority for approving all scripts and budgets. Early in 1933 Selznick resigned in a fury and was replaced by Merian C. Cooper, whom Selznick had brought to the studio as a producer.

Cooper favored action-adventure pictures, winning lasting acclaim for *King Kong* (1933). He worked with enthusiasm, turning films out quickly, cheaply, and for the most part competently. In September 1933, Cooper suffered a mild heart attack and spent a few months recuperating. During that time Pandro S. Berman, Cooper's principal assistant, filled in for him, overseeing production at RKO until his boss returned in December. Two months later, with his health again failing, Cooper was forced to take another extended leave, throwing the studio into turmoil. He resigned in 1934, later resuming work as an independent producer. He was replaced as RKO's production chief by Pandro Berman.

Intelligent, a keen judge of talent, and knowledgeable about the filmmaking process, Berman proved himself an excellent studio head. He guided the Fred Astaire–Ginger Rogers musicals, acquiring for them the best songwriters in the business—Irving Berlin, Jerome Kern, George Gershwin, and Cole Porter. He hired good directors—George Stevens, Mark Sandrich, Gregory LaCava, Leo McCarey—and gave them the freedom to do quality work. "He only wanted the best," choreographer Hermes Pan remembered, "and he spared nothing."

Berman, however, was never comfortable running the studio, preferring to concentrate on his own pictures. In mid-1934 he returned to the producer ranks, surrendering overall production responsibility to B. B. Kahane, the company's president. Once Kahane had approved a project, he left creative decisions to the team in charge. When another administrative shake-up found Kahane replaced as president by Leo Spitz, Spitz appointed Sam Briskin, a veteran from Columbia Pictures, as RKO's studio head. From an artistic standpoint Briskin's two years at the helm constituted a step backward; although RKO released more films than ever before, most of them were second-rate comedies and melodramas. In 1937, when Briskin returned to Columbia, Spitz asked Pandro Berman to take charge of production at RKO once again, which he reluctantly did. Two years later, Berman left the studio over a dispute with the new corporate president, George Schaefer. "RKO was an amazing place," Berman said. "I was there for seventeen years, and I'm sure I worked for seventeen different administrations. The atmosphere there was bewildering."

When Schaefer became president, he resolved to convert RKO into a prestige operation, emphasizing quality productions based on acclaimed literary and theatrical properties. Schaefer's goal was to hire promising talent and provide an environment conducive to making superior films. Formerly general manager of United Artists, Schaefer at first had the support of the Rockefellers, who owned considerable RKO stock, but he caused alarm in 1941 by replacing Harry Edington as the studio's production head with Joseph Breen, the industry's chief censor. Since Breen had never produced a picture and had no experience running a studio, his appointment was questionable. Schaefer's insistence on quality did entice the brilliant, young Orson Welles and his Mercury Theatre group to the studio, where *Citizen Kane* was made in 1940, but Schaefer's policies nearly bankrupted the company. He resigned in 1942, a few months after RKO's board directed that Charles Koerner replace Breen as production boss.

Earlier Koerner had been in charge of exhibition for RKO, but his instincts for gauging public taste soon proved a major asset. Under Koerner the studio abandoned its emphasis on prestige films in favor of smaller-budget pictures designed to make money. By then the studio's reputation for rapid turnover in administration had become infamous in Hollywood. According to legend, when Koerner visited Edgar Bergen and Charlie McCarthy on the set of *Here We Go Again* (1942), Bergen's impudent dummy remarked, "Hello, Mr. Koerner. I'm here for six weeks. How long are you here for?" (B. Lasky, *RKO*)

But Koerner was well-liked by studio employees, and his program of detective yarns, horror pictures, war stories, comedies, topical movies, and musicals soon made the company prosperous. Once RKO was restored financially, Koerner began to approve projects with more limited box-office appeal, and RKO became one of the best-run studios in the business. "Charlie Koerner really pulled that studio out of receivership and made it one of the top," director Edward Dmytryk said. "By the time he died it was probably third in the industry."

Koerner's death from leukemia in 1946 forced N. Peter Rathvon, the new corporate president, to take over production for a year. In 1947 Rathvon hired writer-producer Dore Schary as RKO's vice president in charge of production. Schary had earned a reputation as a realistic picturemaker, and at RKO he demonstrated his philosophy that film should be a means for social commentary. But in 1948 billionaire Howard Hughes bought the studio, as well as RKO's theater chain, for a reported $9 million. Schary and Hughes clashed, with Schary returning to MGM, where he had previously been a producer. Hughes ordered all production at RKO temporarily stopped. During the summer of 1948 Peter Rathvon, still the company's president, was fired, along with seven hundred other employees. The chaos under Hughes proved fatal, bringing to a close RKO's days as a major movie studio. "Howard Hughes . . . abused the studio like the women in his life," Betty Lasky observed.

The irony is that RKO's extraordinarily unstable history was partly responsible for the studio's most lasting strength. Since it was the weakest of Hollywood's major studios, RKO welcomed individualistic filmmakers and allowed them the freedom to express themselves. Although the studio had a staff of contract producers (Adrian Scott, Val Lewton), it also hosted a melange of independents, granting them the artistic latitude they demanded. Women, too (notably Joan Harrison and Harriet Parsons) were given opportunities to produce at RKO, unusual during the big studio era.

Later Universal became a haven for independent production, although for years that studio, too, was plagued by poor administration. Carl Laemmle, Jr., tried to lift Universal to major status during the 1920s, encouraging prestige films, but under his leadership the studio nearly went bankrupt. Eventually Charles Rogers was made president in an effort to rescue the studio's faltering finances, but Rogers knew nothing about making movies and was replaced by Nate Blumberg. A theater man, based in New York, Blumberg could sell pictures, but understood little about story. His producers were inclined to let their directors have a free hand, so long as they turned out a successful product.

In the mid-1930s Joseph Pasternak emerged as the most powerful producer on the Universal lot, since his musicals featuring Deanna Durbin had saved the company from financial ruin. "Pasternak had a great sense of what people wanted," producer Felix Jackson recalled. "Sometimes he came close to being a genius." An uneducated man, Pasternak had a reliable instinct and was always decisive. "I learned one thing from him which I used in my own career later on," Jackson maintained: "It is more important to make a decision than to make the right decision. A wrong decision you can reverse, but if you don't make any decision, everybody is up in the air."

During the early 1940s Universal continued to struggle as a minor studio, producing a series of budget films. Around 1943 Edward Dmytryk observed at Universal the kind of behavior the public generally associates with Hollywood: "I found producers on the make for actresses," said Dmytryk. "I'd never run into that before. Surprisingly little of that goes on in Hollywood, or did in those days. Universal was the first place where I saw it."

A major organizational change occurred in 1946 when Universal merged with International Pictures to become Universal-International, with William Goetz, formerly second in command to Darryl Zanuck at 20th Century-Fox, the head of production. Goetz wanted to make fine films, yet inherited Universal's tradition of jungle pictures, cheap comedies, horror films, and inexpensive musicals. The merger resulted in a strange dichotomy—prestige films like *Another Part of the Forest* (1948) along with westerns, Abbott and Costello comedies, adventure pictures, and popular programmers. Top producers at Universal-International included Leonard Goldstein, Aaron Rosenberg, Robert Arthur, and Ross Hunter, as well as such independents as Walter Wanger and Jerry Bresler.

At Columbia, studio boss Harry Cohn supervised production, while Sam Briskin served as studio manager, in charge of signing contracts and generally executing Cohn's orders. Briskin, according to director John Brahm, "was a bastard, a real son-of-a-bitch. Cohn wasn't that; Cohn was a *charming* bastard. But Sam Briskin was a horrible guy." (Brahm int., AFI) Ben Kahane, on the other hand, another of Harry Cohn's subordinates, was a kindly man, who acted as confidant to actors and directors in conflicts with their boss and represented Columbia in dealing with the various craft unions.

In 1942 Sidney Buchman was named vice president of Columbia and Cohn's assistant production chief. Later in the decade Cohn made writer Virginia Van Upp executive producer, sharing a common reception room with her. Van Upp was an unattractive woman, with buck teeth, in her mid-fifties. "She had a triangle figure," actor Cornel Wilde remembered, "upside down, no shoulders and a wide butt. She wore thick glasses, so you often couldn't see what she was thinking or feeling. She had her hair dyed in the worst blazing henna I've ever seen." Yet Van Upp proved an able administrator. Still later Lillian Burns became Cohn's executive assistant, going over scripts, buying plays, and working closely with producers and directors.

Irving Briskin headed Columbia's B pictures, an intelligent though crude man who hammered at those under him until they often sank to his level. Producers on the lot ranged from Sam Katzman, who specialized in cheap programmers, to producer-director Frank Capra, who was responsible for the studio's biggest hits during the 1930s and early 1940s. In 1951 Stanley Kramer brought his independent unit to Columbia, producing excellent films. Kramer considered Cohn a ruffian and seldom spoke to him. Their agreement stipulated that Kramer couldn't make a picture that cost over $2 million or ran more than two hours in length. If he did, the studio had a right to take over.

Trem Carr was Republic's initial production chief, but when Carr left the studio in less than a year, his position was taken by Nat Levine. Levine established the serial format, which helped make Republic a successful action studio. In conjunction with executive producer Sol Siegel, Levine cast Gene Autry in a series of singing cowboy westerns that proved immensely popular at box offices, particularly in small towns. By the time Roy Rogers replaced Autry, Armand Schaeffer was production head for most of Republic's B westerns, rural comedies, and inexpensive country musicals.

Monogram operated mainly on a freelance basis with writers and directors, but did make contractual arrangements with producers. A different producer would be in charge of filming Monogram's *Joe Palooka*, *Bowery Boys*, and *Charlie Chan* series, each responsible for seeing that a certain number of pictures were made every year. Each producer would secure writers and directors to put his program together, and at the same time would look for subjects that could become the top half of a double bill. Producers at Monogram noted what the major studios were turning out and tried to copy their successes. Occasionally they succeeded, producing a tentative A that made it into the first-run theaters. Steve Broidy, who had been administrative assistant at Monogram, became the company's president in 1945; six years later Broidy put Walter Mirisch in charge of production for the studio. Mirisch had produced *Bomba, the Jungle Boy* (1949) and subsequently upgraded the quality of his Monogram films. Later he would become one of Hollywood's most esteemed independent producers.

Often a conflict arose between the producers and executives on the one hand and the actors, director, and crew on the other. "The boys in the trenches down on the set wanted to make a fine picture," actor Gregory Peck insisted. "Very often the producer or the boys up in the head office were less interested in the quality of an individual film, and of course they were less involved in the creative process. Their real interest lay in the financial welfare of the studio and the balance sheet at the end of the year. Anything that threatened their job meant a good deal more to them than the day-to-day work of trying to create a fine film."

If by personality, talent, or temperament the producer was less dominant than a powerful director, he might willingly accept a subordinate position, content to serve as a catalyst in putting together the necessary ingredients for a picture and then stand aside when the time came for actual filming. Some of Hollywood's most successful producers were organizers rather than creative forces. Others—Selznick, Goldwyn, and Wallis most notably—shaped their films as much as any of the great directors, and a strong director often felt restricted working for them. But regardless of the individual producer's role, throughout Hollywood's Golden Era the producer system was the foundation of the big studio's order of operation, assuring the product necessary to fill the theaters and achieve a profit.

4

Directors

Legendary director Cecil B. DeMille. Courtesy of the Academy of Motion Picture Arts and Sciences.

Film has often been called the director's medium, since a director puts his stamp on the finished product far more than in the live theater. The motion picture director blends various arts and techniques into the texture of a film. The director is a problem solver, and while this work can be immensely satisfying, it is also demanding. Directors in the big studio era came from varied backgrounds, but the successful ones had a visual sense and an ability to grasp the cinematic qualities of a screenplay. Some of the directors of silent films continued to think of dialogue as titles long after the arrival of sound. "Did he say that title right?" Al Green asked when he later directed screen musicals.

Although most written dialogue retained a literary flavor acceptable on the stage, in films it became essential for actors to speak realistically. Skillful directors simplified dialogue, telling as much of the story as possible with the camera, so that much of the director's time was spent with the cinematographer, who embellished or heightened the director's concept. Actors complained that this collaboration prevented them from receiving sufficient attention, but the great directors encouraged performers to take chances, to go beyond the standard interpretation. They instilled in them confidence that they would not be permitted to look foolish.

Besides conferring with writers on the script, the director also helped choose the cast and assemble the crew. He, or in rare cases she, approved sets and costumes and helped with the selection of locations. To save time for working with actors on scenes, the director tried to minimize moving a company and the equipment involved from one location to another or even from one set to another on the same stage, which meant shooting scenes out of sequence. On big productions involving masses of extras, an assistant director normally supervised rehearsal, although the first assistant was mainly a foreman, who made certain everything the director needed was at hand. Sometimes a dialogue coach would be called in to work with actors on intimate scenes, freeing the director to devote more time to logistics. For historical films or those with foreign locations, extensive research sometimes demanded much of the director's attention before filming began.

Each of the major studios maintained a team of directors under contract, as they did actors, producers, and writers. If the director was well established, he rarely suffered interference from studio executives; producers visited the set only occasionally, even though there were constant meetings and private discussions. Top directors could command top designers, cameramen, editors, and the best of the studio's company of actors, although the producer and distributors would normally insist on recognizable names in the leading roles. For lesser directors the control of management was more frequently exercised, and even a powerful actor could present obstacles. With independent producers like Goldwyn, Selznick, and Wallis the domination of directors might result in serious clashes and perhaps dismissal. George Cukor had done extensive preparation for *Gone With the Wind* (1939), made hundreds of tests, and shot for months before David O. Selznick replaced him with Victor Fleming. On his remake of *A Farewell to Arms* (1957), Selznick fired John Huston the day before shooting started. For a week or more the cast worked with a second unit director; then Charles Vidor was brought in, although it was evident that the producer intended to remain in command. "Making *A Farewell to Arms* wasn't a pleasant experience," remembered Rock Hudson, who starred in the film opposite Selznick's wife, Jennifer Jones, "and there were lots of arguments. David exhibited an air of confusion. There was nothing clear and set." Rouben Mamoulian worked on *Porgy and Bess* (1959) for more than eight months, then had a fight with Sam Goldwyn and quit. The director considered the picture ruined.

Executives mindful of budgets often grew impatient with leisurely directors, insisting they stay on schedule and keep costs down. "Like all artists," said Eugene Zukor, son of the Paramount mogul, "directors will work on the biggest canvas you can give them." Consequently compromises had to be made. Bryan Foy, who knew how to stay within a modest budget, having produced scores of inexpensive pictures, once advised, "Keep it moving. You can't see the teeth on a buzzsaw." (Bare, *Film Director*)

During the Golden Era most of the great directors were keen individualists, their personalities reflected in the films they made. For all directors the screenplay was vital, but with secondary directors there were fewer distinctive touches in the finished film. "If you've got a good script, let the office boys direct it," veteran director Andrew Stone remarked. "With good actors and a good script, anybody can direct." For the creative mind the experience was exhilarating, the possibilities for nuances endless. As Orson Welles commented upon his arrival in Hollywood, "This is the greatest electric train any boy ever had to play with." (Boller and Davis, *Hollywood Anecdotes*)

Cecil B. DeMille emerged as the prototype of the Hollywood director, a showman with a flair for publicity who dressed in riding breeches and boots and gave the public the glamour it craved. Steeped in silent filmmaking, DeMille was convinced that visual images carry more impact than dialogue. "The camera has no ears," he used to say. "If you want to say it, get it on the screen." (Leisen int., Chierichetti) Convinced that audiences wanted pretty women and handsome men rather than searing social commentary, DeMille flaunted as much sex in his movies as censorship would permit. His pictures were commercial, aimed at the widest possible distribution, and although critics deprecated his spectacles, the multitude flocked to see *The King of Kings* (1927), *Cleopatra* (1934), *Samson and Delilah* (1949), and both versions of *The Ten Commandments* (1923 and 1956).

DeMille was not adept at directing intimate scenes, preferring to focus on the broader sweep, and often turned his actors over to a dialogue coach. On the set he issued commands to his throng of extras through a megaphone. Most days he'd begin with, "Now I don't want any extras on my set. I want *actors*. I want every one of you to know what you're doing." (Wilcoxon int.) He could be cruel, bullying actors and crew members, yet he had a warmth that endeared him to Paramount regulars. Always he had a twinkle in his eye for a pretty girl.

DeMille researched every detail for historical authenticity. An artist sketched each scene in advance, and some scenes were even painted in oil. In preparing the remake of *The Ten Commandments*, the director asked choreographer LeRoy Prinz to stage a game that young women in ancient Egypt might play for the scene where Moses is found in the bulrushes. "That required research," Prinz said, "and I studied. I have more books on ancient games and ancient music. We found out that the Egyptians only had four tones of music instead of five like we do. So everything had to be conceived with a four-tone scale." DeMille liked to insert small scenes, what he called gem scenes, in with spectacle, like the old woman getting her clothing caught under the moving rock in *The Ten Commandments*, which had to be choreographed with extras and stand-ins. Big scenes, like the orgy at the foot of Mt. Sinai, might take weeks or even months of planning. "DeMille always had to have an orgy in his pictures," Prinz remembered, "and I have staged more of them than any human being around. I was always dreaming up orgies. How many times can you squeeze grapes down a girl's throat upside down?"

Perhaps the greatest of all camera directors was John Ford, a master at capturing terrain through imaginative composition, particularly in Monument Valley, which he repeatedly photographed in his westerns. Ford seldom adhered to a script, retaining the concept in his head and writing lines as he went along. Having achieved success during the silent era, he continued to think in terms of pictures rather than dialogue. "The more you can tell with a camera the better it is," he frequently said. "A lot of words just waste film." His method was to gather his cast around a long table. He would come in, lift the patch off his bad eye, thumb to a page in the script, and ask, "Now what would you say here?" The actors would have to improvise. "He hated actors," Shirley Jones concluded after working with Ford on *Two Rode Together* (1961). "If he could have made a film without them, he would have loved to do that."

Gruff at best, Ford could be a tyrant during filming. On every picture he chose somebody to needle, often character actor Ward Bond. "The more he loved you, the tougher he'd be on the set," observed Harry Carey, Jr., a member of the John Ford stock company. "Here I take you out of eight-day westerns and put you in big movies," Ford fumed at John Wayne during the making of *The Man Who Shot Liberty Valance* (1962), "and you give me stupid suggestions."

Despite Ford's fearsome demeanor, there was camaraderie on his set. "Pappy" Ford, as regulars called him, loved pranks and after working hours liked people who drank. He kept Danny Borzage on the set to create a mood by playing "Bringing in the Sheaves" and "Red River Valley" on the accordion. Ford spent a great deal of time working with the cinematographer, chewing on a corner of his handkerchief when he was nervous or in serious thought. He swore, mumbled with a pipe between his teeth, and yelled impatiently. No one doubted that Ford was in charge. "You had to make up your mind in the first few days either to play it his way or go home," Harry Carey, Jr., maintained.

An astute technician, Ford became known as a one-take director, preferring freshness over precision. He reserved close-ups for emphasis, and unlike DeMille, drew rich performances from his actors, seeming more comfortable working with men than women. "He would incorporate tiny things that sounded like nothing when he told you," actor Pat O'Brien remarked. "Then when you went in and saw the dailies, something came off the screen that was revelatory." Ford used a lot of body language in directing and exaggeratedly acted out dialogue to show what he wanted.

He had entered the movie business as an assistant propman and eventually worked for all the major studios, earning four Academy Awards for best direction: *The Informer* (1935), *The Grapes of Wrath* (1940), *How Green Was My Valley* (1941), *The Quiet Man* (1952), plus two for World War II documentaries. Ford refused to allow a producer on his set. "Don't you have an office?" he'd ask if one appeared. At 20th Century-Fox he worked out an arrangement with Darryl Zanuck whereby he'd direct two pictures the studio chose in exchange for one of his own selection. Zanuck didn't like Ford, but respected his talent and gave him major assignments.

"John Ford was a unique and complex man," Harry Carey, Jr., said. "He wanted to be one of the guys, but never could be because he was the boss. He was intrigued with machoism, yet was a delicate, artistic man, tender and loving. You could see that his hands and eyes were so gentle. Yet somehow there was a part of him that John Wayne physically exemplified, something he always wanted to be. He wanted to be a two-fisted, brawling, heavy-drinking Irishman—to do what Wayne did and clean up a barroom all by himself, which Ford couldn't do. So he created that on the screen; that was part of his genius." When someone called Ford the greatest poet of the western saga, the director retorted, "I am not a poet, and I don't know what a western saga is. I would say that is horseshit. I'm just a hard-nosed, hardworking, run-of-the-mill director." (Ford int., Wagner)

He was widely imitated both in his heyday and after. Howard Hawks, another of Hollywood's great directors from the silent era, copied Ford whenever he could. "Anybody can stand still and talk," Hawks said, "and even in motion I've tried to make my dialogue go fast." (Hawks int., Schickel) Hawks believed in telling a story as simply as possible, modifying dialogue as he went along so that the personality of the actor became fused with the character. (Bogart and Bacall in *To Have and Have Not* and *The Big Sleep* serve as classic examples.) Hawks's persistent rewriting irritated producers, and he had an iron will. "The problem with Howard," said Hal Wallis, "was that he would come into the studio in the morning and just sit there and think. Nothing would happen for two or three hours. Then he would rewrite the day's dialogue and try to make it conversational, which he succeeded in doing. His characters were usually pretty down-to-earth, and the exchange of dialogue was realistic." Hawks didn't rehearse much, talking with actors in his laconic way, taking whatever time he needed. On *Red River* (1948), his classic western, Hawks worked on location in Arizona with a group of young actors with practically no experience. "If they didn't give him what he wanted," actress Coleen Gray remembered, "everybody would shut down, and he and the actor would go walking off in the desert. They'd be gone for half a day, just talking. They'd come back and shoot the scene, and it would be fine." But Hawks denied that he spent much time analyzing the story he was telling, and sometimes he improvised with no script at all. "We just made scenes that were fun to do," he claimed. "I think our job is to make entertainment. . . . I just imagine the way the story should be told, and I do it." (Hawks int., McBride)

Most of the old silent directors were tough characters who had learned their profession by trial and error. Allan Dwan had, for he launched his career as an independent at a time when Edison's Trust came onto his set with guns to disrupt filming. Dwan kept a pistol in a holster and reputedly ran off more than one Trust representative. Raoul Walsh, who played John Wilkes Booth in Griffith's *The Birth of a Nation* before becoming a director, was another from the old school. Walsh could be salty and profane, and had an earthy sense of humor. "Let's get the goddamned thing made!" he'd say. Walsh would read through a script and complain, "Dialogue, dialogue, dialogue," until he came to an exterior sequence he thought could be turned into engrossing pictures. "Now here we have a *scene*," he'd exclaim. (Peck int.) Sometimes, like Ford, he'd toss the script aside. Often when directing a sequence, he'd roll a cigarette, not seeming to watch at all. But he had a frame in mind and knew whether or not it added up to something worth watching. "He had almost a balletic sense of what should occupy that frame," Gregory Peck said, "whether people were in the right position, whether they made a composition, or whether they moved in an interesting way in relation to each other. It was truly picturemaking. Raoul was a visual storyteller with an enormous instinct for keeping the screen alive with movement and conflict."

Walsh was at his best as an action director. "He got more guts and drive and fire into a movie," actress Virginia Mayo remarked. "His films lived because he gave them an earthy quality." Walsh kept his cast and crew laughing. "I never met a director that had such a great sense of humor," Mayo said. He seldom seemed to listen to dialogue, unaware at times that an actor had forgotten or garbled a line. That could always be dubbed later. Subtlety was never his strength. Jack Warner once claimed that a tender love scene to Raoul Walsh was burning down a whorehouse. (Walsh int., Schickel) But Warner, to whom Walsh was under contract, seldom interfered once work began. Frequently the mogul would plead with Walsh to direct a film the studio needed to meet a specific deadline, a picture already sold to distributors but not made. A good company man, Walsh usually acceded, sometimes with the lure of a bonus for bringing the picture in under budget. Once when Warner Bros. was stuck for a release and had nothing coming up, Walsh suggested remaking *High Sierra* (1941) as a western, calling it *Colorado Territory* (1949). "All right," Warner barked. "Start tomorrow." (Sherman, *Directing the Film*)

During the big studio era directors were assigned daily schedules. For short A pictures thirty-two days would be allowed, for extra-long pictures forty-two days. "I would work sometimes until one o'clock in the morning to finish a picture down on what we called the New York street and go home," recalled Walsh. "At eight o'clock the next script was thrown on my lawn like the *Los Angeles Times*. I'd go to the studio and meet the people that were in it." Although he made movies that were personally satisfying, Walsh accepted the fact that he was working in a commercial medium. "Some scenes I put in for my own gratification," he said, "but I generally decided to play for the public, because that's what kept us alive." He made some hits, some near hits, and several failures. "It's like raising children," the director claimed. "Some go out and make good, and some don't." (Walsh int., Schickel)

William Wellman's approach to movie directing was even more visceral. A high school dropout, Wellman joined the Lafayette Escadrille during World War I and by 1927 was sufficiently established in Hollywood to have Paramount's now-classic aviation picture *Wings* assigned to him, which won the first Academy Award for best picture. Like his colleagues from the silent days, Wellman thought primarily in terms of images. If a writer handed him a speech with twenty words, he'd ask if it could be briefer. He constantly looked for shots that could take the place of dialogue. Known throughout the industry as "Wild Bill," Wellman so wound up his performers emotionally that they gave excellent performances. Actors admired him, even though he drove them relentlessly. When he directed *Battleground* (1949) at Metro-Goldwyn-Mayer, Wellman turned his cast into soldiers, putting them through a modified basic training program. On *The High and the Mighty* (1954) he used a different method with every member of the cast. Sometimes he yelled at people to get the performance he wanted, or he belittled them until they decided to prove him wrong. Other times he was gentle and understanding. "Somehow he knew how to approach each one of us and get what he wanted," actress Ann Doran said.

King Vidor's technique was more subtle. Brilliant in his judgment but not always explicit in his direction, Vidor viewed silent films as superior to sound. "Silent pictures were more poetic," he said, "and more of an art. They were not so literal." He felt silent filmmakers had discovered a technique of pantomime articulate in itself. Vidor planned his films carefully, yet left room for the extemporaneous, shaping his pictures so that they emotionally rose and fell like musical compositions. He called the rhythm he achieved "silent music." Influenced by D. W. Griffith, Vidor used a metronome and a bass drum while filming *The Big Parade* (1925) to keep forty soldiers moving to the tempo he set. *The Big Parade* helped establish Metro-Goldwyn-Mayer as a studio of quality and won King Vidor a major reputation, enabling him to make important films like *The Crowd* (1928), *Hallelujah* (1929), and *Our Daily Bread* (1934).

Influenced by the German Expressionists, Vidor experimented with the perambulator, boom shots, and forced perspective, filling the screen with subconscious visual metaphors that grew out of his own experience. He preferred to allow actors to discover their characterizations, talking with them as a psychiatrist would. He wanted spontaneity and felt cameramen, electricians, and technical crew received the greatest benefit from rehearsals. Even after sound Vidor remained primarily a visualist, despite the initial restrictions on movement caused by cameras in stationary boxes. "It took a long time to get back the flexibility and mobility," the director explained, "and then get back to leaving out words." Sound meant the script had to be more specific, including details that earlier had been generalized. The first scene Vidor directed with sound was a crap game in *Hallelujah*. "A bunch of blacks were shooting craps on a dock in the Mississippi River, and I had to decide whether they were shooting for twenty dollars or one dollar," the director said. "Before it was just a crap game; now it had to be pinned down."

Vidor found big studios helpful rather than restrictive, particularly Metro during the Thalberg era. Thalberg felt that the studio could afford an occasional experimental film, since it made fifty pictures a year and the deficit from a few could be covered by the rest. Later Vidor sometimes had to raise the money himself for less commercial films or put up his salary as backing. His most frustrating experience with a producer came during the making of *Duel in the Sun* (1947), when David Selznick persistently interfered. "I thought I was going to make an intense, intimate western," the director claimed. "Selznick was trying to make another *Gone With the Wind*." Eventually Vidor walked off the picture. "David's zeal got the best of him," recalled Gregory Peck, the film's star, "and King, with his pride, really had no choice." Vidor left *Duel in the Sun* when the picture was about 60 percent completed.

Filmmaking to King Vidor was a personal experience that could not be accomplished by a committee or with a domineering producer. Vidor was involved in the writing, laying out the film, shooting, and supervising the music and editing. "By the time you get ready to start moving actors around, you should have done about half your work," said Vidor. He estimated that the actual shooting of a picture took about 50 percent of his time; the rest was planning. Each filmmaker, he felt, had something unique to say, something growing out of his personal philosophy.

Most of the great Hollywood directors shared Vidor's insistence on full control. "One man, one film," was Frank Capra's motto, and Capra forced Harry Cohn to accept his dictum. Born in Sicily, Capra grew up poor, but fell in love with America while still a young boy. He began in the movies as a gagman for Hal Roach and Mack Sennett and always preferred comedy. "I don't believe in tragedies," said Capra. "I'm an optimist, and comedy to me is victory. Tragedy is failure." (Capra int., Schickel) He solidified his creative autonomy by sweeping the Academy Awards in 1934 with *It Happened One Night.*

A charming, quiet man, Capra gave his actors freedom. "Mean it" was all he'd normally tell them. Actors respected him, knowing he respected them in return. Capra never belittled performers; he helped them become their characters. "Don't act, just be that person," he'd say, showing infinite patience. He'd sit in a chair under the camera's lens, rehearsing as long as anybody wanted to rehearse. He'd listen and laugh, an attentive audience of one. "You worked for his response," actress Carolyn Jones remembered. Meticulous in his work, Capra showed no fear of Harry Cohn. If Cohn came on the set, Capra would talk to him but refused to work until the studio boss had left.

"The magic of these movie directors," said stage actress Helen Hayes, "was that they could do it right then and there, without testing it out on an audience." Once films acquired sound, Hollywood raided the New York theater to find directors who could help actors speak the "titles." Among those summoned from the stage was George Cukor, engaged by Paramount in 1929 as a dialogue director, a position created with the coming of sound. Cukor worked as the dialogue coach on Lewis Milestone's *All Quiet on the Western Front,* then was assigned by Paramount to co-direct *The Royal Family of Broadway* (1930) with Cyril Gardner, who had been a film editor and knew camera technique. Cukor later said he entered movies as a novice, but watched and learned. Gradually the studios adjusted to sound, while Cukor became a major director, never viewing himself as the ultimate creator of a film as many of the old silent filmmakers did. He considered his first obligation was to the script, which provided the director with a foundation from which to probe motivation of characters. Cukor initially knew little about the camera and never felt completely comfortable with it, taking for granted that cameramen knew their business.

A man of taste, Cukor was kind and gentle; he was considered a better director for women than men. He took pains with actors, calling them aside to whisper suggestions. "You've got to make a climate in which people can work," he said, "and you've got to know how to get the best out of actors." He felt that in films it was the director's job to keep the excitement going, the element the audience supplies in the theater. On a movie set, however, actors wait for hours while the crew shoots a scene and then stops to reload the camera. There's a letdown, and Cukor found it important to talk to actors.

"They don't have to listen to what you say," he acknowledged, "but you must keep their excitement going." Intelligent and witty, Cukor was meticulous about sets and costumes, sensitive to the decor of his films. "But the main thing is the story," he insisted. "If the story is good, the director is halfway there, and it's easy. The scene plays itself. If the story isn't right, you're beating a dead horse, even if you try every trick in the world. The text tells you where the camera should be, what the emphasis should be, and I don't think that's a question of judgment. It is the truth of the scene that guides you." Like the silent directors, Cukor realized that too much rehearsal could kill spontaneity.

Cukor found the old studios gave talent every chance to develop. Directors were assigned the best stories available, the best leading ladies or leading men, the best designers; they were surrounded by experts in every field. A director's prestige depended on how well he got on with executives and how talented the studio thought he was. "If you've made some flops," Cukor acknowledged, "then things get rather cool." He soon became recognized as a major director and often filmed versions of stage plays: *Camille* (1937), *The Philadelphia Story* (1940), *Gaslight* (1944), *Born Yesterday* (1950), *My Fair Lady* (1964). Coming from the theater, Cukor believed in adhering faithfully to the text, building on the foundation provided by a solid script. He could talk endlessly about character and motivation. "Cukor could motivate every word of every line of every sentence of every speech," actor Lew Ayres remembered.

Rouben Mamoulian, whose stage successes included O'Neill's *Marco Millions* and Gershwin's *Porgy and Bess*, was more experimental than George Cukor. "What fascinated me," Mamoulian said, "was the things we could do with the camera. If the camera can be used metaphorically and stylistically, I saw no reason why sound shouldn't be used the same way. It didn't have to be realistic." Studio heads were sometimes skeptical of Mamoulian's innovations, but his reputation was high enough that they gave in, letting him record Sylvia Sidney's audible thoughts for a close-up in *City Streets* (1931) and merge contrasting sounds on two strips of film in the lab for *Applause* (1929). On *Love Me Tonight* (1932) he used rhythmic dialogue and rhyming verse, stylizing the satire. For *Blood and Sand* (1941) he used the palette of the Spanish artists, approaching the film as a painting. To Mamoulian, color had to serve the drama. "Each color is an actor," he said, "making entrances and exits and conveying certain emotions. There isn't an inch of color in *Blood and Sand* that wasn't put there on purpose."

Some actors didn't grasp Mamoulian's concept. "I never could understand his so-called artistry," actress Lynn Bari confessed. Bari felt he treated her like part of the scenery. Sometimes the director was testy with stubborn executives who tried to hold him back. But he was amazed at the facilities at his command; nothing was impossible, for the studios had masterminds for everything. "A director has to be sensitive on one side, and on the other he has to be like the Rock of Gibralter," Mamoulian claimed. Irene Dunne was astonished at Mamoulian's attention to detail when she worked with him on *High, Wide and Handsome* (1937). "I can't imagine his approach ever happening today," the actress said, "with such terrifically expensive overhead." On *High, Wide and Handsome* the crew might wait an hour for a cloud formation to form the way the director wanted it.

Mamoulian realized the sensuous pleasure of movement in filmmaking: the movement of the camera, the movement of actors, the movement of film cutting—how one shot follows another. Life is movement, he insisted, and visual motion is more powerful than anything the mind perceives through sound or reading. "There's nothing as strong as imagery," Mamoulian said. "It's like translating a thing from one language to another."

Other directors from the theater were less inventive. Columbia brought Michael Gordon to Hollywood as a dialogue director, assigning him twenty films within the first year and a half. Gordon, a member of the Group Theatre in New York, arrived with an admittedly snobbish attitude toward movies, but soon found himself genuinely interested in the medium. He started out directing low-budget crime films, gradually learning about the camera, the sound track, and what went on in the cutting room. When he asked for a better picture, studio executives said, "We can't give you a good picture because you don't make enough money." He returned to Broadway, but was later allowed to direct prestigious films like *Another Part of the Forest* (1948) and *Cyrano de Bergerac* (1950), as well as the successful Doris Day–Rock Hudson comedy *Pillow Talk* (1959).

Musicals brought another group of directors from Broadway, several of whom became cinematically imaginative. Vincente Minnelli, who started as an art director but later directed the Jerome Kern show *Very Warm for May* on the stage, was invited to Hollywood by Metro producer Arthur Freed and given a year to observe before receiving his first picture assignment, *Cabin in the Sky* (1943). Minnelli soon became one of MGM's specialist in musicals; he was skilled at creating atmosphere and was strongly influenced by the world of art. A visual genius, Minnelli used a realistic framework for fantasy; he demonstrated a keen sense of pacing, and emerged without peer in fashion and decor. A perfectionist, he often had trouble verbalizing what he wanted from actors and spent a great deal of time placing people. Minnelli was sometimes referred to by crew members as "the gargoyle director," since he insisted on framing every setup like an artist. "He just wanted actors to walk to a point and when they got there, tilt their head in such a way that there was a gargoyle over their left earlobe," songwriter Ralph Blane remembered. On *Yolanda and the Thief* (1945) he kept Lucille Bremer in a bubble bath from eight until eleven in the morning trying to line up the best angle through the crook in her arm when she answered the telephone. Finally the actress was in tears. Soft-spoken but fidgety, Minnelli could also be thoughtful and generous. "I drive everybody crazy," he admitted to Fred Astaire one day, after changing his mind so many times the dancer was becoming irritable. But Minnelli's results were stunning; he directed some of the most beautiful pictures Hollywood ever produced—*Meet Me in St. Louis* (1944), *An American in Paris* (1951), *Gigi* (1958), as well as such dramas as *Lust for Life* (1956).

When a psychiatrist suggested that Minnelli and his wife, Judy Garland, not work together again, choreographer Charles Walters took over the direction of Garland's movie *Easter Parade* (1948), reinforcing the already powerful team of musical directors within the Freed unit at Metro. Originally a dancer on Broadway, Walters had choreographed *DuBarry Was a Lady* and *Meet Me in St. Louis* for Freed, then was assigned *Good News* (1947) by the producer as his introduction to directing. Awed by the task, Walters asked Freed to let him start by filming a musical number, though he realized that most of his time as director would be devoted to the principals. "Technically," he said, "those big dance numbers were more difficult than the dramatic scenes." Slick in his approach and full of energy, Walters considered himself a company director, adept at keeping within his budget without sacrificing quality. Later he ventured into comedy, but his reputation rested on musicals like *The Barkleys of Broadway* (1949), *Lili* (1953), and *High Society* (1956).

Many of Hollywood's great directors came from within the industry, acquiring their mastery of filmmaking elsewhere before being assigned their first film to direct. Joseph L. Mankiewicz had been working as a screenwriter since 1929 and had been a producer for ten years before assuming his first directorial job on *Dragonwyck* (1946) as a last-minute replacement for the ailing Ernst Lubitsch. Four years later he scored the unprecedented coup of winning Academy Awards for both writing and directing in two successive years, *A Letter to Three Wives* (1949) and *All About Eve* (1950). Brilliant, witty, and ambitious, Mankiewicz made his job seem simple. He enjoyed his work and made other people enjoy theirs. Interested in details and realistic performances, Mankiewicz asked his cast on *The Late George Apley* (1947) to lunch together and play games like Scrabble while the crew was lighting the set. "He wanted to establish a family relationship, and that was his way of attempting to achieve it," actress Vanessa Brown explained.

George Seaton also had established himself as a screenwriter before he directed some of his own scripts—*Diamond Horseshoe* (1945) and *The Miracle on 34th Street* (1947), both at Twentieth Century-Fox. John Huston had worked at Warner Bros. as a scriptwriter for a decade prior to being allowed to direct *The Maltese Falcon* (1941); Preston Sturges wrote for Paramount seven years before convincing studio executives to let him direct his screenplay *The Great McGinty* (1940). Billy Wilder launched an extraordinary collaboration with Charles Brackett in 1938, extending the partnership into a producer-director-writer relationship at Paramount four years later. Wilder became an excellent director, the acknowledged heir to Ernst Lubitsch, although the mechanics of filmmaking were never his strength. Like many writers, Wilder quipped that he became a director because he grew tired of others "fouling up my screen stories." (Madsen, *Billy Wilder*)

Edward Dmytryk and Robert Wise both came to directing from film editing. Dmytryk got his directorial start on B pictures at Columbia, winning acclaim later at RKO for his anti-Fascist picture *Hitler's Children* (1943), which led to better assignments. Eventually he directed *Crossfire* (1947), *The Caine Mutiny* (1954), and *Raintree County* (1957). "Dmytryk was like a truck driver," character actor Walter Abel said; "he went straight ahead. He knew the script, he served the script, and he knew what he wanted to get on film—and he got it." Dmytryk acknowledged the intensity of a director's job: "There's no work in the world that occupies as much of your time, intelligence, and feelings," he said. "It is miserable to make a picture; you suffer tremendously. The camera was never any problem to me. I'd always had a kind of scientific attitude from the time I was a kid. The problem was drama and understanding people and emotions." Dmytryk admitted he was purely a motion picture man, but felt his background enabled him to create realistic films in which characters talked like normal people. "In pictures things have to be spontaneous and immediate, coming out of the director's mind and not out of a writer's mind," Dmytryk claimed.

Robert Wise also considered himself a man of the cinema, more comfortable working with technical setups than with actors. "I found it was one thing to be working in the cutting room on a movieola and having the actors in a nice frame of celluloid," Wise noted. "It was another to go up on the set and have them all look at you, hands on hips, saying, 'Well, boss, what do we do?'" Every director works differently, and Wise had to find his own technique. "I give the actors a kind of general movement and general thrust," he said. "Then I let them rehearse it and bring to the scene what they discover in doing it rather than trying to be didactic." Performers found his method frustrating at times and wanted stronger guidelines.

George Stevens, who directed *Shane* (1953) and *Giant* (1956), started in movies as a cameraman. Later he became a comedy director for Hal Roach and was eventually assigned *Alice Adams* (1935), his first important dramatic story. Brilliant and humorous, Stevens nevertheless had difficulty in expressing himself with actors. Nina Foch, who coached Millie Perkins in the title role of *The Diary of Anne Frank* (1959), remembered that Stevens "was marvelously smart in the cutting room, but he didn't like to talk to actors at all. He would rather not speak to them." His method was to shoot thousands of feet of film, then watch hours of footage at night. "He loved that part of it," said Foch. "The wonderful thing was he'd sit there and watch the film, and then he'd say, 'That's the moment.' It was uncanny; he could just tell. His cutter would be right next to him and Stevens would say, 'Use that.'"

Behind the scenes Stevens ran a command post, shouting orders like a brigadier general. A chart in his office designated each job and the person assigned it. He held people responsible and didn't tolerate mistakes. In preparation for *Giant* Stevens read everything on Texas he could find. His office had bulletin boards on rolling stands with articles and pictures on Texas. "He did most of his directing with me before the picture began and said hardly a word during shooting," Rock Hudson claimed. "He had me so rich and powerful in my thinking and so bigoted that by the time we were shooting I *was* Bick Benedict. I didn't have to play it. Stevens was a hypnotic man; I followed him around like a puppy."

Paramount's Mitch Leisen (*Hold Back the Dawn* in 1941 and *To Each His Own* in 1946) began his film career as a costume designer for Cecil B. De-Mille, later becoming DeMille's set designer. A visualist and a man of taste, Leisen might spend two hours adjusting a drape on a window, rather than rehearsing his actors. If he had a banquet scene, he ordered what went on the trays. He relied on the studio's dialogue coach, Phyllis Loughton, to work with actors in putting scenes together, then Leisen would take over and film them. If Loughton didn't like a shot, she would whisper her objections to him. "There is nothing more invisible than a dialogue director, if you're smart," Loughton said. "The director is the director, and his is the final word."

A prime example of a great director who came up through the studio ranks is three-time Academy Award winner William Wyler, who launched his career writing for Universal's publicity office in New York. Then he transferred to Hollywood, where he worked as a propman, grip, script boy, casting director, and second assistant before making his directorial debut in

1925. He began at Universal directing two-reel westerns and program comedies, but graduated to more important pictures with the coming of sound,
winning Oscars for *Mrs. Miniver* (1942), *The Best Years of Our Lives* (1946),
and *Ben-Hur* (1959). "You had to learn the business, like any other business," said Wyler. "My training in westerns stood me in good stead, because
you really learned about movement. That's what motion pictures are all
about. There's quite a science to photographing movement effectively."

Wyler enjoyed making different kinds of films, from dark tragedy to romantic comedy to musicals. As a director under contract, he often accepted
pictures simply because it was time for another assignment or because a particular script seemed the best available at the moment. Thorough in his
work, Wyler often ran over schedule, but he seldom faced trouble from the
studio bosses since his reputation for quality was impeccable.

Actors complained that Wyler never knew what he wanted, asking them to
repeat a scene over and over until he saw something he liked. While making
The Heiress (1949), the director had Olivia de Havilland descend a staircase
thirty-seven times, never telling her what she was doing wrong. Finally the
actress dropped her fan from exhaustion. "Great!" Wyler yelled. "That's it."
Wyler's instincts told him what was right, and actors learned he'd not allow
them to go wrong. Above all he was concerned about the way characters
would react in certain situations, sometimes drawing from his own experience. During a production he was totally absorbed in his work. "When
shooting started," his wife said, "it was as if he dove under thirty feet of water
and stayed there until it was finished. While he was working, social life
ended completely. The children and everything else became my job."

During the 1930s important directors came to Hollywood from England,
among them James Whale, John Brahm, Robert Stevenson, and Edmund
Goulding. But the greatest of the British imports was Alfred Hitchcock,
whom François Truffaut compared with such artists of anxiety as Kafka, Dos-
toyevsky, and Poe. Part humorous fat man and part sadist, Hitchcock was extremely possessive, particularly with actresses. "He liked to take someone
who felt she was in control of her life and slowly break her down," said Tippi
Hedren, who starred in Hitchcock's *The Birds* (1963). "Hitchcock was a difficult person and had all kinds of frustrations and neuroses. He thought of
himself as looking like Cary Grant. That's tough — to think of yourself one
way and look another."

Hitchcock wanted actors to be puppets he could move like animated
props, molding every expression. "Your performance was Hitchcock's performance," Laraine Day said. "You read lines the way Hitchcock read the
line. You added nothing and you took away nothing. You did *exactly* as he

told you. He had drawn it out on paper. You'd see the scene like a comic strip, and that's the way you played it. You'd bring nothing but the body to the set." Camera angles were planned before the picture began. Hitchcock was a master at letting viewers use their imaginations, never sure of what they saw, as in the famous shower scene in *Psycho* (1960). The script to Hitchcock was law, and he tolerated no transgressions. Actress Carolyn Jones had a tendency to transpose lines. When she worked for Hitchcock on *The Man Who Knew Too Much* (1956), the director would say, "Cut," and in his wicked English accent comment, "Miss Jones, I've spent *so* much money paying the writer to prepare the script. Would you please say it the way he wrote it?"

Hitchcock loved pulling gags on the set. But rather than face an argument, he would simply walk away. He often called on his own experiences, as well as other people's, to help develop a character, a scene, or one of the myriad details of a film. And he was so organized that everything fell into place. "This business just gets one Hitchcock in its lifetime," character actor Norman Lloyd remarked. "He was wonderful to work with, the epitome of the international director to me."

Foreign directors made an early appearance in Hollywood, adding to the industry's growth. Erich von Stroheim arrived from Vienna in 1914, established himself as a regular member of D. W. Griffith's company, and in 1924 made the silent classic *Greed*, a seven-hour indictment of avarice and human degradation. A dreamer, indifferent to costs, von Stroheim wasted money and soon found himself at odds with the studio system. Disillusioned, he left Hollywood without making a sound film. Ernst Lubitsch came from Germany in 1923 to direct Mary Pickford and stayed in Hollywood until his death in 1947. Known for his sophisticated comedy (*Ninotchka* in 1939, *To Be or Not to Be* in 1942), Lubitsch was remarkably successful, considering that his pictures rarely made money. While the big studios were factories employing methods of mass production, at the same time Hollywood executives showed a naive respect for artists, and Lubitsch was considered an artist with genuine taste. The Lubitsch touch was full of humanity, graceful in style. With Lubitsch nothing was improvised; he knew how to elicit the best from actors and was always precise about what he wanted. "Working with Lubitsch was like going to three years of college," actress Signe Hasso said. "He was a master."

Michael Curtiz, a Hungarian who had directed pictures in Germany and Austria, was brought to Hollywood by Harry Warner in 1926; he completed more than a hundred films for Warner Bros. over the next quarter century, including *Casablanca* (1943). Curtiz spoke with a thick accent and mangled the English language worse than Sam Goldwyn did. "The next time I send an idiot," he supposedly said in disgust, "I'll go myself." He directed *The Adventures of Robin Hood* (1938) and *The Sea Hawk* (1940) with Errol Flynn, persistently calling the actor "Earl Flint." Curtiz worked fast and was at his finest moving masses of people. He thought lunch a waste of time, and barked obscenities at actors who slowed him down. He stormed about the set like a martinet, yet nothing escaped him—a stocking that needed straightening, a color that wasn't right. His expertise was with the camera; he was concerned always with the panoramic effect, and he had little to say on intimate scenes, assuming actors knew what they were doing. "You're standing like a stick," he might observe. "Be alive. Be like a fox, be a lion, be like a tiger." (Randall int.) A demon for work, Curtiz hated to go home at the end of a day, although he relied on his wife, writer Bess Meredyth, for story insights.

With the rise of Fascism in Europe, the swell of emigrant artists to the United States increased sharply—in architecture and music, as well as film. Several of Europe's top filmmakers suddenly found themselves in Hollywood, where they banded together with a colony of intellectuals for discussion and cultural nourishment. Fritz Lang had been offered the leading post in the German film industry by Joseph Goebbels, Hitler's Minister of Propaganda, but went to Paris instead. He arrived in the United States in 1934, where he spent the rest of his career making films dealing with mankind's inner conflicts. Strongly influenced by the German Expressionists, Lang was a cinematic psychologist. His films are full of violence, dark struggles, and nightmares; he was fascinated with stories of criminals, spies, and war. At odds with the conformity demanded by the big studio system, Lang made a series of anti-Nazi movies, including *Man Hunt* (1941), *Hangmen Also Die!* (1943), and *Cloak and Dagger* (1946), each reflecting his strong personal involvement and keen awareness of the Fascist mind. On the set he could be a bully, stomping about in Prussian boots and brusquely issuing orders. He had everything worked out in advance, proving himself imaginative as well as a master of detail. "The stage hands threatened to drop sand bags on him," character actor Walter Abel said, "but what Fritz could do with actors in front of the camera was incredible." Lang had wanted to be a painter, but became convinced that film was the art of the twentieth century.

Not all foreign directors forced such dark hues. Jean Negulesco, a former painter from Rumania, demonstrated a light touch, notably in box office hits like *How to Marry a Millionaire* (1953) and *Three Coins in the Fountain* (1954). Henry Koster, born in Germany, came to Hollywood from France in 1936, winning acclaim at Universal with the early Deanna Durbin pictures.

Quiet and gentle, Koster created a relaxed set, although he found directing and dealing with temperamental personalities difficult. "I consider myself very fortunate," Koster admitted, "because I know my talents were limited and my personality not quite what a director's should be." He worked hard and welcomed the security of a regular paycheck. Later he directed major pictures at Metro and Twentieth Century-Fox, including *The Robe* (1953), the first production in CinemaScope.

Each studio had a team of program directors, who were workmanlike and could be relied on to bring pictures in on time, within budget. Metro-Goldwyn-Mayer, for example, had Jack Conway, Robert Z. Leonard, Richard Thorpe, Roy Rowland, Richard Whorf, Tay Garnett, and Norman Taurog. Each was able and efficient, not particularly demanding; each wanted to get the job done as quickly as possible. They rarely made any major contribution to the script.

The Warner Bros. contract directors included Lloyd Bacon, Frederick DeCordova (who in 1971 became producer of "The Tonight Show" on television), David Butler, Roy Del Ruth, Delmer Daves, and Gordon Douglas, all fine technicians, though their creative accomplishments were modest. The best scripts at Warners went to Michael Curtiz or Raoul Walsh; lesser directors would be assigned spoofs or musicals featuring young leading men and women. Stars rarely criticized stock directors, although performers didn't get excited about working with them, since they rarely told actors what to do. Practical, sometimes uninspired, these program directors were reliably commercial.

Vincent Sherman proved more innovative, although he directed numerous B movies for Warner Bros. Frequently he'd be assigned a picture to put certain actors the studio had under contract to work. Sherman accepted such assignments, extracting Jack Warner's promise for a better film afterwards. When option time came around, Jack Warner would inquire what pictures the director made the past year and whether or not they made money. If they had, the option was automatically picked up. "I'd finish a picture on a Friday," Sherman said, "and many times I'd come in Monday morning and be handed another script." He felt a studio director needed two eyes: one toward artistic merit and credibility of material and the other toward entertainment and box office success. "You knew that if the pictures did no business, there would be no studio," he said.

Most of the directors at Republic were action directors who spent little time trying to get real performances out of actors. Schedules on Republic's westerns were short, budgets small, and profit the primary motive. Gene Autry remembered that on his first serial B. Reeves ("Breezy") Eason, a Hollywood veteran, and Otto Brower alternated chapters—Eason directing one, Brower the next. By 1950 Joseph Kane had become Republic's foremost house director, making most of the Roy Rogers pictures. Kane was outstanding on action, but paid little attention to nuances in the script, content with broad performances. John English was better on script and more sensitive; "oater" crews referred to him as a "boudoir man," since English preferred indoor scenes over the sweat and manure of the outdoors.

Under the big studio system the director had a stipulated number of weeks to deliver his cut of a film before it was turned over to the producer, who had the final say. Repeatedly directors claimed their work was ruined by capricious editing, for the finished product was ultimately decided by the producer and the studio heads.

While labor conditions for studio directors remained strenuous, the formation of the Screen Directors Guild in 1938 eliminated working on Saturday and Sunday without extra compensation, set the number of hours members could work, and established intervals at which meals must be served. Despite these new standards, directors arrived on the set at six A.M. when they were shooting a picture and didn't finish until seven or eight that evening, spending most of the day on their feet. What the big studio system offered in return was a constant supply of properties, the equipment and facilities, a guaranteed income, and the opportunity to make an occasional film that was superior.

5

Creating Stars

*The Paramount commissary during a Golden Circle dinner
to introduce the studio's future stars to members of the press.
Courtesy of the Academy of Motion Picture Arts and Sciences.*

All the major studios had a talent department, whose purpose was the discovery and grooming of young performers. Every year a team of talent scouts scoured the country for promising youngsters with potential. Warner Bros. had Solly Baiano as its principal scout, Metro had Lucille Ryman and Al Trescony, Paramount had Milton Lewis, and 20th Century-Fox had Ivan Kahn, all of whom attended the same shows in Los Angeles and New York. Most studio scouts made an annual trip to Broadway, where they surveyed not only leading performers in new shows, but secondary actors and choristers as well. Often they attended tryouts in Philadelphia, Boston, or New Haven and most of the plays that opened in Los Angeles, Hollywood, and Pasadena, making appointments with people who looked promising. Studio representatives also attended regional theater productions, performances at colleges and universities, beauty contests, pageants, and nightclub acts. Solly Baiano of Warner Bros. and Al Trescony of Metro frequently covered events in Los Angeles together, at times accompanied by their wives, usually arriving in the same car. "If we both were interested in the same talent," Trescony remembered, "we had an agreement that whoever's car we were in would have first call."

If the scout was impressed by someone out-of-town, he or she would usually be invited to California for a screen test, all expenses paid. Others were instructed to contact a studio representative if they were ever in Los Angeles. Those who came at studio expense usually arrived on the Santa Fe Super Chief, were met in Pasadena by a limousine and housed in the Garden of Allah, the Marmont, the Beverly Hills, or the Beverly Wilshire hotel, where they were treated royally.

Talent also came from vaudeville, radio, jazz bands, even from burlesque. Promising youngsters were brought to Hollywood to see how they would photograph. Local talent had an obvious advantage. Singer John Raitt, walking out of the MGM commissary one Friday afternoon where he was visiting his friend dancer Dan Dailey, was accosted by a test director and asked to make a test for a picture the next day. The studio's wardrobe department gave Raitt Clark Gable's boots for the test, and he sang unaccompanied a song he had learned in half an hour. Solly Baiano discovered Lana Turner in a lingerie shop, not Schwab's Drugstore as has often been reported. He was Christmas shopping for his wife and daughter at a clothing shop on Hollywood Boulevard when he noticed an exceptionally pretty clerk. He started talking to her and asked if she had done any acting. The girl didn't seem interested, but he gave her his card and told her to come and see him tomorrow morning at Warner Bros. She took a long lunch break the next day, during which Baiano introduced her to Mervyn LeRoy, who directed her first picture.

Talent scouts at all the studios were besieged by mothers writing about precocious children or invading their offices with photographs. Some hopeful mothers made trips to California, accompanied by their children, which was generally a mistake, since thousands of talented youngsters in Los Angeles competed for a limited number of roles. Studio representatives encouraged young people to complete their education and acquire professional training back home. One woman kept writing Solly Baiano about her child and eventually included a snapshot, which she said wasn't very good. Baiano thanked her, and rather than hurt the woman's feelings, he told her the child was too pretty for the kinds of roles the studio was casting.

Many hopefuls brought to the studio for tests were ultimately rejected, and some of those selected were photographed under less than ideal circumstances. Arlene Dahl, one of the screen's great beauties, made her test at Warner Bros. in an ill-fitting outfit Alexis Smith had worn in *San Antonio* (1945), but she did receive a contract. Broadway actress Carol Bruce was treated well when Universal signed her to a contract, but was miscast in the two pictures she did there. "They bought gold in New York City," Bruce said, "and turned it into tin." Jan Clayton, who created the role of Julie in *Carousel* on the stage, fared even worse. Clayton was signed to almost every studio in town over a period of ten years and never survived an option.

Usually there would be a silent test first, essentially a photographic test, followed by a sound test, a short scene from a play or an earlier picture. With inexperienced actors and actresses, there would usually be a reading before the silent test in the office of the studio's drama coach, with representatives from the casting department present. Frequently the coach read the supporting role. Then discussion would follow about whether to proceed with a test. Paramount had a room with a one-way mirror along one side, variously called the "Glass Cage," the "Fishbowl," or the "Snake Pit," with chairs outside where producers and members of the casting office could watch young people read scenes. "The room was intimidating to say the least," actress Yvonne DeCarlo recalled. "You never knew who was on the other side." (De Carlo, *Yvonne*)

Lillian Burns, MGM's drama coach, made reports on everybody who read for her, sending copies to Benny Thau, Louis B. Mayer, and the casting department. "So-and-so came in," her evaluation might state. "She will *never* be an actress." Or the coach might insist, "She *will be* a star." Mayer and the Metro producers greatly respected Burns's instinct for talent. "She could scare the living daylights out of the guy or gal reading," talent scout Al Trescony commented. "But there's always a little light that pops out when you know someone has talent, even though they're nervous. Lillian had that intuitive flair, she would always know. She'd say, 'All right, that's fine. Now let's do it all over again. Just forget about us. You're wonderful, my darling, wonderful.' Relaxed, they'd come back and do the scene, and I'd be out getting ready to negotiate for a test option."

Occasionally an actor or actress would read for a director, especially if a specific part were involved. Catherine McLeod related her experience reading for Frank Borzage, who would direct her in Republic's big-budget *I've Always Loved You* (1946): "He had a white satin couch in his office, and it was hot. We didn't have stockings in those days, because it was wartime, and we painted our legs with makeup. My legs just melted all over the white satin couch."

Tests were generally made in black and white, since color was expensive. Sometimes the test would take place in New York, but more often it would be shot at the studio in California on a prepared set. The studio's drama coach selected a scene that would enhance the abilities of the person under consideration and occasionally might direct the test herself. One of the studio's contract players would rehearse with the newcomer and play the scene they prepared before the camera. A crew not working on a particular morning or afternoon would be assigned to the test, along with one of the studio's regular cameramen, perhaps even a top cinematographer like Metro's Joseph Ruttenberg or George Folsey. "To be able to have someone like a Joe Ruttenberg test a beginner," Al Trescony noted, "was like having Picasso paint their portrait." Someone new in the business, a second unit director, or someone who made short subjects usually directed the test.

Within a few days, studio executives and producers and the appropriate department heads viewed the test, sometimes with the person tested present. Seated in a small screening room, the newcomer was subjected to a barrage of critical comments. "What'll we do with her nose?" someone from makeup might ask. "Well, her hair's got to be fixed," a stylist might add. "Well, we'll have to pad her fanny." "We've got to pad her bosom" was often the consensus. Actress Jeff Donnell remembered such a humiliating experience at Columbia: "The only things that were right were my ears and my teeth."

Studios were searching for beauty, acting ability, and a unique personality. "I always looked for the actor with personality first," Al Trescony said. "I've found that a person with a fantastic personality will attract the attention of everyone in a room, not unlike a beautiful young woman. But people will look at the beauty and after a moment return to their conversation. The one with personality will hold their attention." Talent scout Solly Baiano felt the most important ingredient in acting was to "make people believe you. There has to be something coming out of the eyes." Trescony agreed that in screen acting the eyes were all-important. "When I directed an interview test," he said, "I would go to a head close-up and watch the eyes, because the eyes show everything the person is thinking. It comes through the celluloid right out to the audience." An ability to recognize talent, Lillian Burns insisted, was a "gift from God." Those who had that gift could sense when "there was something like buried gold."

But sorting through the chaff was not easy. Talent scouts spent most mornings in their offices booked solid with appointments. They discussed clients with agents, interviewed talent they or a studio representative had seen in plays or nightclubs, and talked with ambitious youngsters who appeared at the studio gate and demanded to be seen. The scout might have the person read a two- or three-minute scene. If the scout liked what he heard, he would set up a longer reading within the next two or three days with the studio's drama coach and casting director. Should they agree the person had possibilities, a test was recommended, which meant negotiating a test option with the performer's agent. Scouts normally devoted two or three hours twice a week in the afternoon to readings, then spent evenings covering theater and nightclub performances, sometimes two or three a night. The next morning they read quickly through *Variety* and *The Hollywood Reporter*, then dictated reports on what they had seen the night before. These reports would be sent to all the studio executives, producers, and directors. "I would try to make my critiques a little humorous to hold my readers," Al Trescony explained. "At one time we tested at least one unknown a week at MGM. I'd set up the stages, get the cameramen, and arrange for the crews."

An actor whose test was approved was signed to a seven-year contract with options, starting at a salary between $75 and $250 a week, depending on experience. "With options" meant that the studio could drop the newcomer at the end of each six-month interval of the contract period, or raise the actor to the next salary level. Usually the casting director decided who received contracts and who didn't, although the studio mogul had final approval. Warner, Cohn, Mayer, Zanuck, and Goldwyn all recognized talent and shared an ability to make decisions quickly. Occasionally they miscalculated. Jack

Warner, for instance, heard Mario Lanza, but decided not to sign him. Neither did he sign Clark Gable nor Burt Lancaster. Metro had Deanna Durbin under contract, but dropped her in favor of Judy Garland, allowing Durbin to go to Universal, where she became queen of the lot. Ann Sheridan, Susan Hayward, Evelyn Keyes, and Yvonne DeCarlo were all stock girls at Paramount, but none became stars until they went to another studio. Studios seldom raided each other for talent, yet they were aware of a dropped option and mindful of what that individual could do. Sometimes a beginner fit into the films one studio was making better than those made by another. Debbie Reynolds signed with Warner Bros. at a time when the studio had nothing for ingenues. Talented though she was, Jack Warner dropped Reynolds in favor of potential leading men and women. MGM immediately signed her and wisely cast her in musicals. Dan Dailey, on the other hand, had to compete with Gene Kelly and Fred Astaire at Metro. When *Mother Wore Tights* (1947) came up at 20th Century-Fox, Louis B. Mayer, upon request, generously released Dailey from his MGM contract, enabling the dancer to achieve stardom opposite Betty Grable.

David Selznick brought Gene Kelly to Hollywood on the strength of Kelly's success in *Pal Joey* on Broadway. Selznick considered using the dancer as the priest in *Keys of the Kingdom* (made in 1945 by 20th Century-Fox), since Selznick International wasn't making musicals. In the meantime the producer loaned Kelly to Metro-Goldwyn-Mayer, where he made *For Me and My Gal* (1942) opposite Judy Garland and won immediate success. Eventually the dancer's contract was sold to Metro, and Kelly's career in film musicals continued with legendary success.

Sometimes a newly signed contract player was told to wander around the lot for a few weeks to learn how motion pictures were made. Robert Alda recalled watching Bette Davis, Ann Sheridan, and Ida Lupino at work as he went from set to set at Warner Bros. Since he was clothes conscious, Alda never went to the studio in the same outfit. That began to catch people's eyes, and studio executives became interested, casting him as George Gershwin in *Rhapsody in Blue* (1945).

Betty Garrett, however, after big success in *Call Me Mister* on Broadway, roamed the Metro lot for a year without working. "I was under contract to the studio without anybody's realizing what I did," she claimed later. Her big break came during one of L. B. Mayer's birthday parties. Ida Koverman, Mayer's secretary, asked Garrett to entertain, putting her on the bill with the child stars. She wore a white dress with a huge poppy on the skirt and a stem that wound around the bodice. She wore a big black hat with more red poppies and wore long black gloves. She stood up to sing at Mayer's party and said, "I think I'm a little old for this show." The audience laughed and applauded everything she did. Garrett was soon featured opposite Frank Sinatra and Gene Kelly in *Take Me Out to the Ball Game* and *On the Town* (both 1949) and was even considered for the title role of *Annie Get Your Gun* (1950).

Young contract players with less experience than Garrett or Alda were put into an extensive apprenticeship program to prepare for stardom. Having signed them, the studio gambled further on the youngsters by investing in elaborate grooming to teach them the craft of moviemaking and how to conduct themselves. The studio's drama coach, who also served as head of the talent department, supervised this training program. Lillian Burns at Metro was a power on the lot. Burns could end careers, but nurtured those she thought had star potential. "I always told the truth," Burns maintained. "I know I had a reputation for being tough in my evaluations, but being less than honest could have meant the end of my job at MGM." Paramount had Phyllis Loughton, later wife of director George Seaton, as drama coach; 20th Century-Fox had Helena Sorrell, and RKO had Lela Rogers, Ginger's mother. Actress Josephine Hutchinson was coach for a time at Columbia, succeeded by Natasha Lytess, who eventually worked privately with Marilyn Monroe. Warner Bros. had the respected Sophie Rosenstein until she left to go to Universal, where she worked with Rock Hudson, Tony Curtis, Piper Laurie, and Clint Eastwood.

Lillian Burns believed in the development of the total persona. "If Ava Gardner didn't know how to hold a champagne glass, she had to learn," Burns said. "Many of these youngsters had not come from the background Katie Hepburn did. They had to learn about antiques, about music, about culture. That was part of their training. By developing the person, you were developing their talent." Besides talent, there was craft. Newcomers to motion pictures had to learn an entirely different technique from stage acting. "In motion pictures there is a camera, what I have termed a 'truth machine,'" Lillian Burns explained. "You cannot say 'dog' and think 'cat,' because 'meow' will come out if you do."

Burns, whom Mayer called Lilly, had studied with Dame Lilian Baylis in England. Before coming to MGM in 1936, she had worked briefly at Republic, where she said her office was next to the stables, since the horses were all-important at Republic. At Metro, Burns had a big room over Stage 5 set up like a living room, with two offices adjoining it and an outer office. Most of her studio contact was with Benny Thau and the producers and directors of specific pictures, although she conferred regularly with talent scouts and the casting director, Billy Grady. Burns did not believe in classes and never ran a school. Hers was an individual approach. "I don't believe you can teach acting," she said. "You can teach voice, you can teach diction, you can teach body movement. You can help develop talent, or even a great personality. That's where the big studios could make stars." She worked with her young contract players for weeks on specific roles, preparing them more thoroughly if she knew they would face a weak director on the set. She would take a young person through an entire script, working as she would in a rehearsal. But the major part of her task was developing the players as people. "An actor or actress brings to life another living, breathing human being. They must have enough dimension to bring another person to life."

On Friday afternoons, if she wasn't busy on a production, Burns called together everyone she was working with in her central room and had them run through scenes they were preparing. These sessions didn't occur often because of other commitments, but they were not treated lightly when they did. Sitting around the room might be Betty Garrett, Janet Leigh, Ricardo Montalban, Arlene Dahl, Peter Lawford, Gloria DeHaven, Barry Nelson, Marshall Thompson, or Cyd Charisse. On special occasions producers and directors were invited to attend. Scenes were carefully studied, presented with players teaming up, and critiqued, while others learned by observation.

Lillian Burns's goal was to help her young people understand a script and a character, sharpen their tools, and let go of their inhibitions. They should never try to "act," but be honest and simple, projecting from the eyes instead of just with the voice. Above all they had to be individuals and not copy another's interpretation. Burns often enticed surprising performances out of people. "I really worked my tail off to make a good showing in front of the Friday afternoon group," said actor Marshall Thompson, "because the others were sitting there criticizing."

Not everyone loved Burns; many feared her. "If you had talent, she'd work with you," Al Trescony said. "If not, she would tell you." Detractors claimed she taught everyone to be clones of herself. Some observers said Lana Turner copied her coach's facial expressions, movements, mannerisms, everything, until the blonde actress became like a younger twin sister.

If Burns felt players had little potential, she paid them slight attention. One day she told young Lola Albright, "Lola, you're such a nice girl. Why don't you go back home?" Albright agreed she was dreadful. "I hadn't the vaguest idea of what I was doing," the actress said after a successful career in films and television. "I was always terrified, for Lillian was a very terrifying woman. Having no confidence to start with, I was destroyed every time I walked into her office. But she was probably right, I should have gone somewhere else."

Louis B. Mayer occasionally suggested that Burns use Stage 5 to put on plays, but she always declined, arguing that the proscenium arch was at odds with screen acting. Phyllis Loughton at Paramount, on the other hand, did mount plays, as she had done at Metro earlier. Although Loughton recognized that film acting was different from stage acting, she maintained that young people had to learn to use their voices, their bodies, and how to be characters other than themselves. To help accomplish these objectives, she put on about three plays a year with Paramount newcomers, to which the studio's producers and directors were invited. Loughton realized, however, that personality, looks, and a good voice could compensate on the screen for lack of acting ability. Dorothy Lamour, for instance, she found "a fascinating young woman, although no great actress." With the coach's help Lamour learned her job, starred in *The Jungle Princess* (1936) and contributed to the success of the Crosby-Hope *Road* pictures. Loughton also made tests at Paramount, read all the scripts about to be filmed to find places for her young people who were ready, and coached them in their roles once they were cast.

Lela Rogers ran a workshop for novices at RKO, which for a time included comedienne Lucille Ball. Sophie Rosenstein, a lovely, brilliant woman from the University of Washington, was talent coordinator, teacher, and Mother Confessor to young contract players first at Warner Bros., then at Universal. Sometimes she worked with them at her house at night. When Rosenstein died, Estelle Harman took her place at Universal. "I think it's possible to teach the craft and art of acting," Harman said, "especially of film acting. There are a lot of craft things to learn. Certainly the development of a persona becomes part of that, helping young people open up and have presence and vitality, the energy, the magic. But filming a long shot is one kind of communication; when they pull in for a tight close-up, it really has to stay just in the eyes. The smallest mannerism is five-feet big. If you move your mouth a little, that's five feet of movement. So that communication is different."

Most studios also had speech and diction coaches. Gertrude Fogler, a charming woman admired by everyone on the lot, was the speech teacher at Metro. A small, plump, white-haired lady, Fogler had her charges read aloud classics and poems as exercises. She was concerned with placing the voice, since a resonant voice records better than a thin one, but she encouraged actors to enhance their speech without assuming a stilted tone. Although she exorcised southern accents, any newcomer who spoke with perfect stage diction had to learn to drop G's and slur words, since speaking for the movies must be natural. Foreigners like Hedy Lamarr, Ricardo Montalban, and Fernando

Lamas took speech lessons to get rid of their accents, or at least make them easier to understand.

As part of their training, studios also offered singing and dancing lessons. At Metro, Arthur Rosenstein was the vocal coach. Known affectionately as "Rosie," he worked with most of the performers featured in MGM musicals, many of whom had never sung before. Maestro Spadoni concentrated on the serious singers at Metro, such as Mario Lanza, and there were dance lessons available from Janet Bates. Elizabeth Taylor, Marie McDonald, Janet Leigh, Margaret O'Brien, and Marie Windsor were all required to take dance lessons at Metro on regular schedules.

At Columbia, Harry Cohn decided in the mid-1940s that all of his contract players should take dancing and called them together on a stage across from the main lot. They all got down on the floor, and choreographer Jack Cole had them go through a series of exercises. Cole's assistant, Gwen Verdon, brought her baby along. Occasionally at Warner Bros. young performers took ballet lessons. "Off in a corner of a stage they had an enormous mirror down one side," Joan Leslie recalled. "They'd have a barre there and maybe a couple of old, deserted dressing rooms, and we'd go in and work out to a portable phonograph. They had an excellent ballet man, and Alexis Smith, Susan Peters, and Delores Moran all worked out with him." Astute young people realized they were being given a unique opportunity to develop into versatile professionals free of charge and took advantage of all they could.

Studios also offered lessons in etiquette, movement, fencing, horseback riding, swimming, boxing, languages, whatever might be needed in a motion picture. Helen Rose helped young people at Metro pick out their clothes, while Della Owens Rice advised starlets at 20th Century-Fox on how to dress. Performers learned about taking care of the body, taking care of their clothes, how to come across in interviews, and how to appear in public. They were taught about makeup and skin care and how to keep trim. "The studio owned you, and they wanted their property in great shape," MGM contract player Jean Porter said. There were exercise classes, guest lecturers, and a barber shop and dentist right on the lot. "The studio decided to straighten my teeth," Joan Leslie said. "I was fifteen years old when I went to Warner Bros. and didn't know how to dress. I remember Steve Trilling, the Warners casting director, saying to me, 'You really must let your nails grow longer, Joan.'" Later Trilling arranged her first date.

Most of the lessons were optional, yet ambitious contract players attended classes four or five hours a day when they weren't making a picture. Saturday was usually a half day, which maybe included just horseback riding. If there was nothing scheduled, the young performer might visit the publicity office or one of the other departments to find out how it functioned. Youngsters with a little experience frequently tested with those just coming in, reading lines with their backs to the camera. Sometimes a group of newcomers would view and discuss a film classic or would sit on various sets and watch veterans work.

Smaller studios and independent producers customarily sent their new players to a drama coach off the lot; some actors preferred that to the studio's instruction. Since studio coaches had a voice in whether or not a performer's option was renewed, many newcomers looked upon them with suspicion and sought a teacher outside studio politics. Florence Enright and Elsa Schreiber were both distinguished Los Angeles drama teachers who specialized in screen acting. Enright, who frequently worked for Sam Goldwyn, was primarily technique-oriented. "She practically changed my whole speaking voice," actress Julie Adams said. "She taught me how to work with props, she taught me how to pick up cues. And she taught me how to look at someone's downstage eye, so that your eyes are pulled farther around. She also taught me to keep my face still for close-ups and how not to move my eyes too much." Enright advised young performers on the direction they should take with roles, preparing each of their scenes with them. Even established stars occasionally consulted with a coach on roles. For a more dramatic part than usual for him, Tyrone Power studied with Elsa Schreiber during the making of *Nightmare Alley* (1947), while his young co-star, Coleen Gray, worked with Schreiber's husband, George Shdanoff.

Through the renowned Maria Ouspenskaya, echoes of the Stanislavsky method reached Hollywood, in the films she made herself and in what she passed on to pupils. A small woman who always wore high heels, Mme. Ouspenskaya was a dynamo, sometimes intimidating to the uninitiated. "What a lady!" Carol Bruce remembered from her study with the great actress during Bruce's days at Universal. "I was in such awe sitting there watching her that I didn't even know what she was talking about. I was enchanted and mesmerized watching this tiny figure who had such dimension." On their first meeting Ouspenskaya told Bruce to be an apple. The young actress couldn't imagine what she was supposed to do, so she blew her cheeks out round and made a face, holding her breath. Ouspenskaya said, "Magnificent! Wonderful. You have imagination." Unfortunately, when Bruce tried to apply what she'd learned on the set the next day, she became more confused than ever.

Even experienced contract players were dazzled by the stars they suddenly found themselves surrounded by, and newcomers were often as awed as any fan. "I was such a runaway movie wack," singer Mel Tormé confided, "that just to be on those lots was thrilling for me." Actress Barbara Rush said, "It was absolutely awesome to be with all those big stars. It was such a world of make-believe. I'd just never seen people that attractive and so perfect. It was absolutely dazzling." For Janet Leigh, Metro was a fairyland. "You'd walk around the lot and pass Clark Gable in the street, or you'd open a stage door and see Katharine Hepburn and Spencer Tracy shooting a scene, or Judy Garland and Gene Kelly bursting through a screen or something." For producer Ross Hunter, then a young actor, Columbia was no less exciting. "The first time I saw Rita Hayworth," Hunter recalled, "I just couldn't believe I was on the same lot. It was a fantasy come true." And for actress Coleen Gray, a midwestern farm girl, 20th Century-Fox was the realization of the great American dream. "I would see all of those beautiful people," Gray said, "and thought I'd dreamed it. To be a movie star in that era was just the living end."

In such an unreal world it was easy for dreamers to lose their way, and casualties were numerous. A young player was expected to project what the studio considered an appropriate image, often at the expense of personal identity. In many cases a newcomer's real name was stripped away and replaced by a name the studio thought would command attention on a marquee. Jeanette Morrison Reames became Janet Leigh, Roy Fitzgerald became Rock Hudson, Rosetta Jacobs became Piper Laurie. Jack Warner wanted to change Arlene Dahl's name to Christine Winters, but the actress insisted on keeping her family name.

Hollywood moguls had definite ideas about how a star should look and refashioned their new talent accordingly. For the first few years Ava Gardner was in films, Metro's makeup department built a pad for her cleft chin. Warner Bros. put a black wig on Arlene Dahl for a photo session and made her up to look like Hedy Lamarr; then they discovered she had one Eurasian eye and decided to make her up like Madame Butterfly. "They shaved off part of my eyebrows," Dahl said; "they're still growing back."

Despite this image-making, the studios demonstrated a paternal attitude toward young people under contract. "There was a big daddy in charge," actor Robert Stack said. The old studios offered security, even though their approach produced more personalities than finished actors.

While the big studios nurtured, they also exploited their young contract players, demanding long hours. At the end of a day on the lot, it was not unusual for a starlet to be sent to wardrobe to select a gown and told she would be attending a premiere or publicity function that evening. A limousine would pick her up, and generally a date would be supplied by the studio. "I never had any days off," Virginia Mayo said. "If I was making a film, I was working; if I was not, I was studying."

Once the novice was ready, the studio drama coach was responsible for bringing him or her to the attention of producers. That meant being aware of every script the studio was preparing to shoot and breaking them down with the various contract players in mind. "I almost felt like an agent," Estelle Harman said. At Metro, Lillian Burns even went to see Louis B. Mayer about placing youngsters in particular parts, although producers frequently consulted her on who was doing well. Ambitious newcomers made sure they got a list of upcoming pictures and eagerly sought out the choice parts, often approaching the producers themselves. Paramount had a special table called "the Golden Circle" in the center of the commissary to bring budding young talent to the attention of producers and directors.

The most common method of placing talent was through the studio casting director, who had likely been involved with the player's progress from the beginning. In many instances the head of casting had been responsible for the person's being signed to a contract. Billy Grady, casting director at Metro, worked closely with Lillian Burns and the various studio talent scouts and for a time occupied an office next to the publicity department. Grady made frequent trips to New York to look over fresh talent, claiming he did more business with actors and agents dining at Chasen's in Hollywood than he did in his studio office. "Carnation Billy," as he was known, cast many MGM pictures in conjunction with Benny Thau, who in turn consulted Louis B. Mayer. Lew Schreiber was Grady's equivalent at 20th Century-Fox; Max Arnow and later Steve Trilling served in a similar capacity at Warner Bros. Arnow eventually headed casting at Columbia; Rufus Le Maire, who had been at Metro, held a corresponding post at Universal.

Most newcomers started out playing small parts in B pictures, where they learned to stand on designated marks to keep the camera in focus, match close-ups with long shots, and other essentials of film technique. At Republic hopefuls were put into B westerns even before they signed contracts, as a less expensive alternative to screen testing. At the larger studios new arrivals were often used as extras, with little or no dialogue, sometimes in two or three pictures at the same time. Promising novices might be given a few lines, perhaps an occasional close-up, eventually working their way into A pictures. If the studio felt youngsters had potential, they were given star build-up by the publicity office out of proportion to the tiny parts they were playing.

Assignments in pictures with big stars were highly coveted, since those meant the best exposure. Lana Turner, Judy Garland, Ava Gardner, Donna Reed, and Esther Williams were all introduced in Metro's *Andy Hardy* series with Mickey Rooney, although a regular part in a series, particularly a B series, was considered death for a young actor. If the career progressed on schedule, a contract player would normally alternate a good role with an inconsequential one. Sometimes a player would be starred in a B picture while filming small roles in major productions, to test public reaction. Later, a young player might be cast opposite an established star in a picture, then teamed with someone of equal rank to see whether the two young people could carry the picture alone. Next each of the two youngsters would likely be paired again with a big name to enhance marquee value.

Competition was brutal, with a swarm of candidates jockeying for the same parts. A player might have a role one day and lose it to a rival just before shooting began. "We never had a moment's security," Coleen Gray said of her contract days at 20th Century-Fox. Boy-next-door types frequently wanted to play romantic leads, while romantic leads fought for character parts. Type casting was a perennial danger; the studios tended to repeat previous successes, assigning players virtually the same role over and over again. Even after a performance was filmed, there was no assurance it wouldn't be cut. "There had to be a kind of strange loving god smiling down on you," Mel Tormé said of his Metro period. "Talent really had very little to do with it. You just had to have a great amount of luck. There was a lot of politicking."

Young actors might get one starring role, but nothing else. Others sat around for six months without an assignment, waiting for their option to run out. The lucky ones, like Janet Leigh, were assigned a major starring role within a few weeks after signing their contract and were guided into parts they were ready for, surrounded by big stars. Unlucky ones discovered something had gone wrong—they looked too much like a star already established on the lot, their talent was too similar, or their scenes were eliminated so that producers at the studio never saw what they could do. Many lasted only six months; the percentage that went on to stardom was low. "I once figured out that if we interviewed 1,000 people a month, 500 of them we *might* have read," MGM talent scout Al Trescony declared; "150 *might* be called back for a second reading, maybe five would be contracted to test, and maybe one would be signed." Of those signed only a fraction made it to the top.

Those who did have their options renewed continued their studies, learned from watching professionals on the set, and sharpened their skills by playing a variety of roles with directors employing different techniques. "I had two lines to say in *Camille* with Greta Garbo," Joan Leslie recalled from her childhood stint at MGM. "One line was, 'So you did come, all the way from Paris.' I had two coaches to coach me on how to say this."

Underage performers like Leslie took their academic schooling on the lot.

California law required that children have three hours of schooling five days a week, except for summer vacation. When not shooting a picture, child actors at Metro attended the studio's "little red schoolhouse," a one-room structure with one teacher who taught kindergarten through twelfth grade. Elizabeth Taylor, Margaret O'Brien, Mickey Rooney, Judy Garland, Freddie Bartholomew, Jane Powell, and Roddy McDowall all took classes there. In a portable dressing room on the set, a tutor offered instruction to all the children working on a particular picture. Since teachers carefully recorded the number of hours each child devoted to study, pupils would be whisked off anytime there was a break in production to put in their required time at school. Each studio had a teacher, and dancer Dante Di Paolo remembered playing football during recess with Donald O'Connor and the other kids enrolled in the school at Paramount. Before a student completed high school, final examinations were administered by the Board of Education in Los Angeles, and arrangements were made for the candidate to graduate with the University High School senior class.

Eventually performers realized that working for a Hollywood studio was not unlike going to a factory for other employees. "My father had a job with Ford Motor Company back home in Detroit," said actress Adele Mara. "He knew that if he couldn't work there he'd have to go someplace else. You didn't try to take over Henry Ford's factory, you just worked there. So that's the way I looked at it when I began my movie career." Many young actors and actresses rode the streetcar to work or walked.

Most actors quickly found that film work was hard and that developing their craft would take time. Actress Julie Adams took a job with the Lippert company, where she made six movies in five weeks. "I was 'the girl' in all of them," Adams said. "I never could remember whether I was the schoolteacher or the banker's daughter. We shot all the stagecoach scenes at once, all the horseback riding scenes at once, and I had three changes of wardrobe for all six pictures."

Despite the competition young players at the same studio developed a rapport and provided each other with a support system. Sometimes two or more contract players would live together to share expenses and would develop lasting friendships. Young women often knew each other from the Studio Club, where they frequently stayed after arriving in Hollywood, since parents knew they would be chaperoned there. "It was run like a great big sorority house," actress Barbara Rush said.

Studios kept a group of girls on hand for the pleasure of film exhibitors and visiting dignitaries, but these seldom stayed longer than the initial six-month option, and rarely did they advance beyond the stock girl level. Sometimes they had been runners-up in Miss Universe contests or cigarette girls in a local nightclub. Generally the casting couch route to stardom, despite much talk to the contrary, was unsuccessful and at best brought fleeting benefits.

Players with talent began to feel the camera's lens and to sense when they were hidden in a scene or part of their face was covered. "I fell in love with the camera," actress Lizabeth Scott declared. "Nothing made me happier than to perform for it. I was mesmerized by the lens."

Hard work though it was, most performers came to love moviemaking. "Those were the happiest fourteen years of my life," Lynn Bari concluded about her time at 20th Century-Fox. Others groused that the studio was exploiting them. Meetings were held regularly to evaluate players, and some insisted the only good parts they got were those when other studios borrowed them. In some unfortunate instances performers discovered that their studio was grooming them to keep an established star from becoming too powerful and demanding. Warner Bros. viewed Ida Lupino, Faye Emerson, and Martha Vickers as emergency replacements for Bette Davis, while Dane Clark served as a potential substitute for John Garfield. At 20th Century-Fox, William Eythe became a cover for Tyrone Power, and June Haver was kept available to replace Betty Grable. Universal initially saw John Saxon as that studio's answer to James Dean. Tony Curtis was becoming a teenage idol about the same time, and within six months Universal had hired five Tony Curtis look-alikes, most of whom remained unknown. Singer Allan Jones realized he would always be in the shadow of Nelson Eddy at MGM, and dancer Gene Nelson's option was dropped by 20th Century-Fox shortly after Dan Dailey was brought over from Metro to star in *Mother Wore Tights*.

Heartbreak was part of the package. During her early days in Hollywood a studio executive told singer Anne Jeffreys, later a star of Broadway musicals known for her beauty: "You aren't particularly pretty, you aren't particularly voluptuous, you haven't got a particularly good figure, your legs aren't particularly good, and your singing is so-so. I really think you should give this whole thing up." Jeffreys was destroyed for a time. Bryan Foy, in charge of B pictures at Warner Bros., was seldom reluctant to hurt people's feelings. Once when an agent brought a cowboy actor to Foy for an interview, the Warners executive turned to the agent and said, "It's too bad about this guy." The agent wanted to know what was wrong. "He suffers from a very severe affliction," Foy remarked caustically. "No talent." (Baiano int.)

While contract players fumed about their working conditions, those who stayed at a studio long enough came to feel part of a family. As with all families, there was bickering and internal dissension, but intimate relationships and shared experiences grew and were nurtured. The police officer on the gate called performers by name, and crews looked after them on the set. Employees could go to wardrobe if they had an evening event and find something to wear.

Having a powerful agent remained crucial for young players, and making contacts on the lot could help. Knowing how to deal with egos and being seen were important, but above all a newcomer had to be liked by the public. Ultimately the public made stars, though the big studios built unknowns into names whose careers frequently extended over a lifetime. Discerning actors realized that much of what they had learned in drama school didn't work in pictures, that for a performer to succeed in movie acting, the personality needed to shine through.

For those who survived, the big studios offered an exciting life, despite the hard work and potential entrapment in stereotyped roles. The studios paced their contract players and built careers. Those who yearned to stretch their talent were often frustrated; those content to work within the system were generally rewarded, monetarily if not with roles that satisfied and challenged them. The exposure was constant, the opportunity to acquire experience limitless, and the publicity invaluable.

6

Actors and Actresses

Metro-Goldwyn-Mayer's twenty-fifth anniversary assemblage of stars and contract players, 1949. Front row (left to right): Lionel Barrymore, June Allyson, Leon Ames, Fred Astaire, Edward Arnold, Lassie, Mary Astor, Ethel Barrymore, Spring Byington, James Craig, Arlene Dahl. Second row (left to right): Gloria DeHaven, Tom Drake, Jimmy Durante, Vera-Ellen, Errol Flynn (in costume for That Forsyte Woman)*, Clark Gable, Ava Gardner, Judy Garland, Betty Garrett, Edmund Gwenn, Kathryn Grayson, Van Heflin (in costume for* Madame Bovary)*. Third row (left to right): Katharine Hepburn, John Hodiak, Claude Jarman, Jr., Van Johnson, Jennifer Jones (in costume for* Madame Bovary)*, Louis Jourdan, Howard Keel, Gene Kelly,*

Christopher Kent, Angela Lansbury, Mario Lanza, Janet Leigh. *Fourth row (left to right)*: Peter Lawford, Jeanette MacDonald, Ann Miller, Ricardo Montalban, Jules Munshin, George Murphy, Reginald Owen, Walter Pidgeon, Jane Powell, Ginger Rogers, Frank Sinatra, Red Skelton. *Fifth row (left to right)*: Alexis Smith, Ann Sothern, J. Carroll Naish, Dean Stockwell, Lewis Stone, Clinton Sundberg, Robert Taylor, Audrey Totter, Spencer Tracy, Esther Williams, Keenan Wynn. *Courtesy of the Academy of Motion Picture Arts and Sciences.*

Once there were no film stars. In the first movies there weren't even screen credits. In the early days of silent pictures studio heads didn't advertise the names of actors, realizing that fame would bring pressure for higher salaries. For the first year or two Mary Pickford was known simply as "The Girl with the Golden Curls." Finally the public demanded to know her name, and the first film superstar was born. But the power of stars grew quickly. Pickford's husband, Douglas Fairbanks, was the boss on his pictures, even telling directors what to do.

Charlie Chaplin worked on both sides of the camera. "Charlie was the most complete man this business has ever known," his friend and later associate Norman Lloyd declared. "He was the greatest actor in the world, he was his own writer, his own director, his own financier. He was the greatest combination of artist, genius, and businessman that I've known. He was a complete individualist; he had the guts to live by it and in the case of his art do it. Charlie was a man of fire, describable but indescribable." Chaplin devised the character of the Little Tramp to convey his views and held to that character into the sound era, aware that his strength was pantomime. "Charlie created a world," Norman Lloyd said. "I always thought of it as the world of the immigrant. He came out of poverty and believed that you had to work every day or you didn't deserve your dinner."

While much of silent screen acting was posturing aided by a pianist or a small ensemble playing mood music, at its best silent picture acting became an art. All the dramatic values had to be conveyed in action and facial expressions, and the story had to be told with the camera. "I think the people who worked in silent pictures really perfected the great techniques of motion pictures," director Joseph Newman said, a view many who rose to prominence after the introduction of sound shared. Douglas Fairbanks, Jr., agreed: "The best of the sound films are those which come closest to being silent. Otherwise we're just photographing stage plays."

Stars became an integral part of movie magic early, and none made a greater contribution than Lillian Gish, who enhanced most of the D. W. Griffith classics. "I felt the screen was invented for a few faces," actress Dolly Haas said, "only a few, and Lillian's face belongs to them." *The Birth of a Nation* (1915), *Way Down East* (1920), and *The Wind* (1928) established Gish as a silent actress without peer, a fact Griffith himself acknowledged. "I never had a day's luck after Lillian left me," the celebrated director admitted. (St. Johns, *Love, Laughter and Tears*) Gish remained the consummate professional throughout a career that spanned seven decades.

But it was Greta Garbo who most completely captivated the public and the industry with her presence and glamour, successfully making the transition to sound. Born Greta Gustafsson in Stockholm, Garbo arrived in Hollywood from Sweden the protégé of director Mauritz Stiller. She proved a sensation in her first American picture, MGM's *The Torrent* (1926). "I remember the day she came into the studio," Joe Newman recalled. "I was in the front office at that time, and in marched this tall, awkward, large-boned young girl. I ushered her and Stiller into Mr. Mayer's office. Six months later, when I saw her on the set, she was transformed. I've never seen such a change. She was magnificent." Actor Gale Gordon, later a fixture on Lucille Ball's television show, was an extra on Garbo's *Flesh and the Devil* (1927). "She was the most beautiful thing I ever saw in my life," remembered Gordon. "She didn't speak English then and wore these beautiful gowns. I'd sit offstage and watch. She was absolutely ethereal." Young Lew Ayres, who made *The Kiss* with Garbo in 1929, was no less affected. "She was something you stood in awe of," Ayres said. "She was quite a recluse even then. Garbo was the first actress I knew to have a private dressing room on the set, just four pieces of canvas siding." Neighboring actors on the Metro lot used to come and peek at Garbo or watch her work whenever they could sneak on the set. Yet she rarely was temperamental or became angry. On *Conquest* (1937) the director kept her sitting in an old whale boat in the Catalina Channel for hours, waiting for the fog to lift so he could begin filming. "Garbo sat out there and talked to us about our lives, our wives, and our children," remembered Gil Perkins, who had been hired as a stuntman on the picture. Earlier, on *Queen Christina* (1933), Perkins heard Garbo threaten to become ill and go home one Saturday morning if the crew wasn't permitted to attend the Notre Dame–USC football game for which several of them had tickets. She gave in only when the assistant director informed her they'd shoot around her if she played sick.

By 1926 sound began to filter into the industry, and by 1928 it was essential whether filmmakers liked it or not. Sound produced an upheaval, with many screen personalities no longer acceptable, most notably John Gilbert and Corrine Griffith. Legend has it that Gilbert's career ended because of his high voice, but director King Vidor denied that. "That's just an excuse. He ended it because he didn't know how to adapt his aggressive lover to sound," Vidor said. "It would have happened to Valentino had he lived. They were both intense lovers. In silent pictures audiences would wonder what they were saying and would fit their own words to the action. But suddenly they had to speak and say things like 'I adore you, I worship you, wait till I get you in bed tonight.' And it became laughable." Others, like Emil Jannings, never fully conquered the English language and went back to Europe.

To fill the void, movie studios began to raid the Broadway stage. Such legitimate actors as Fredric March, Paul Muni, James Cagney, Spencer Tracy, and Pat O'Brien decided to give Hollywood a try. Once they had made a hit in a play, studios were eager to test them. Lloyd Nolan signed with Paramount because the studio was the only one that would give him six months off to go back to Broadway. Anyone with stage training was in demand, including drama students, and youngsters with theatrical experience—Bette Davis, Joan Bennett, Katharine Hepburn—suddenly found themselves in an advantageous position.

Paramount's Astoria studio in New York was quickly renovated; its facilities were convenient for Broadway actors who knew how to handle dialogue and wanted to moonlight during the daytime hours. Some were consciously testing their effectiveness in pictures, hoping the talkies might offer them a chance to experience fame, fortune, and California sunshine. Ginger Rogers was "discovered" in the Gershwin musical *Girl Crazy*, made her initial appearance before movie cameras at the Astoria studio in Queens, and in 1931 went to Hollywood under contract to Pathe.

The second gold rush was on, as actors and would-be actors flocked to California. In New York the theatrical grapevine insisted, "It won't last," but the westward migration continued. Charles Starrett, a handsome college athlete with minimal Broadway experience, soon found himself a cowboy star, famed as the Durango Kid, while Clark Gable, with slightly more stage exposure, became a matinee idol, eventually known as the "King of Hollywood." Others considered movies beneath them; New York actors looked down on film personalities. "There was an ingrained snobbism among actors like myself that would vent itself on purely picture people," Norman Lloyd, a veteran of the Mercury Theatre, confessed. "It was ill-founded and silly." Yet most Hollywood producers and directors gave preferential treatment to anyone who had worked in the theater, even though studio moguls complained that Broadway types didn't photograph well or weren't sexy enough for the screen.

During the early 1930s Warner Bros. billed the august stage actor Paul Muni as "Mr. Paul Muni." Screen co-workers frequently found the actor haughty and difficult and failed to understand his introspective nature. "He was a creep," said Cornel Wilde, who worked with Muni on *A Song to Remember* (1945), "a very insecure, confused, constantly upset, jealous man. All through the picture he never looked me in the eyes; he would look between my brows or at my forehead. He never allowed any kind of real communication between us, which destroys ensemble playing. I played my role sort of low-key. He played his very high and full of all kinds of tricks—blowing his nose, taking out his glasses and breathing on them, constantly moving." Muni depended on his wife, Bella, and many felt she dominated him. Bella would stand behind the camera, and if she didn't like the way a scene had gone, she'd shake her head. Muni would demand that the scene be reshot.

Most stage actors took film work seriously and were disposed to delve into characterization, exploring the historical context of scripts as well as mannerisms and technical points. Spencer Tracy emerged as the supreme screen actor, perfecting the art of reacting. Always prepared, Tracy not only had a way of phrasing that made dialogue sound natural, he also made even the slightest gesture seem spontaneous. "I would see the dailies," MGM contract player Barry Nelson remembered, "and what I didn't learn sitting on the set watching Tracy's expressions, I could see in the close-ups and two-shots. I saw how much he had added of what that character was thinking when someone else was speaking. Tracy never wasted a movement or a thought. There was great economy in his playing."

Tracy's credo was "Just learn the lines and don't bump into the furniture." His first take was usually his best, although in successive shots he would rearrange props slightly to sustain spontaneity. During the 1950s Katharine Hepburn, with whom he collaborated on screen and off, talked him into going back to Broadway for a play. After six weeks he quit. "I couldn't say the same goddamn lines over and over again every night," Tracy complained. "I'd forgotten how boring that can be. I wasn't creating anything. In films every day is a new day for me. But in the theater once you've done opening night, that's it. There's very little more creating you can do." (Dmytryk int.)

Other stage actors—Henry Fonda, John Garfield, Robert Ryan, and later Montgomery Clift, Marlon Brando, and James Dean—became commendable screen actors, but none was more natural than Spencer Tracy. "Screen acting," Gregory Peck explained, "is as close as you can get to the literal truth. If you're thinking straight, if you're drawing on honest emotions and you've examined your inner self to find the truth of the situation, you are virtually living the role, and the camera will pick it up. It will also pick up falsity and phoniness and concealment. Anything that isn't true will register, and something then falls between you and the audience."

Some actors never understood camera technique; they didn't realize that the camera picks up everything. Intimate scenes in particular require great concentration. "In a two-shot," observed veteran stage actor Walter Abel, "if you're not hearing what is said to promote or invoke the answers to the dialogue you've just heard, you're out. Charles Boyer was one of the greatest listeners in the business. He would *pin* you, he would fascinate you like a snake, so that you couldn't get out of his grasp." Above all, screen acting must be realistic, so subtle and honest that it appears not to be acting at all. "A lot of people work from the outside in; I work from the inside out," Academy Award–winning actress Joan Fontaine said. "I don't use false noses and artificial gestures. Once you become the person you're playing, these things happen automatically."

Solely a screen actor, Gary Cooper mastered his craft so that he knew exactly how to react for the camera. "He learned how to use his eyes so subtly and so well," actress Barbara Rush explained. "You had to get up close to him before you could see that he was giving something, and it was all with his eyes. Cooper knew how to make things small." Though he was capable of more than his "Yep" stereotype, Cooper told his writers repeatedly, "Keep the dialogue short and I'll be able to handle it."

Cooper took acting seriously, yet was not as confident as many of his colleagues. Ronald Colman, for instance, came from the stage, had a magnificent speaking voice, but like Cooper, entered films during the silent era. "You felt Ronnie wasn't doing a damn thing," said actress Jane Wyatt, who worked with Colman on *Lost Horizon* (1937). "It all happened in his eyes." Employing a different approach, James Stewart emerged as a no less brilliant screen actor. "I remember one day I was doing a scene with him," said Julie Adams, who appeared with Stewart in *Bend of the River* (1952). "I was sitting in the back of this covered wagon reading the lines with him when I thought, 'How does he do that? He isn't doing anything.' He had mastered that absolute ease, that absolute communication of everything he wished to get across. Only he wasn't working at it at all."

The camera was particularly kind to a few people. "It either loves you or it doesn't," dancer Julie Newmar said. "It's a strange phenomenon." During the big studio era most of the leading men and women were physically beautiful people, with photogenic faces. Tyrone Power and Robert Taylor were both handsome and meticulously groomed. Power had a chiseled profile, as well as a sonorous speaking voice, and while he worked hard at becoming a fine actor, his good looks won him a mass following. Charming and sensitive, Power was a gentleman, admired by those who worked with him. "He had an aura about him that set him apart from everybody else," actress Coleen Gray recalled. "I had an enormous crush on him and felt his feet never touched the earth." Robert Taylor, although not as accomplished an actor, was no less admired. Quiet and shy, Taylor never adjusted to being a matinee idol and regarded acting as an unworthy occupation. "Stardom didn't sit very happily on his shoulders," producer Armand Deutsch said.

Hedy Lamarr, among the screen's beauties, never developed into a great actress, but hers was an extraordinary presence. A private person, Lamarr was difficult to deal with, a prima donna on the set. Yet many fell in love with her. Other glamour queens demonstrated more equanimity. Rhonda Fleming—gorgeous, especially in Technicolor—was a better actress than most people thought, and Arlene Dahl—with her red hair, blue eyes, and alabaster skin—took her work seriously, returning to the stage and stretching her talents. Elizabeth Taylor, an instinctive actress and a renowned beauty, proved her intrinsic abilities over and over again. "In my book," Metro drama coach Lillian Burns said, "Elizabeth is a definitive motion picture actress."

"I came to realize very early in films," actor Charlton Heston said, "that the actor is the servant of the camera. The camera is what tells the story, and the actor, in a sense, is merely the most important prop." (Heston int., McBride) But the actor must be groomed to meet the camera, and in the golden days of Hollywood, perfection became the goal. Even serious actresses were concerned with physical appearance—keeping trim, having caps put on their teeth, adding false bosoms and eyelashes. Claudette Colbert hated her nose, so she had the makeup department apply green paste along it to reduce it for the camera. Clark Gable's ears had to be altered, and Rita Hayworth's hairline was raised. Kirk Douglas, playing an Indian fighter, would do push-ups right before the cameras rolled, so he would come into the scene exuding vigor and athleticism. If an actress had a blemish, she had to be photographed carefully. If she was exceedingly beautiful, the camera might luxuriate on her in prolonged close-ups, particularly early in her career when the public was discovering her. Even in dramatic scenes realism had its limits. During the making of *The Sea of Grass* (1947), Louis B. Mayer complained that Katharine Hepburn cried too much in the film and that the channel of her tears was wrong. When director Elia Kazan asked for an explanation, Mayer replied, "The channel of her tears goes too close to the nostril; it looks like it is coming out of her nose like snot." Kazan insisted he couldn't do anything about the channel of Hepburn's tears. "Young man," Mayer answered, "you have one thing to learn. We are in the business of making beautiful pictures of beautiful people, and anybody who does not acknowledge that should not be in this business." (Kazan int., Ciment)

While stars sometimes felt like puppets in the hands of producers and directors, they became gods and goddesses to a public around the globe. At their height Douglas Fairbanks and Mary Pickford were probably the best-known couple in the world because of the wide distribution of silent pictures. John Gilbert and Rudolph Valentino were admired by millions, including a number of youngsters who later became Hollywood luminaries themselves. By the 1930s women stood in line to see what Joan Crawford was wearing in her current film, and designers took into account the public's interest. In the early 1940s beauty shops all over America advertised Veronica Lake's "peekaboo" hair style, and girls across the country cut their hair to look like June Allyson. The story was told of an American woman who visited Westminster Abbey some years after actor George Arliss won an Academy Award for his portrayal of the title role in *Disraeli* (1929). Coming upon a bronze figure of the great British statesman, she exclaimed, "My, what a wonderful statue of George Arliss!" (Warner, *My First Hundred Years*)

Few were exempt from Hollywood's magic. Actresses Helen Hayes and Ruth Gordon frequently went into New York City together when they weren't rehearsing a play, would catch the noon movie at the Capitol, have lunch, see the four o'clock show somewhere else, then maybe end with a third feature before driving back to Hayes's home in Nyack. "We were that nutty about movies," Helen Hayes said. Girls everywhere dreamed of finding their celluloid prince. After Anne Frank had died in a concentration camp, the world learned that among the possessions this child of the Holocaust had treasured during her family's last days together was a photograph of Hollywood actor Robert Stack.

The gloss and hyperbole of movies may have helped successive generations through depressions, wars, and global crises, but seldom did anyone outside the business understand the responsibility placed on Hollywood stars. A picture's success or failure often rested on their shoulders. Most actors and actresses were perfectionists, determined to do their best even with a bad script. "The trick is to make a film of integrity and style and still recognize that you're borrowing somebody's money that has to be returned," Gregory Peck said. "It's art and commerce blended, and the challenge is to overcome the commercial aspect and come out with something that's honest and fresh and creative." Some stars claimed they were assigned three or four mediocre scripts, or more, for every good one. Others admitted there were times they took on an assignment solely for the money. Even more than the director, the star was accountable for a picture's draw at the box office. Although stars might be treated like royalty, theirs was a precarious position, dependent on popular appeal.

For those at the top a bevy of experts—press agents to hairdressers, fashion designers to studio limousine drivers—catered to their needs and enhanced their images as movie stars. The size of the fortunes amassed during the old studio days has been exaggerated, however; most stars did earn large sums, but only a few managed to save enough to survive comfortably in later life. In 1945 Lana Turner became one of the most highly compensated actresses in the world when MGM paid her $4,000 a week. But even the biggest stars worked twelve to fourteen hours a day when they were making a picture. Throughout most of the big studio era crews worked six days a week, and actors spent most of Sunday resting or learning lines. Many went to bed on Saturday night and spent all the next day recuperating. Robert Cummings remembered starting nine films in one year, and Dale Evans made eight pictures her first year at Republic. There were also constant interviews, picture layouts for fan magazines, and personal appearances. Seasoned movie actress Virginia Mayo recalled working weekends with young Paul Newman, fresh from the stage, preparing him for his first screen role in *The Silver Chalice* (1955). "Many times somebody did the same thing for me," Mayo said.

The standard studio contract was for seven years, divided into options. During the first year, if executives thought an actor or actress could draw at the box office, they took up the six-month option. From then on, if they were convinced a performer was on his or her way, each succeeding option was for one year, except the last, which was for two years. A raise accompanied each picked-up option. Contracts were for forty weeks a year; players went twelve weeks without pay. During this "vacation" time they couldn't work elsewhere without special permission. Rarely was the three months a single span; it was divided between picture assignments. If an actor finished a picture on the first day of the month and started the next on the eighth, seven days without pay were counted as part of the vacation period. Players had little or no choice over which roles they were assigned or when. If they refused a picture, they were suspended, banned from working for anyone else, and the time was added to the term of their contract.

Although the studios insisted on allegiance from their contract players, they in turn took care of nearly everything for them, both personally and professionally. "If you didn't like the cook you had," Rock Hudson said, "you told the studio, and somebody else was hired. The house you lived in, the shopping, the gardener, your meals were all managed for you so that the only thing you had to be concerned about was your performance." If actors wanted an airplane ticket, they called the studio travel department, and they were met at their destination by somebody to take care of details. "MGM treated us like we were a mixture of jewel and retarded child," singer Fernando Lamas declared, "because they kept us in cotton. Everything was done for us. If we had a headache, three doctors came on the set with three pills and three shots. If we needed a driver's license, some inspector from the department of motor vehicles came to the studio. We never went anyplace." Such pampering was not healthy for some in the business, and they got into trouble emotionally. Many of those who became caught up in the myth didn't survive.

By the 1930s powerful agents had appeared in Hollywood—Myron Selznick, Minna Wallis, Leland Hayward—who negotiated advantageous clauses for their top clients, sometimes splitting their contracts between studios, securing directorial and casting approval for them, making sure they had dressing rooms and accommodations commensurate with their salaries. Lana Turner by 1941 already made $1,500 a week at Metro; on the other hand, five years later, as a star at low-budget Republic, Catherine McLeod earned $450 a week. "They were good to us," 20th Century-Fox contract player Don Ameche acknowledged, "but they were also exploitative. All the studios were. I don't really think they cared a hell of a lot about what our future was going to be. The only studio I know that had a retirement program was Metro."

The old studios were huge repertory companies shuffling their employees around in order to keep them working. MGM boasted "more stars than there are in the heavens," but the studio was in reality a gigantic factory with top talent turning out a steady stream of product, all with Metro's special trademark. Warner Bros. had a smaller contract list, while RKO and Universal had even smaller rosters, but each churned out a distinguishable product in sufficient quantities to fill their theaters. "It was all a way of making a great deal of money for everyone," Myrna Loy said; "it was a factory. You walked onto the set and people were still hammering and putting things together. There was actually no glamour, just damned hard work." (Kotsilibas-Davis and Loy, *Myrna Loy*) Stars who survived came to regard moviemaking as a way of earning a living. "Acting was a basic element of my life," Virginia Mayo said. "It was something I did that I enjoyed, and the studios paid me well."

Casting offices tried to place actors and actresses as strategically as possible, teaming them so their "chemistry" caught the public's attention. Once a winning combination was established, vehicles were developed for the stars involved. No sooner had Alan Ladd and Veronica Lake proved a sensation in *This Gun for Hire* (1942) than Paramount paired them again in *The Glass Key* (also 1942). If an actor and a director made a highly successful picture together, they were usually teamed again and again, as were Gregory Peck and Henry King at 20th Century-Fox. Casting always entailed box office considerations. When Hal Wallis decided to hire Broadway actress Shirley Booth to re-create the role of Lola in the film version of *Come Back, Little Sheba* (1952), the producer protected his picture at the box office by casting popular screen star Burt Lancaster opposite Booth as Doc, even though Lancaster was too young for the part. In other instances casting was determined for dramatic impact. Alfred Hitchcock wanted an established personality killed in the now-famous shower scene in *Psycho* (1960), so that it would come as a shock to audiences. Had he used an unknown actress, the scene (positioned as it is early in the picture) would have had far less impact. Often actors with stage experience were chosen for supporting roles to give a picture depth.

Assignments for promising contract stars were meticulously orchestrated to insure public acceptance. After Jennifer Jones won an Academy Award for playing a French saint in *The Song of Bernadette* (1943), David Selznick felt it important for her career to make the transition to an American girl-next-door type in *Since You Went Away* (1944). He then cast her as the sexy half-breed in *Duel in the Sun* (1947), linking her romantically on the screen with Gregory Peck, who played an amoral Texas cowboy and a devil with women, shortly after Peck had portrayed Father Chisholm in *The Keys of the Kingdom* (1945). Offbeat casting was a frequent device for adding interest to a picture, and performers generally welcomed such departures.

Having a major star in a film assured its producer a level of quality regarding budget, wardrobe, sets, and the like, because the studio had an investment to protect. Stars protected their own interests by not permitting actors or actresses who resembled them in appearance to be cast in their pictures and sometimes tried to have them fired from the lot. If a young rival was cast against a star's wishes, his or her part likely was reduced during shooting.

Promising new stars had their assignments paced so that roles matched their developing talents. Rock Hudson was cast in Universal's remake of *Magnificent Obsession* (1954) four years before the picture was actually shot, allowing the handsome young actor time to mature into a solid dramatic part. "I give the studio credit," Hudson said later. "They put me in all of those westerns and Indian pictures to get some training. By the time *Magnificent Obsession* came around, I was ready."

Agent Leland Hayward wisely maneuvered newcomer Gregory Peck into a series of pictures in which the actor played opposite famous leading ladies—Greer Garson, Ingrid Bergman, Jane Wyman, Joan Bennett, Jennifer Jones—which introduced him to their fans. "I had an extraordinary good run of pictures," Peck observed, "probably because my contract had been split between David Selznick and 20th Century-Fox. Neither felt they owned me."

In some cases performers with talent simply didn't make the impression on the public that producers judged essential. Sometimes they were misused or cast in one good part, then ignored. Robert Alda won success at Warners playing George Gershwin in his first picture, *Rhapsody in Blue* (1945). "After that they weren't quite sure what to do with me," Alda recalled. Finally the studio put him in a number of B pictures. "They started me in a Tiffany production and then started putting me in Woolworth shows," the actor claimed. Meanwhile Warners refused to loan him to other studios, promising him better parts that never came. Alda left Hollywood for Broadway, where he created the part of Sky Masterson in the musical *Guys and Dolls*.

Many actors and actresses enjoyed playing the variety of roles the old studios assigned to them. If they were cast in a picture they didn't like, they sometimes used that film to bargain for another part they really wanted. As in all factories, there was griping within the Hollywood studios. "Everyone complains," Doris Day said. "No matter what you're doing, you complain, especially when you're a contract player." Many felt stifled by their long-term contracts and envied those actors, such as Charles Boyer, who never committed themselves to any one studio. Ricardo Montalban once insisted that in all his years at MGM he never got a single good role. (Tormé int.) Actress Audrey Totter claimed she accepted a mediocre assignment in *Any Number Can Play* (1949) at Metro because she was promised the part of Jo in the studio's upcoming remake of *Little Women*. Later the executives involved denied having promised her the part, which went instead to June Allyson. Totter left MGM right after that. Others complained they were forced to make film after film they despised. James Cagney claimed that most of the scripts he was forced to do were dreadful. Cagney once received a script at Warner Bros. and asked the producer if he had read it. When the producer assured him he had, the actor incredulously asked why he wanted to make the film. "We've got to do it," the producer explained. "The picture is already sold." (Cagney, *Cagney By Cagney*)

At Warner Bros. the actors' attitude seemed to be "What can we do to thwart Jack Warner?" The production head fought with nearly everyone on the lot—Bette Davis, Humphrey Bogart, Olivia de Havilland, James Cagney, and Errol Flynn. "Everybody under contract to Warner Bros. couldn't wait to get the hell out," actor Robert Stack said. "They always felt it was penal servitude."

Studio executives not only dictated the roles their actors played, they tried to control their social life as well. Louis B. Mayer encouraged contract player June Allyson to date studio heartthrobs Van Johnson and Peter Lawford and would have been delighted had she married either. But when the actress announced her plans to wed actor Dick Powell, whom he had forbidden her to see, "Papa" Mayer called Allyson into his office. "Dick Powell is too old," the mogul hotly informed her. "He's a has-been. And he's a divorced man. What are you doing to yourself, child? Your career will be over. This will ruin you." (Allyson, *June Allyson*) Not only did Allyson and Powell marry, they later made two pictures together at Metro.

Although a few contract players cheerfully accepted whatever roles were offered to them, including those other actors had turned down, or frankly admitted they were not ambitious, arguments between studios and their contract personnel were endless. "The casting director giveth and the casting director taketh away" became a familiar Hollywood motto. When Lana Turner refused the part of Lady de Winter in *The Three Musketeers* (1948) because it wasn't a starring role, Metro-Goldwyn-Mayer suspended her. After meetings and negotiations, the script was rewritten enlarging Turner's part, and she finally accepted it. Laraine Day had been featured in an earlier Turner film, *Keep Your Powder Dry* (1945), which she hated. She wanted to do another script called *Undercurrent*. Day agreed to make the Turner picture if the studio would let her star in the other film afterwards. Then Katharine Hepburn read the *Undercurrent* script and wanted the part. The studio chose Katharine Hepburn. Day complained to Benny Thau and asked for a release from her contract, which Metro willingly granted.

Joan Fontaine consented to make *From This Day Forward* (1946) for RKO but detested the script, complaining about writer Clifford Odets's realistic approach. For contractual reasons she had to do the film. Eddie Bracken spent three miserable years at Paramount, where he played Norval Jones in Preston Sturges's fine comedy *The Miracle of Morgan's Creek* (1944), under the screenwriter's direction. After that the actor Eddie Bracken was forgotten: "The writers were writing scripts for Norval Jones," Bracken said, "and I didn't want to play Norval Jones anymore unless Preston Sturges wrote it. The others were writing about a jerk who did stupid things. There was no substance behind it." But actors and actresses under contract were expected to accept whatever roles they were assigned. While their names and faces were kept before the public, professional growth often suffered.

Frequently the studio would star a budding contract player in an inexpensive picture to test his or her box office draw. The problem was that young actors could become so identified with B pictures that they were trapped there. When actors applied for jobs at one of the major studios, they rarely mentioned their work at Columbia or Republic. "If a big company like MGM heard you had been working for those Poverty Row companies," Douglas Fairbanks, Jr., noted, "that would be a black mark against you and you wouldn't be hired."

Overnight success could be a curse as well as a blessing. Lew Ayres enjoyed early success with *All Quiet on the Western Front* (1930), followed by two other popular films. "After that began my period of struggle," said Ayres, "trying to maintain what had been achieved so quickly and simply. Having no knowledge of what it meant to try to sustain a career, I was in for some great shocks." Studios tried to keep their stars young, creating an image that was at odds with personal maturity. "I wanted to grow up and not be the girl next door," Jane Powell remarked about her MGM years. To mature as an actress, she had to leave the studio.

Performers complained when their studio loaned them to other companies at a profit. If a studio had nothing scheduled for an actor and another company asked to borrow that person, an arrangement might be worked out whereby the home studio loaned the actor for $1,000 a week, paying him his usual $450 and pocketing the difference. Actors had little say in such negotiations, and they frequently objected to the parts or grumbled that they were being sold like cattle. On other occasions they seemed flattered that another studio wanted them and claimed they received better treatment away from their home lot. Sometimes they were given excellent roles. Grace Kelly's best pictures were always made away from Metro; Rock Hudson's Academy Award nomination for *Giant* (1956) came as a result of a loan-out to Warner Bros. "What *Giant* did for my career also increased Universal's investment in me," Hudson pointed out.

Conversely stars were often angered when their studios refused to loan them for a part they wanted. Rock Hudson had an offer to make *Ben-Hur* (1959) at Metro, but Universal wouldn't let him go unless he added more years to his contract, which Hudson refused to do. "I wanted out of Universal as soon as possible," the actor said, "but I was wrong. I should have given them whatever they wanted." Arlene Dahl was set for the movie version of *Cyrano de Bergerac* (1950) opposite Jose Ferrer, until Metro made such exorbitant demands for her services that producer Stanley Kramer felt he couldn't meet them. Dahl had performed the role on Broadway for a while and was crushed to lose such a choice screen part, convinced her career suffered from the loss.

The big studios could be spiteful, blackballing actors who incurred their wrath. If performers became restive or threatened to quit, no other major studio would hire them without their home studio's permission. Sometimes deals were arranged that posed problems for directors and cast members during production. In order for RKO to acquire Dorothy McGuire from David Selznick for the lead in *Till the End of Time* (1946), director Edward Dmytryk also had to take Guy Madison. "Madison was a beautiful-looking guy," Dmytryk said; "he just couldn't act." Fan magazines had made Madison popular before he ever made a film, and Selznick thought he could make a star out of him, which he did. Earlier Selznick, the prime wheeler-dealer, who had more stars under contract than he had parts for them, made a two-picture commitment for Ingrid Bergman with Warner Bros. in order to secure the services of Olivia de Havilland for *Gone With the Wind* (1939). Bergman, a newcomer then, was included as part of the package; she later made *Casablanca* (1943) and *Saratoga Trunk* (1944) for Warners, both of which were good roles for her. "It was like a butcher with a whole lot of dressed lambs in the window," said Joan Fontaine, who was also under contract to David Selznick, but made only one picture for him. "You were the lamb on the end. There was no consultation about plans for me, and I received terrible, insulting telegrams if I balked at a film."

Warner Bros. became notorious for suspensions, although actor Pat O'Brien claimed that much of Jack Warner's toughness was an act. "When you went on suspension," O'Brien said, "you didn't get a salary, but Warner gave it to you later as a bonus. It looked good in print: 'Disciplinary action has been exercised.'" The threat of suspension was enough to bring most stars to heel. Swedish actress Viveca Lindfors, under contract to Warners, refused to appear in *Backfire* (1950) at first because of the film's violence. When she was placed on suspension, she changed her mind. "I sold out," the actress later said.

In 1937 Bette Davis, Warner Bros.'s biggest money-maker, determined to test the studio's suspension powers by refusing to make *God's Country and the Woman*. Intelligent and strong-willed, Davis decided the time for action had come. "If I never acted again in my life, I was not going to play in *God's Country*," she avowed. "It was . . . a matter of my own self-respect." The actress was put on a three-months suspension without pay as punishment, while the press made her out to be greedy and obstinate, reporting that she simply wanted more money. After *God's Country and the Woman* went into production with a lesser actress, Davis departed for England, planning to make two films in Europe. Warners served her with an injunction, prohibiting her from working anywhere. "It was evident that the entire Motion Picture Industry was backing Warner Brothers," the actress wrote, "for were I to win the case, every major star would rush for the nearest exit and follow me to freedom. Not one other film company in Hollywood would touch me with a ten-foot pole." Davis sued Warners from London, but failed in a heated court battle. Although she had lost the skirmish, in effect she won the war. To win Davis back, Jack Warner not only agreed to pay her legal fees, but began assigning her better roles. "I won after all, Davis concluded. "Jack Warner now offered me the excellent *Marked Woman*, and I settled down to real work." (B. Davis, *The Lonely Life*)

There was a feeling of unrest brewing in Hollywood. James Cagney already had taken Warner Bros. to court over his contract; Carole Lombard was fighting with Paramount, Katharine Hepburn with RKO, Margaret Sullavan with Universal, and Eddie Cantor with Samuel Goldwyn. Stars were beginning to realize their strength. When Warner Bros. handed Olivia de Havilland a series of mediocre scripts, following her triumph in *Gone With the Wind* for Selznick, the actress rebelled. Placed on suspension, de Havilland took her fight to the Superior Court of California. Over the years the studio had suspended her six times, and in each instance her contract had been extended the length of the layoff period. De Havilland filed suit against Warners, invoking California's antipeonage law that limited to seven years the time an employer could enforce a contract on an employee. Hearings dragged on for months, while an angry Jack Warner blacklisted the actress with every studio in Hollywood. For three years she was absent from the screen. Then in March 1944, the court handed down the landmark de Havilland decision, whereby actors were released from serving out time added to their contracts through suspensions. "Every contract player owed Olivia a great debt of gratitude," her sister, Joan Fontaine, wrote. (Fontaine, *No Bed of Roses*)

While stars grumbled and often couldn't wait to terminate their contracts, most knew it was their studio's publicity machinery that had obtained their celebrity status. Some performers were too independent or too undisciplined for the big studio structure and fought the system. All agreed, however, that the system provided a comfortable life, one in which stars were spoiled outrageously. "If you got to be the super echelon star," Eddie Bracken said, "you were the king of the walk."

There were always gradations in acting ability, but movie stars of the Golden Era fell into two basic categories—actors and personalities. "One kind really becomes the part, and you don't even know that he is an actor," director Henry Koster explained. "The other is always an actor, but is so great that no matter what he does you admire it and believe every line he says." Most of the old studio stars had created a persona, and they acted that personality no matter what role they played. Audiences flocked to the theaters more to see their favorite stars than to watch realistic performances. Much of the fascination and excitement came from the beauty and personality of the stars, and if they spoke simply and truly, as themselves, audiences were satisfied. "Sometimes it really [was] an intensity inside," director George Cukor said. "Often it [was] the way the light hit the eyes. They [were] born to be creatures of the camera." (Sherman, *Directing the Film*)

Even Spencer Tracy, while he delved deeply into the parts he played, was always Spencer Tracy, and Bette Davis remained Bette Davis, although her technique was honed to perfection. Both were genuine actors with strong screen personalities. So were Katharine Hepburn, James Cagney, Rosalind Russell, Gregory Peck, Greer Garson, Edward G. Robinson, Ingrid Bergman, Barbara Stanwyck, Humphrey Bogart, and Claudette Colbert. Others, such as Dorothy McGuire, Alexis Smith, Eleanor Parker, Anne Baxter, and Patricia Neal, were probably more complete actresses, in part because their personalities lacked the thrust of bigger stars. "The problem with every great star is holding them down," film director Vincent Sherman observed. Bette Davis and Katharine Hepburn, for instance, were inclined to press things to the limit, largely because theirs were such strong personalities.

Most of the great Hollywood stars were almost pure personality: Clark Gable, who didn't much like acting; Jean Harlow, sweeping out of her dressing room followed by her maids and retinue; Marlene Dietrich, disciplined, but reserved; Errol Flynn, a charming rogue with an irreverent, wicked humor, whom everybody adored, men as well as women; Carole Lombard, Hollywood's darling, with beauty, generosity, and a raucous wit; Mae West, who was her own writer, director, photographer, art director, and star; John Wayne, who realized that playing John Wayne was his most saleable asset and seldom tried to "act"; Lana Turner, a great presence who developed into a skillful actress; Alan Ladd, created by his agent-wife, Sue Carol, who possessively looked after his needs; Joan Crawford, dedicated to her craft, who dressed and behaved like a movie star at all times; and, of course, Marilyn Monroe, who became the tragic icon of Hollywood's demise.

Many film stars came from the music field, popular through their recordings and radio appearances before they entered movies. Gregarious, mellow-voiced Bing Crosby emerged during the Depression as the leading "crooner," enjoyed huge success with Paramount's *Road* pictures, and won an Academy Award as best actor in 1944 for playing a Catholic priest in *Going My Way*. Frank Sinatra, the bobby-soxers' singing heartthrob of the World War II era, also won an Oscar, cast against type as tough little Maggio in *From Here to Eternity* (1953). Sinatra was a natural actor who disliked extensive rehearsals, preferring to keep his performance spontaneous. Elvis Presley, the rock idol of the 1950s, became a competent actor, with a natural dramatic instinct like Sinatra and Crosby. "I shall never forget his eyes," said actress Lizabeth Scott, who made *Loving You* with Presley in 1957, "so beautiful, so exquisite. And he was so talented. Elvis was one of the most polite young men I have ever met. He was a simple person, a little boy really."

Of the highbrow singers, Jeanette MacDonald, Nelson Eddy, and Mario Lanza, all of whom at one time sang in opera, enjoyed popularity in films. Young Jane Powell had had extensive radio experience before becoming Metro's "baby soprano," while Paramount's Dorothy Lamour and Betty Hutton, Warners' Doris Day, and Republic's Dale Evans had all been big band singers. Irene Dunne, who became a fine actress and a sophisticated comedienne, entered movies from the musical comedy stage, and Judy Garland, a formidable actress as well as a peerless vocal talent, began in vaudeville. Revered by those who worked with her, even after she became difficult, bitter, and emotionally unstable, Garland remains one of Hollywood's haunting tragedies.

Every major studio had its personality actors under contract, performers the camera loved but whose dramatic talents ranged from solid to competent to painfully inadequate. MGM boasted June Allyson, Van Johnson, Ava Gardner, Ricardo Montalban, Esther Williams, Peter Lawford, and later Debbie Reynolds. Paramount had Paulette Goddard, Bob Hope, Veronica Lake, Sonny Tufts, and Gail Russell. Twentieth Century-Fox made much of Alice Faye, Betty Grable, Victor Mature, Gene Tierney, Cornel Wilde, Jeanne Crain, Don Ameche, and Linda Darnell. Warner Bros. claimed Joan Blondell, Ann Sheridan, Ronald Reagan, Lauren Bacall, and Virginia Mayo, while RKO capitalized on Ginger Rogers and later Jane Russell. Universal went in for exotic types, notably Yvonne DeCarlo, Maria Montez, and Maureen O'Hara; Rita Hayworth became Columbia's major asset after director Frank Capra left. Skater Vera Ralston, the *inamorata* of studio head Herbert Yates, may have been queen of the lot at Republic, but cowboy personalities Gene Autry, Roy Rogers, and Bill Elliott brought the company its profits. These performers projected a charisma the camera picked up. "What makes a star," actress Jane Wyatt declared, "is something to do with the whole person, not just the way they read a line."

Some stars were not fundamentally interested in acting, but saw movie-making as a means of earning a living. Twentieth Century-Fox leading man Dale Robertson was one. "After World War II, I needed something to make enough money to get back in the horse business," said Robertson. "That doesn't mean I took my job any less seriously than some so-called dedicated actor." Those from the legitimate theater deplored such casual attitudes. Actress Nina Foch remembered Paul Muni calling her into his dressing room on the first day of shooting her second picture, *A Song to Remember*. "Do you want to be an actress or a whore?" demanded Muni, who had known Foch's father. The young actress was shocked and assured him she wanted to be an actress. "Well, work, work, work," Muni told her. "You have your choice here, but you must make that decision right now." Foch conceded he was right. "That was the decision," she acknowledged later, "not only in Hollywood, but in any big industry where the stakes are terribly high. There was nothing to recommend this town at all; there was nothing cultivated about it anywhere. It was filled with brown bodies and brown minds."

Yet not only were most Hollywood actors and actresses hardworking, the successful ones were obsessed with their careers. Joan Crawford's preoccupation with her work was so intense she wasn't interested in anything outside Culver City. After Crawford and Douglas Fairbanks, Jr., married in 1929, they went to Europe on a delayed honeymoon. "She hated it," Fairbanks said, "hated every moment of being in London and Paris, where she'd never been before, because she felt insecure away from the movie studios. She was the hardest worker I ever saw. Her only excess I can remember was an excess of ambition. She was completely absorbed with her career and with work and intensely jealous of her competitors at MGM."

Actress Irene Dunne once remarked, "If a picture of mine is good, people will praise the director and the writer and the cameraman, and they will say I was very good in it. But if it's bad, they will say, 'Irene Dunne's latest picture was bad.'" (Doran int.) Stars therefore felt a heavy burden, and while most were dedicated professionals, they could also be temperamental and demanding, concerned with their appearance and with perfection in every detail surrounding a production. "There are only two times when I like actors and actresses," Hollywood costume designer Orry-Kelly supposedly said, "the first three months of their career or the last three months of their career. In between they're impossible to handle." (Koster int.) The difficulties stemmed from the insecurities stars felt from knowing they had a reputation to maintain, aware that if they slipped, their career could be finished. "Looking back on Hollywood," Joan Fontaine wrote, "I realize that one outstanding quality it possesses is . . . *fear*. Fear stalks the sound stages, the publicity departments, the executive offices. Since careers often begin by chance, by the hunch of a producer or casting director, a casual meeting with an agent or publicist, they can evaporate just as quixotically." (Fontaine, *No Bed of Roses*)

Hollywood could be savage, and the expression "You're only as good as your last picture" proved true. Men claimed actresses were more terrified than actors, since they depended on beauty, but even Gable evidenced insecurity about his work. Many performers admitted to being shy, and most were sensitive and vulnerable. "I think there are two things you have to have in acting," said Pamela Mason, former wife of British actor James Mason. "You have to have enormous determination, resilience, and the ability to be hurt. And you have to want it enough to suffer for it. All actors, no matter how important they are, are always out of work, and they always think they'll never work again after each job. It's a cruel business." Although the big studios offered steady employment to contract players, they were feudal courts, with a well-defined pecking order that could change overnight. Several actors who left to serve in World War II returned home to find audiences had forgotten them. When Joan Crawford's contract with MGM was terminated after eighteen years, the star went to Warner Bros., where she revamped her image and won an Academy Award for *Mildred Pierce* (1945). But the trappings in Burbank appeared second-rate after the pampering Crawford had received at Metro, and playing suffering heroines in pulp melodramas didn't compare with the glitter at Culver City.

Underneath the hyperbole and glamour, many stars were painfully aware of their humble origins. Joan Crawford's lowly beginnings explained in part her inordinate need for success. "Nobody could be any more down-to-earth than I was born," Nebraska-bred Fred Astaire said. "I was a plain, middle-western child. The fact that I did shows with a blasted dress-suit on tagged me as a swell. That really amuses me, because it's so silly." Twentieth Century-Fox contract player Coleen Gray, another midwesterner, admitted, "I frequently felt I was a clod from Minnesota. It's something I had to overcome."

Veronica Lake, Paramount's blonde siren of the 1940s, became one of many casualties of Hollywood. Known to millions during the World War II period and treated like a valuable piece of jewelry by her studio, Lake was a sensitive woman who experienced sudden fame but didn't want to be an actress. She was tiny, possessing a delicate beauty. With her makeup on and her hair done, "she was one of the most breathtakingly beautiful girls you've ever seen," publicist Teet Carle said. "You would see her on the set and you stared in amazement. But I went out to her house to do a couple of interviews, and when I looked at her there, I said to myself, 'Who dragged her in from the backyard?'" (Carle int., AFI) Co-stars found Lake difficult, weird, self-contained. "I work at loving people," Eddie Bracken said, "and I hope they love me back. The toughest of all was Veronica Lake. She had a chip on her shoulder from the word go—with everybody. She was insulting to everybody. She was known as 'the bitch' and deserved the title."

Lake would have preferred to run a ranch and came to hate the picture business and everything it stood for. "I think she is one of the greatest tragedies of Hollywood," said her ex-husband, director Andre DeToth. "She didn't want to be in pictures at all." Once discovered, Lake was torn between Hollywood glory and a desire to return to the farm and raise children. She became a split personality. "I saw her going down the drain," DeToth said, "and there was nothing I could do about it. It was dope and liquor and liquor and dope. That's where all the money went. Later, she became unreliable, when earlier she had been a pro. That was her tragedy."

Alan Ladd, Lake's perennial leading man, also suffered chronic bouts of alcoholism. Simple and reserved, Ladd portrayed tough guy roles, very different from his own nature. "He was such a sensitive man," said Lizabeth Scott, who appeared with the actor in *Red Mountain* (1952). "I think in *Shane* you saw some of those qualities that he tried to submerge in his personal life." On the screen Ladd became the macho gunman, despite the fact that he was less than five feet eight inches tall. During close-ups with actresses taller than Veronica Lake he stood on a box. In other shots he walked on a platform, or his leading lady worked in a trench. His screen image and reality were at such variance that embarrassment and shame followed him throughout his career, intensifying insecurities present since childhood. "By the time I knew him," recalled Carolyn Jones, who made *The Man in the Net* with Ladd in 1959, "he was so into alcohol and drugs to try to keep off of alcohol that he was pickled most of the time—to the point where he was unable to function well." A possible suicide, Ladd died in 1964 from an overdose of sedatives mixed with alcohol. He had nearly killed himself fourteen months earlier by a self-inflicted gunshot wound, officially pronounced "accidental."

Marilyn Monroe was temperamentally a child—naive, sweet, far from ignorant, but a slow study. She became difficult to work with and invariably arrived on the set late because she was so unsure of herself, terrified of the stardom she had fought to achieve. "She had this childish charm connected with a great sexual attraction," director Joe Newman observed. "It was a natural talent that she had, but I don't think she ever realized it. And instead of just being satisfied with her native talent, she was trying to develop into a great dramatic actress. Basically she was a nice girl." Fellow actors found it agonizing to work with her. They'd be at the studio ready to begin by nine in the morning, and Monroe wouldn't show up until five o'clock that afternoon, unable to retain more than a few words of dialogue. She was frightened, incapable of supporting the reputation suddenly heaped upon her, yet something miraculous happened between her and the camera.

Other performers insulated themselves by withdrawing into aloofness when fear actually was the driving force. Some stars, like Alice Faye, Deanna Durbin, and Kim Novak, left the business at the height of their success, weary of its unreality and pressures and increasingly paralyzed by the weight of celebrity. Durbin never liked the movie industry and, although she enjoyed singing, had no ambition to become an actress. Soon she felt Universal was exploiting her and left without a qualm.

"You never escape being a film star," Veronica Lake complained. (Lake, *Veronica*) Fans waited for their idols everywhere, hordes of swooning, screaming teenagers invading a star's privacy. At the height of his popularity, Van Johnson could go nowhere without hundreds besieging him, turning a casual outing into mass hysteria. "Some nights were so bad we didn't dare leave the studio," MGM actress June Allyson remembered; "there was that kind of unruly crowd outside the gates at Culver City. When they started really getting rough—trying to tear off clothes and yank hair—Van simply went into hiding." (Allyson, *June Allyson*) One of the heaviest prices stars paid for fame was losing their anonymity. No longer could they shop at a grocery store, go out on dates, or engage in any other public activities without being recognized and mobbed. This frenzied attention added to an insecure actor's confusion and unhappiness. "Once a person has achieved stardom," Dale Robertson pointed out, "he's going to find out that the tires on his car go flat just like they do on somebody else's that isn't a star. And he's still got to shave every morning, he's still got to put his britches on one leg at a time. All of a sudden this person realizes that nothing has basically changed. He thought all of a sudden he was going to start flying instead of walking and that life was going to be different, but really it isn't."

During the making of a film, actors seldom indulged in social life because they had lines to memorize and had to get up early to go to work. "I don't think you ever really get out of character," Lizabeth Scott said, "because there's so little time during the shooting of a picture." Those who sustained long careers came to view the business realistically. "I just did my three or three and a half pictures a year," Barbara Stanwyck once said. "Maybe one out of six would be good, but I kept going." (Lloyd int.) For many the important thing was to keep going. "There was a great belief in professionalism," actor Norman Lloyd said. "You got your work done. But there was a great anti-arty feeling and no nonsense. That was true of any of the major studios."

Stars knew when their studio no longer produced special material for them that their time there was coming to an end. Either they left voluntarily, or they simply drowned in mediocre assignments. "I wanted to get out before they finished me off," Myrna Loy said of her MGM contract. "They used to do that in the studios—they'd either get very careless or do it deliberately. Somebody new comes along, they get all excited, and all their interest goes there. Even if you're still bringing in shekels at the box office, they have a tendency to ignore you." (Kotsilibas-Davis and Loy, *Myrna Loy*) Ultimately stars realized they could make more money freelancing. With mixed emotions Alan Ladd left Paramount, where he had been under contract for twelve years, to sign with Warner Bros. when Warners offered him more money.

Freedom from a studio contract brought the right to say no to roles an actor or actress judged unsuitable. But freedom also meant making judgments on scripts and deciding on countless details the old studios had taken care of; in retrospect the system often looked more attractive than freelancing. Even Bette Davis, who had fought Warner Bros. vigorously when she was under contract, found leaving the financial security the studio provided difficult after nineteen years. For her it was "like leaving home and going out in the world for the first time to seek your fortune," she said. (B. Davis, *The Lonely Life*)

Some performers felt that they didn't develop as individuals until after their careers were over. "There comes a time," Lizabeth Scott said, "when living in the limelight, the impersonalization of your personality, becomes so offensive that you finally say, 'I've had it.'" Viveca Lindfors declared, "The whole system of acting in front of the camera, being a personality instead of being an artist in front of a live audience, is dangerous if you're determined to be a professional actor. It takes a while for an actor or an artist to know what he or she wants to do with his or her life. You're so fed by myths that it's hard to say, 'This is not for me.' Finally I realized that what Jack Warner wanted me to do was not what I wanted to do."

Despite problems and complaints, however, most studio players looked back on their contract period with a sense of fulfillment, remembering the camaraderie. "I had such a great time," Virginia Mayo said of her years at Warner Bros., "and learned so much. I felt I never wanted to stop making pictures. Somehow I knew it was my moment." Nearly everyone agreed that

working in the picture business was a lesson in human nature. And all conceded that the big studio system represented a glamorous moment in American history, one that will never be seen again.

7

Supporting Players

Character actor Walter Brennan in 20th Century-Fox's Banjo on My Knee, *1936. Courtesy of Larry Edmunds Cinema Bookshop.*

Each studio had its stable of character actors, supplying producers and directors with talent for minor roles. They gave a picture cohesion and sometimes provided interest when the stars seemed to fail. Metro-Goldwyn-Mayer boasted the largest array of contract players; such veterans as Lewis Stone, C. Aubrey Smith, Dame May Whitty, Frank Morgan, Leon Ames, Marjorie Main, Louis Calhern, and Selena Royale added to the luster of Metro pictures, although youngsters like Marsha Hunt, James Whitmore, Keenan Wynn, and Angela Lansbury were also important. Warner Bros. had on its roster Sydney Greenstreet, Peter Lorre, Claude Rains, Frank McHugh, Guy Kibbee, Glenda Farrell, and Alan Hale, each with multiple facets. Paramount featured William Demarest, Cecil Kellaway, Walter Abel, Gail Patrick, Jack Oakie, Brian Donlevy, Virginia Fields, Barry Fitzgerald, and Mona Freeman. Twentieth Century-Fox's supporting roster included Clifton Webb, Celeste Holm, Edmund Gwenn, Thelma Ritter, Cesar Romero, John Carradine, and Jane Darwell.

Many actors preferred not to be a star, aware of stardom's pitfalls and confident that character acting offered greater breadth of roles. "A character actor is all I ever wanted to be," Columbia's Jeff Donnell explained, "because they last longer, it's steady employment, and you get different parts." Natalie Schafer, who specialized in playing society women and later was television's Mrs. Howell on "Gilligan's Island," admitted to being privately shy. "I wanted parts that were unlike me," Schafer confessed, "parts I could hide behind. I never wanted to be a star. I don't like being 'on' all the time." Others insisted that a supporting actor has more latitude in the characterizations he can invent. "He is like a painter with a very large palette of colors from which to paint an interesting picture," actress Agnes Moorehead observed. "It can be a subtle performance or an eccentric one." (Moorehead int., Steen) Moorehead herself played a Chinese peasant in *Dragon Seed* (1944), a chic society lady in *Mrs. Parkington* (also 1944), and a 104-year-old woman in *The Lost Moment* (1947), roles she enjoyed immensely.

Director Frank Capra once commented that making a motion picture is like building a table: "On the top of my table, which is bright and shiny, I have these lovely dolls that are my leading actors and actresses. But it is not a table until I put legs under it, and those are my character people. That's what holds my picture up." (Doran int.) Like many top directors, Capra used certain character actors in nearly every picture. Audiences knew that if certain players were in a movie, they were seeing a Warner Bros. film, or a Metro film, or a Paramount film. Supporting actors were recognized as old friends, and audiences loved seeing them change guises again and again. "For me," producer Martin Jurow remembered, "the character people provided the core and substance. Their interpretations gave flesh to a movie."

Not only did these contract actors constitute an efficient stock company, they became a vital part of the studio family. They were developed, nurtured, publicized, well-paid, and most in turn were loyal, recognizing that their colleagues knew their craft. Most considered it a privilege to work among talented people.

Hours were customarily long during the making of a picture, but a workhorse usually was rewarded, having his or her option picked up at the end of the contract period. Marsha Hunt averaged six pictures a year at MGM during the early 1940s, with most of her time in between devoted to publicity assignments or war work. She enjoyed the variety of roles Metro offered her: a neurotic suicide, an unwed mother, a Brooklyn chorus girl, a Southern belle, a society snob, a farm girl, a young schoolteacher, a nightclub singer, and her first aging part, in which she went from sixteen to sixty-five when she herself was twenty-one. "It was heaven," Hunt declared. "This wasn't the road to stardom, but I didn't care. I wanted to be the best actress I could become."

If the studio didn't have a part for a player at the moment, executives loaned that actor at a substantial profit for the company. Contracts with character people, like those with stars, guaranteed a salary for forty weeks a year, and it was essential to the balance sheet that everyone on the lot be kept working. If no one under contract proved suitable for a role, another player would be borrowed or brought in from Broadway. Metro summoned Natalie Schafer to play Lana Turner's mother in *Marriage Is a Private Affair* (1944), after her success in *The Doughgirls* onstage in New York. The actress was housed at the Beverly-Wilshire Hotel, and a limousine picked her up and drove her to the studio. "I had the most divine costumes," Schafer remembered. "The studio built up your ego, because everything was done for you." Later she regretted not signing the long-term contract Louis B. Mayer offered her.

For unknowns, landing a studio contract generally was difficult. Character actress Ann Doran, who eventually played some two hundred supporting film roles, began looking for jobs during the late 1920s, going from studio to studio. She lived at the time with her mother, a silent movie actress, at the Barbara Worth Hotel near downtown Los Angeles and took the streetcar every morning out to Hollywood. She'd either get off at Sunset and Gower to prospect at Columbia or stay on and ride over Cahuenga Pass to Universal in the San Fernando Valley. "In those days you could just go in a studio, and if they needed somebody, they'd come out to the front office and say, 'You— come with me.' And you went to work." Gradually Doran met people and day after day pestered them for jobs. Eventually she developed a routine, taking the streetcar to Paramount, then walking around the corner to RKO. From there she'd walk down to Sunset Boulevard and check things out at Columbia, before crossing the street to Darmour. If there was no work at either, she'd catch a streetcar out to the Valley, stopping first at Warner Bros. From Warners, she'd walk over to Universal. Other days she'd go out to Fox, not as accessible, and finally she made her way to Culver City, where she checked out prospects at MGM and the Hal Roach studio. Meanwhile she wrote letters and kept lists of people she received answers from. Between acting jobs she worked as a stand-in. Eventually she started making westerns, employable in part because she could cry so easily. "My father always said that my bladder was just behind my eyes," she claimed later. "I could cry very easily."

By the mid-1930s Doran was a low-paid contract actress at Columbia. Gradually she moved up to more money and featured-player status at Paramount and other major studios, in 1955 portraying James Dean's mother in *Rebel Without a Cause*. "I have never been a star," she said. "I have never had that responsibility on my shoulders and didn't have to stay twenty-five or twenty-six years old. I've made my living in a business that is fun, exciting, exasperating, rewarding, soul-searing, and soul-satisfying. I performed to the best of my ability and tried to be a good supporting player."

In the golden age of glamour, Hollywood studios insisted that their character people as well as their stars appear flawless. Not only were the leading ladies and leading men attractive, but performers playing old men and old ladies were, too. Marsha Hunt had a birthmark on her right wrist, so Metro insisted she cover it with bracelets or long sleeves. Makeup, they told her, might come off on the leading man's coat in an embrace. Even in relatively realistic films an aura of glamour was maintained. Hunt worked in a World War II picture in which two actresses wore GI fatigues. "Because the coveralls were male garments," she said, "they allowed a certain division of the rear cheeks to show in movement. If you look closely, you will see that the two actresses are wearing girdles, the better to look slim-hipped, and there was slight shoulder padding in the outfits." Hunt and Robert Sterling played a scene in which they got drenched in a downpour and ran for cover, beginning a love scene once they reached cover. But the studio wouldn't permit them to appear unflatteringly wet, allowing only strategic little trickles of water that the makeup man applied to catch the light and glisten effectively.

A character actress might be allowed to wear steel-rimmed glasses if she was supposed to be awkward or homely, but that was the extent of realism during the big studio era. Audiences wanted to see their favorite character actors project variations of a personality they had established over the years, not immerse themselves so deeply in a role that enchantment was destroyed. Movies provided fantasy, and character actors, like stars, were expected to project a mysterious aura that fed the public's dreams. Rarely was the formula broken, not until the end of the big studios's heyday. Celeste Holm fought for a part in 20th Century-Fox's *The Snake Pit* (1948), a hard-hitting drama set in an insane asylum, in which Holm played an inmate. To prepare herself for the role, the actress wore a straitjacket for eight hours. In her scenes the studio put a skinhead on her rather than shave her head, since she was working on another picture simultaneously, and blacked out some of her teeth. But such concern for reality was exceptional, and even in *The Snake Pit* much of Holm's footage was cut.

Jeff Donnell insisted that unless a young actress had a thirty-eight or forty-inch bust, she was destined to play the ingenue's friend. At Columbia Jeff herself became typed in gullible roles. "When you were dumb in the old days," Donnell said, "you were *dumb*, and you said dumb things that were funny. They just don't write characters like that anymore." Actors with unusual or irregular features were equally handicapped. Lloyd Bridges remained largely in supporting roles at Columbia because his eyes were deep set. After seeing Clark Gable in *It Happened One Night* (1934), Jack Warner decided that contract player Lyle Talbot should grow a mustache to approximate Gable's sophisticated good looks.

Most actors and actresses accepted the fact that the studios were profit oriented and needed to give mass audiences what they wanted. Generally performers were impressed by the care with which their pictures were mounted, the attention to detail, and the courtesy and amenities shown them; they were aware that they were well paid and were content to be part of a glamorous operation. Most character actors, like the stars, privately recognized that many of the pictures they made were not of high quality, but were confident that a better assignment would soon follow. Others balked at the edicts of studio executives and found the shallowness of Hollywood and its institutionalization of glamour either frustrating or infuriating. These actors often became cynical and mechanical in their work. "I just make funny faces," claimed actor Peter Lorre, who after his success in Germany in *M* (1931) became typecast as the slippery character with a trick voice. (M. T. Wyler int.) Eventually Lorre was reluctant to deviate from that type, fearful that his popularity would suffer, and toward the end of his career he mocked himself on the screen.

Leon Ames, who made over 110 pictures, was under contract for eight years to MGM, where he specialized in father roles. "I did some good things," the actor admitted, "but nothing sparkling. I didn't have the guts to quit because the money was so good." Often his better roles came as loan-outs to other studios, as in *The Velvet Touch* (1948), in which he played opposite Rosalind Russell. Ames was respected at Metro, where he was rightly viewed as a company man, although he later served as president of the Screen Actors Guild. When he arrived on the set of *They Were Expendable* (1945) for a tiny vignette, director John Ford introduced him to the assembled cast and crew: "New actor. Played his whole part in *Meet Me in St. Louis* a foot out of focus and damn near stole the picture." Ames admitted he became lazy, working mainly for the money.

Mary Astor, who played the mother to Leon Ames's father in *Meet Me in St. Louis* (1944), felt that being under contract to a studio meant becoming permanently typed. In her case at MGM it was mother roles. "Metro's Mothers," she wrote, "never did anything but mothering. They never had a thought in their heads except their children. They sacrificed everything; they were domineering or else the 'Eat up all your spinach' type. Clucking like hens. Eventually every actor on the Metro lot called me Mom. I was in my late thirties and it played hell with my image of myself." (Astor, *Life on Film*) Astor felt trapped and, like Ames, worked mainly for her paycheck.

Many of the actors's complaints were routine. Nearly everyone at Metro, as elsewhere, complained about decisions handed down from the "Iron Lung," as the MGM administration building was called. Most actors and actresses performed their share of parts they didn't like and later fumed that they should have refused and taken a suspension. A few were simply grateful for the work. Sometimes better assignments could be won by changing studios. As a newcomer at Paramount, Marsha Hunt played nothing but ingenues—"sweet young drips in twelve nearly identical roles, with a change of wardrobe and a change of leading men," as she described them. Since these were all female leads, Paramount thought the budding actress should be thankful. "They didn't understand that I wanted a challenge," said Hunt. "I wanted range and to grow in my craft." After two years of stalemate she left Paramount, finding opportunities for growth at Metro.

Most actors felt that interesting material was the main criterion for judging parts. "To hell with the money," insisted Milburn Stone, who after portraying dozens of character roles in movies won fame as Doc Adams on television's "Gunsmoke." "Money comes as a consequence of being a good actor." Other Hollywood veterans said their experience as contract players during the old studio era compensated for their mediocre roles. When Ann Doran started at Columbia in the mid-1930s, the studio had five men and five women under contract as character actors, earning $75 a week. They worked in every picture made on the lot, and Doran said that was where she received her best training.

Yet some felt exploited. "I made seventy pictures at MGM, and I acted nine times," Keenan Wynn said. "If Gene Kelly got in a fight, I held his coat." Keenan's father, comedian Ed Wynn, used to jest, "For those of you who don't know who Keenan Wynn is, when Esther Williams dives in the pool, he's the fellow who gets splashed." Wynn loathed most of the films he made at Metro, although he recognized that financially they were necessary for the studio. Still he felt like a phony playing the parts he was assigned. "The conviction was growing stronger all the time," he said, "but I didn't have the strength of purpose to do anything about it." Instead, he became "a professional griper."

According to Wynn, the studio's practice was to feature an actor in two good pictures and follow those with twenty bad ones, on the assumption that by then the player's name was established with the public. Wynn often felt that someone without acting experience could do as well or better. "My contract with MGM expired after seven fat years of frustration," he affirmed. Later in live television, Wynn proved himself an actor of substance.

A few claimed their careers were destroyed by the big studios. "It was the beginning of the end when I came to Hollywood," actress Anna Lee said. In England Lee had been a star, but she never commanded starring parts in the United States, appearing instead as a character actress. Czech refugee Francis Lederer had been a matinee idol in Europe and a smash hit on the London stage before coming to Hollywood, where he made a few important pictures. Then Frank Capra selected him to play Chopin in a film biography the director was scheduled to make at Columbia. Harry Cohn demanded that Lederer sign a contract for three additional pictures in exchange for the part, promising to turn him into the biggest star in movies. The actor agreed and was on salary at Columbia a few months when Capra walked off the Chopin picture. Cohn wanted Lederer to repay the studio what he'd earned and buy out his contract. "I have no use for you," the Columbia boss contended. When the actor said he couldn't afford to do that, Cohn threatened to ruin him in the business by putting him in two terrible movies back to back. Lederer had no choice; Columbia cast him in *The Lone Wolf in Paris* (1938), a B picture, and his career never recovered. A year later he made *Confessions of a Nazi Spy* at Warner Bros., and that role finished him as a star. From then on Lederer played only character parts, frequently Nazis. Even his agent, Charles Feldman, one of the most powerful agents in Hollywood, couldn't fight Harry Cohn without the risk of being barred from the Columbia lot.

When actor Charles Bickford was under contract to MGM, he realized that the studio did not plan to build him into a major player after he was cast in *Anna Christie* (1930) with Garbo, Metro's biggest box office star, and his appearance went unpublicized. The studio shortly loaned Bickford to Universal and Fox and began using him either in roles he considered unsuitable or in B pictures. Many of these assignments he turned down, and at one point he threatened to walk out of the studio rather than play an inferior part. His agent informed him that Metro could secure an injunction to prevent his appearing in any entertainment medium until he had fulfilled the terms of his MGM contract. Hot-tempered, the actor continued his unruly conduct, eventually alienating Louis B. Mayer. Within hours after Bickford and Mayer clashed, the news had spread all over the Metro lot. "In the commissary that noon," the player wrote, "the long table at which I usually ate with five or six companions was strangely empty. Acquaintances with whom I had fraternized in a spirit of camaraderie either avoided me or greeted me with surreptitious nods from across the room. The only friendly faces I saw were those of the waitresses."

No studio in Hollywood would hire Bickford. When the recalcitrant player told his agent he intended to fight, the agent refused to back him, and they discontinued their association. Meanwhile the Hollywood publicity machinery was set into operation against the actor, and Bickford became an outcast. "The whispering campaign was the most effective," he remembered. "The saboteurs spread the word. Overnight I was miraculously transformed into box-office poison. No longer was I a great actor; I had suddenly become a mediocre ham. I was a troublemaker. I was impossible to handle. I had no sex appeal. I dyed my hair. I was a Bolshevic." (Bickford, *Bulls, Balls, Bicycles, and Actors*)

Bickford turned to independent producers, putting aside artistic considerations to earn a living, avoiding the major studios. For the next ten years he was a renegade, averaging about four pictures annually—some good, most not. In 1944 he was cast in 20th Century-Fox's *The Song of Bernadette*, a hit, and regained favor with the studio heads, subsequently appearing in such successes as *The Farmer's Daughter* (1947), *Johnny Belinda* (1948), *A Star Is Born* (1954), and *Days of Wine and Roses* (1962).

While there were countless stage actors of proven reputation the big studios used poorly or didn't know what to do with, others sustained long careers as Hollywood character people. Claude Rains, a thoughtful, intelligent actor, consistently studied parts to find nuances that would make each distinctive. Sydney Greenstreet—congenial, helpful to young performers, with a keen sense of humor—always came to the set prepared, although toward the end of his career he found it difficult to make last-minute changes in lines. "He couldn't unlearn what he already knew," producer Frederick Brisson said. "So he would stutter if there were too many changes, and sometimes that meant twenty or thirty takes or cutting the film in such a way that instead of the whole speech, he would do it piece by piece." Agnes Moorehead, who came to Hollywood with Orson Welles from the Mercury Theatre, epitomized professionalism. She invariably knew the script, knew her contribution, and demanded that her colleagues know theirs. Actors consistently claim that they do their best work when everyone else is good. "It's the people who don't know what they're about and are self-indulgent who are very difficult to play with," Moorehead said. (Moorehead int., Steen)

Boris Karloff, a soft-spoken and gentle man in contrast to his Frankenstein character that won him fame, was a fine actor from the Canadian stage. Articulate and humorous, he often recited poetry on the set. Karloff was cooperative and charming, even when he had to sit in the makeup chair for hours while his monster image was created. Vincent Price, who often played villains, became an authority on fine art, well-read and knowledgeable. Jane Darwell, best known perhaps as Ma Joad in 20th Century-Fox's adaptation of *The Grapes of Wrath* (1940), was a warm, motherly person, far from the stereotype of the temperamental actress. George Sanders, on the other hand, seemed remote and was inclined to be lazy; despite his urbane screen appearance, he seemed tense on the set, sweating profusely so that he'd need to change shirts three or four times. Thelma Ritter, who had been a leading lady in stock and had then retired to raise a family, entered films in the late 1940s, specializing in cynical, wisecracking, outspoken roles that were well suited to her own personality. In twelve years Ritter was nominated six times for an Academy Award in supporting roles, yet never won.

Giants from the theater, particularly in their later years, sometimes found sanctuary in the studios, where the work was easy, the pay good, and the professional exposure worldwide. Ethel Barrymore, with her enormous eyes and imperious voice, was under contract during the 1940s to MGM, where she reigned like a queen. At that time more interested in sports than in discussing the subtleties of acting, Barrymore sat in her wheelchair, her ear close to the radio, listening to a boxing match or baseball game, treating interruptions with regal impatience. She occupied an imperial dressing room, had a driver to meet her at the set and take her wherever she needed to go, and *nobody*, not even those in the front office, called her anything but "Miss Barrymore." Generally she remained aloof, but if a producer should arrive ten minutes late for an appointment, he could expect a reprimand. "You didn't direct Ethel Barrymore," Michael Gordon chuckled. "She was a national monument. You just hoped her scenes came out right and you could say, 'Beautiful, darling.' She was a majestic presence."

Her brother Lionel Barrymore was also at Metro during his last years, confined to a wheelchair, in pain from arthritis and partially paralyzed from a leg injury; he frequently dozed in his chair between takes. Gruff on the surface, he often grew talkative and pleasant with colleagues he respected. Like his sister, Lionel Barrymore was a "presence," and he loved to show off his skill with words. Gregory Peck remembered Barrymore on the set of *The Valley of Decision* (1945) declaiming pages of dialogue over and over, growing more histrionic with each repetition.

Gladys Cooper, who had achieved fame on the London stage in the early 1920s and was nominated for Academy Awards for *Now, Voyager* (1942) and *The Song of Bernadette*, signed a contract in 1944 with MGM, where she first played Irene Dunne's mother in *The White Cliffs of Dover*. A delicate beauty in her day, Cooper remained under contract to Metro for five years, appearing in a variety of lovable mother and grandmother roles not always to her liking. Louis B. Mayer respected dignified character actresses, and Cooper fulfilled his ideal. From her standpoint the Metro contract offered the actress her first financial security in America. Cooper found the studio "just like a factory," although not an unpleasant one. "They have two hundred of us under contract," she said, "and you have to fight to get anything done." (Morley, *Gladys Cooper*) Cooper played a drunken neurotic in *Mrs. Parkington*, but most of her roles held few surprises. Reviews of her work were normally brief but good, and with the MGM publicity department constantly churning out stories, she appeared regularly in the fan magazines, as did all of Hollywood's major character players.

Although a majority of the distinguished character players had extensive stage experience (Ralph Bellamy, Mildred Natwick, Sam Jaffe, Norman Lloyd, Dan Duryea, Jane Wyatt, Arnold Moss, Martha Scott, among countless others), many came from other areas of the entertainment industry. Aside from their theater background, Edward Arnold and Dorothy Gish had both worked extensively in silent pictures, and Harry Carey had established himself as a star of silent westerns. Milburn Stone and Walter Brennan arrived in Hollywood with vaudeville experience, yet Brennan was the first actor to win three Academy Awards—for supporting performances in *Come and Get It* (1936), *Kentucky* (1938), and *The Westerner* (1940). Red Buttons became the first burlesque comedian to win an Oscar (for *Sayonara* in 1957), followed later by Jack Albertson and Art Carney. Rosemary DeCamp, Hans Conried, Betty Garde, Mercedes McCambridge, and Audrey Totter all had years of radio experience before entering films. Marshall Thompson and Jean Porter were among the many who began their careers with a studio contract.

For each of these performers the first obstacle, once a part had been assigned them, was to learn how to act in front of a camera. Few directors devoted much attention to their character people, spending most of their time with the stars. Supporting players were expected to know their craft, and aside from grouping actors before the camera, directors allowed lesser performers to find their own interpretations. The focus in lighting and camera placement was consistently on stars. "Sometimes my best lines, even though I might only have five, would be delivered with the camera showing the reaction of the star," Hans Conried said. "They would shoot me speaking from behind my ear."

Rehearsal for character parts was normally minimal. Actors ran through their lines a few times before a scene was filmed, mainly so that the cameraman could check his moves. A script person might go over a supporting actor's lines with him to make sure he knew them, but that was often the extent of his preparation. Betty Lynn recalled her opening scene with Robert Young in *Sitting Pretty* (1948), in a car with a moving backdrop. The two began the scene, seated side by side, before they had even met. Lynn was new at the time and nervous. She looked over at the star, with the car rocking to simulate movement, and suddenly she thought, "'My gosh, that's Robert Young!' It was a funny feeling, but I kept going." Most youngsters on a stock contract were kept too busy to study extensively with the studio's drama coach. More often they learned from the actors working with them, some of whom they got to know, others they didn't. Mostly supporting players learned by trial and error. "It was a frightening way to learn your craft," Marsha Hunt commented.

Some found watching the dailies helpful; others hated seeing themselves on the screen. "I went to see dailies if I had a limp or a scar or something I wanted to check," said Keenan Wynn, "or if I had an accent. Otherwise I never went." When not busy on the set, ambitious newcomers scoured the lot for their next part, talking with producers and meeting people. Occasionally the studio gave supporting performers a trip to New York for publicity or sent them out of town to attend a premiere at company expense. But mostly contract players spent forty weeks a year hard at work on pictures, sometimes alternating two or three at once.

"There were lots of times when I'd finish one movie on a Saturday and start another one on Monday," Natalie Schafer recalled. Mary Astor had no sooner signed her Metro contract on a Friday than she was told by the casting office to be in the wardrobe department the following Monday morning at eight o'clock. Rosemary DeCamp remembered being fitted for costumes for overlapping parts at Warner Bros. and running from set to set asking, "Who am I and what stage are we playing on?"

Yet featured players were well cared for, and work for them was not without its glamour. Most enjoyed commodious dressing rooms—not in the same building with the stars, but freshly decorated. When they were working, a studio car drove them to the stage door and to the commissary for lunch. "This was before hairspray," Marsha Hunt pointed out, "and they didn't want your hairstyle to be thoroughly ruined in a stiff wind on the way to lunch and back." Each featured player had a portable dressing room on the set, with his or her name printed in a slot on the door, and a personalized canvas chair near the camera. Generally there was a stand-in for each, along with most of the niceties enjoyed by stars. "I can't remember feeling second banana," said Marsha Hunt.

Featured players, like stars, were expected to drive expensive cars, live in impressive houses with servants, and dress stylishly. Status was mandatory in Hollywood on all levels. Studios sometimes encouraged contract personnel to go into debt, assuming that financial obligations would insure subservience. Even when actors weren't busy on a picture, they were not free to leave town without their studio's permission. Since they rarely knew for certain when they'd be working and had little or no control over what roles they played, character actors like stars complained they felt like property. Whether or not an option was picked up depended on the actor's cooperation, the success of his or her recent pictures, the quality of reviews, and the public's reaction to the player, as evidenced in the amount of fan mail received and the attention commanded in the popular press. "Yesterday's newspaper is nothing compared to a dropped option," Milburn Stone said. "There's nothing in the world colder."

For contract players who stayed, as for the stars, the studio became home after several months. Members of the studio fraternity were protected and shown consideration. Dancer Cyd Charisse, who was featured in some of the studio's top musicals, was in her early twenties when she came to MGM, and workers in the various departments recognized how naive she was. "So they would watch over me," Charisse explained, "make sure I was all right, cluck like a bunch of mother hens." (Martin and Charisse, *Two of Us*)

As in most corporate bodies, the stock people at the various Hollywood studios formed cliques. It helped to know the right people. Being a close friend of the star assured a character player better treatment, and sleeping with the producer might at least have temporary advantages. Avoiding factionalism on the set was usually wise, although character actors sometimes served as a steadying influence on temperamental stars who respected their featured players and often learned from them.

Few of the veteran players spent much time in weighty dramatic analysis. Asked what his method was, Claude Rains replied, "I just learn the lines and pray to God." (B. Davis int., McBride) Lloyd Nolan viewed himself as a natural actor. "The simpler it is, the better it is," Nolan said. "That isn't to say that simplicity is easy." Walter Huston, a man in his sixties, preferred to talk about sports or women over the motivation of characters he was playing. Still, Huston was always prepared and generally gave a strong performance with what seemed a minimum of effort.

When Cesar Romero drew the role of Hernando Cortez in 20th Century-Fox's *Captain from Castile* (1948), he decided to seek the help of a noted coach. Romero went to the coach once, concluding that approach wasn't for him. "I just learned my lines and performed naturally," he said. Occasionally character people didn't know when a picture started which role they'd be playing. Leon Ames played the father in MGM's remake of *Little Women* (1949), but he was on standby for the lead ultimately given to Rossano Brazzi, since the studio wasn't sure at first whether they could import Brazzi from Italy or not.

Although rehearsal time in films was limited, character actors might spend weeks on the studio's payroll learning some technical skill. When Arnold Moss played a dragoon in *The Loves of Carmen* (1948) at Columbia, an Olympic gold medal champion taught him how to use a saber. The instruction took six weeks, during which Moss was paid full salary without a foot of film being shot, despite the fact that the duel in the picture lasted only about fifteen seconds. Marsha Hunt was a member of an all-girl orchestra in *Music for Millions* (1945), in which the actress had to learn to fake playing various instruments, particularly the violin. For days the women practiced the passages to be filmed to the playback of a Dvorak symphony prerecorded by the MGM orchestra. Hunt came to hate the *New World Symphony*, but the results looked real.

The whole of a B film would probably be shot in two or at most three weeks. Character actors at the bigger studios made B films between more prestigious assignments. "They were killers!" declared Ann Doran, who was featured in dozens of B movies. "You had so many pages of script and so many days to do it in, and you simply said your words and got to the door."

Not only were many of these character people highly regarded for their talent, but some also became instrumental in labor negotiations with studio executives through the Screen Actors Guild. The Guild was formed in 1933 much against the studios' wishes. Membership was open to all motion pictures performers, and the Guild grew into a powerful union, capable of forcing demands on the studio and winning for its members better working conditions, higher salaries, and minimum employment protection. Hollywood conservatives had launched the Motion Picture Academy of Arts and Sciences in 1927 in an effort to raise the standard of film production, but also to preclude the formation of craft guilds. The Academy not only rewarded outstanding achievement in picturemaking, it also elevated the prestige of the cinema in the United States. Eventually the Academy became a source of international publicity. Most careers were enhanced by winning the Academy's Oscar during a given year, even though studios frequently manipulated such victories.

But the Screen Actors Guild was necessary to improve working conditions. Leon Ames, for several years the Guild's president, was a founding member. At the time the Guild was formed, numbers were drawn out of a hat at actor Jimmy Gleason's house; Ames drew number fifteen. "I've forgotten who drew number one," the actor said, "but whoever it was traded numbers with Ralph Morgan, because Ralph became our first president." Meetings were initially held in secret—at the Masker's Club, somebody's house, or Boris Karloff's garage. The members followed the lead of Actors Equity, which had been formed in 1913 by stage actors in New York, and demanded similar benefits. At first the office of the Screen Actors Guild consisted of two orange crates, an old typewriter, a telephone, a chair, and a secretary, Midge Farrell. Everything was done privately, since any actor identified as a Guild member would immediately have been terminated by his or her studio, which would have meant the end of that career in the motion picture industry. Character actors led the way, although stars gradually began to lend their support—covertly, since they were under surveillance, too. Each Guild member paid annual dues, while SAG went to work on a pension plan, shorter hours, and better amenities. Later a credit union was established with $50,000; within twenty years the credit union had assets of over $20 million.

Ann Doran was talked into joining the Guild by Leon Ames and actor

Walter Pidgeon, who assured her SAG was a good thing long before it was a recognized union. Later she was a member of the Screen Actors Guild board for twenty-three years, served on several negotiating committees, was secretary for a time, and was finally a vice president. "I learned that there was something to this business besides being an actress," she said.

For character players, making movies offered much the same excitement, fulfillment, and glamour as it did for stars. If the pay was less, the security often was better. Leon Ames, an Indiana boy from a farm along the Wabash, turned a career in pictures into a substantial livelihood and retired in comfort. For Tom Ewell, who came later, playing opposite Marilyn Monroe in *The Seven Year Itch* (1955) and Jayne Mansfield in *The Girl Can't Help It* (1956), movies offered an opportunity to live out some fantasies. "I was what you might call a gawky kid from a small town in Kentucky," Ewell said, "skinny, with red hair, certainly unprepossessing. I had a great deal of the rural in me. If I put on a new suit, two hours later it looked like I'd slept in it. To have had the career I've had and to have played opposite the beautiful women I have is to me quite remarkable." Others looked back with satisfaction on their film work, the people they'd met, and the friendships that lasted. More than the plush dressing rooms, the dream parts, and the Hollywood tinsel, what seemed to make working in the big studios memorable to many character players was the camaraderie they experienced, along with the knowledge that the finest craftsmen available were working to make the best possible product.

8

Publicity

*Judy Garland and Mickey Rooney in the courtyard of Grauman's
Chinese Theater in Hollywood. Courtesy of the Academy of Motion
Picture Arts and Sciences.*

The old studios, like governments or American universities, were organized by departments, each with its department head who met regularly with representatives of the central administration, sometimes the studio moguls themselves. Howard Strickling, the long-time publicity head at Metro-Goldwyn-Mayer in charge of the West Coast office, worked in conjunction with Metro's corresponding executive in New York, Howard Dietz. At 20th Century-Fox, Harry Brand, a giant in the field, headed the publicity department. At Warner Bros., Hal Wallis supervised publicity before he became Jack Warner's production head and was replaced in the publicity department by Bob Taplinger, who later went to Columbia. Terry DeLappe headed publicity at Paramount during the late 1930s, and Russell Birdwell served as publicity director for Selznick International, masterminding the advertising campaign for *Gone With the Wind* (1939). Since a studio's stars were often on loan, publicists occasionally worked between studios, but mainly they concentrated on productions nearing completion and events on their home lot. Someone in each publicity department handled fan magazines; somebody else dealt with national magazines like *Life* and *Look*; staff members called "planters" scurried to place items in newspapers throughout the country. Each contract player on the lot and each picture in production was assigned to a studio publicist whose job it was to turn out stories to capture the public's attention. By 1945 a Publicists Guild had been formed in Hollywood, and a system was devised by which members advanced from apprentice to junior to senior levels.

The publicity department at MGM became a model, in which Howard Strickling, the dean of studio publicists, was a stern, knowledgeable, tough disciplinarian who demanded peak performance from his employees. Before World War II, workers put in a six-day week; Strickling believed in keeping his staff busy. He never let anyone feel too secure and rotated assignments to keep employees from stagnating. On the other hand, when staff members worked well together, he left them alone. Strickling was conservative and aloof, alternating the two suits he wore to the office—one a pepper and salt tweed with a vest, the other a navy blue the staff called his funeral suit. "He wasn't easy," Metro publicist Emily Torchia remembered, "but he was fair." He was one of the first to hire women. "Mr. Strickling didn't pay the highest salaries, but neither did he hire and fire," Torchia said. "Many of us were in the department for years, and we still keep in touch even today."

Above all Strickling was loyal. He permitted no one to speak negatively about MGM in his presence; L. B. Mayer was one of his idols. Longtime employee Ann Straus recalled that he was loyal not only to the studio but to the stars as well. In Strickling's accounts MGM actresses didn't drink, they didn't smoke, they didn't even have babies. He wanted perfection from his staff, and he got it. No chore was too menial to Strickling if it publicized MGM, and he expected his employees to perform whatever duties were necessary. On Saturdays most publicists had finished their work by four or four-thirty in the afternoon, but their boss frequently found a preview for them to attend or a star who needed to be picked up at the airport on Saturday night.

Strickling was fond of saying, "Don't forget that the bedrock of publicity is the day-to-day routine." (Chandlee int.) He began most days at nine o'clock with a department meeting, during which everybody was expected to contribute ideas on current projects. At its height the Metro publicity department numbered around sixty people, each of whom was assigned three or four of the studio's stars and four or five national press correspondents who covered Hollywood. It was the publicist's responsibility to match stars and correspondents. In the case of Walter Seltzer, who worked as a Metro publicist for four years, his biggest star was Joan Crawford. In addition, he covered the trade papers and Los Angeles newspapers. "Besides my other duties," Seltzer remembered, "I would make a swing of the town every day (a fifty-mile round trip in a 1928 rattletrap) to each of the eight dailies."

After the morning meeting, publicists covering a production unit normally went to the set, unless the actors were shooting, to glean information that might appeal to columnists like Louella Parsons or Hedda Hopper. After visiting the set, publicists came back to the office and wrote—feature stories on the stars permanently assigned to them, as well as production items. A major goal was to have a story in the *New York Times* with the publicist's byline, but material was also constantly being fed to the smaller newspapers across the country. Biographies of each star had to be updated, and future campaigns were meticulously planned. During the afternoon, unit publicists returned to the set, looking for more stories and perhaps an opportunity to talk with players unavailable earlier.

Located on the corner of Washington Boulevard and Ince Way in Culver City, Metro's publicity office, always busy, was often the scene of intense pressures and strain until six-thirty or seven o'clock at night. "I learned at MGM that I could work with a train going through the room," said veteran publicist Esmé Chandlee. "There was a lot of give and take, and we learned to work as a team. What it taught me in the end was to be flexible." All agreed that being in the department under Howard Strickling meant hard work, but also a lot of fun. There were frequent pranks, which made even the bad times seem bearable. When World War II broke out, staff members arrived on Monday morning to find the following note on Howard Strickling's door from Ralph Jordou, their head planter: "Sorry, Howard, had to leave, covering the war." Jordou became a war correspondent until the war ended. (Torchia int.)

As at other Hollywood studios, MGM's publicity department was organized like a newspaper office. There was a copy editor, Harley Schultz, a distinguished, well-educated man, who reviewed all the publicists' copy. Downstairs were the planters, usually headed by a former newspaper man. Planters took the stories written by reporters and placed them wherever they thought they could get the best coverage. The publicist who wrote a story could suggest a certain trade paper, like *Variety* or *The Hollywood Reporter*, or a specific columnist, such as Harrison Carroll or Jimmy Starr, but it was up to the planter to make the final decision about where an item would go. Often there were disagreements. Esmé Chandlee started at Metro as a planter, feeding stories to Hedda Hopper. Sharing her office was Bill Lyon, a close personal friend, but Lyon was Louella Parsons's planter. "I used to laugh at this myself," Chandlee said, "and so did Bill, but when we were planting, we found another office to make the call. There was a lot of rivalry right within the studio."

Emily Torchia began her thirty-three years in the Metro publicity department operating the switchboard. Eager and bright, Torchia was quickly promoted to roving reporter, but eventually concentrated on unit work. "You killed yourself to get a story into a national magazine," she said, "but our stories went all over the world, since MGM had offices in all the capitals of Europe."

Main title billing—determining the order in which stars were billed in credits and advertising, as well as the size of their names in print—was another concern of the publicity office, which meant checking contracts and frequent consultation with the studio's legal department. Anytime a producer on the lot started a new production or began casting, announcements had to be prepared for the trade papers. Somebody in the publicity office even had to phrase the captions under pictures that went out.

Although Howard Strickling worked closely with Benny Thau and Eddie

Mannix and the rest of the MGM executives, he reported directly to Louis B. Mayer. Much of Strickling's time was taken up in meetings with individual producers, while his two assistants, Ralph Wheelwright and Eddie Lawrence, took care of details. Emily Torchia once did a layout and sent out pictures without showing them to Wheelwright. The department was fined $5,000 because cleavage was showing. "In those days stars couldn't be pictured holding a cocktail or smoking a cigarette," Torchia explained, "and you didn't show cleavage; it had to be airbrushed. Fortunately the resultant publicity overshone the fine, so nothing much was said to me."

There was always friction between the California and New York offices of Metro. Howard Dietz, the noted Broadway lyricist ("Dancing in the Dark") who headed publicity for MGM in New York, was credited with creating the company's trademark, Leo the Lion, and coming up with the slogan "More Stars Than There Are in Heaven." Dietz was bright and a far more sociable man than Howard Strickling, who preferred his home and ranch to hobnobbing with stars. Yet it was Strickling and his staff who worked with the studio personnel on a daily basis and watched pictures being made. Dietz normally didn't see Metro stars except when they came to New York on a promotional tour or visit. Still, since New York was the corporate headquarters, Dietz considered his office at 1540 Broadway to be the company's publicity headquarters. He and his associate, Sy Siedler, came to Los Angeles for meetings two or three times a year, during which they consulted with the Culver City office on specific problems and planned promotional strategy, but it was Howard Strickling who remained close to L. B. Mayer, making himself useful at all hours and attending weekly brunches at Mayer's Santa Monica beach house on Sundays.

Dietz usually met and entertained celebrities when they arrived in New York City. "One of the things that Howard did not like was coping with the stars," Dietz's wife, costume designer Lucinda Ballard, said. Dietz was far less patient with Spencer Tracy's drinking than Howard Strickling was and felt there was a great deal of back-scratching among the top Metro-Goldwyn-Mayer personnel. Shortly after Dietz was placed in charge of publicity in New York, he borrowed a sign that had hung over the desk of Harold Ross, editor of the *New Yorker*, which said: "Don't be famous around here."

The MGM publicity department earned a reputation for cosseting its stars, covering up for them when they behaved less than admirably. The public was shown how the stars lived at home, how they ate, how they dressed, their pets, their boats, their cars, but only under the most ideal circumstances. Stars were beautifully photographed, and all were reported to lead storybook lives, even in the face of tragedy. If Louis B. Mayer worried privately about the sexual exploits of his talent, his studio publicity department pictured every romance as heaven-made, a union of deities who bewitched and sanctified, even when they changed partners seasonally.

Warners' policy was not to assign stars to studio publicists but to have everyone in the department responsible for all the talent on the lot. Publicists at Warners had an assigned number of columnists whom they kept supplied with stories, choosing material from anywhere in the studio. Hal Wallis was twenty when he became publicity director of Warner Bros. in 1923; he was elevated to studio manager within the next five years. During the 1930s Robert Taplinger, a young man brought from New York, served as Warners' publicity chief, working under S. Charles Einfeld. Like Metro, the department at Warner Bros. had reporters, unit writers, publicists who handled magazines, photographers, and even specialists assigned to radio. Still photographer George Hurrell turned Ann Sheridan into the "Oomph Girl" and glamorized Bette Davis at Warners, as he had Joan Crawford at Metro earlier. During the early 1940s Warner Bros. had between twenty-five and thirty publicists, in addition to secretaries and assistants. "In those days," said Bill Hendricks, who eventually headed the department himself, "publicity began when Warner Bros. bought a story; it never stopped until the picture was released."

Jack Warner realized the value of publicity, and his publicity offices, occupying two floors of an entire wing, were located near his own suite. Bill Hendricks's day as department head normally started with an early morning call from Warner, checking on activities Hendricks had scheduled. If a special guest was visiting the lot, Warner was on hand to present the dignitary with a brass key to the studio. On other occasions the mogul himself received medals, plaques, and citations; all of these activities needed to be covered by the press. Fan magazines commanded the full-time attention of a publicist and a couple of assistants. Every Monday morning Hendricks would receive a sheaf of clippings pertaining to Warner Bros. personnel and would check to see if any of the stories were libelous.

"The fan magazines were the greatest star builders that ever existed for motion pictures," declared Teet Carle, a publicist at Paramount. "Then came the newspapers." (Carle int., AFI) If a new actor arrived on the Paramount lot or an interesting scene was about to be shot, a studio planter would call Hubbard Keavy of the Associated Press and suggest he come out and do an interview. When Dorothy Lamour made her first dramatic picture after a string of sarong roles, Paramount publicists decided it would be a great stunt to have Lamour burn her sarong, with the press gathered around taking pictures. A publicist selected a sarong from the wardrobe department, but when the press arrived, the sarong wouldn't burn! It had been fireproofed, which made an even better story.

Ben Schulberg had been Adolph Zukor's first publicist long before the formation of Paramount, shortly after Zukor formed his Famous Players in Famous Plays. Schulberg had helped launch the publicity drive for *Queen Elizabeth* (1912), featuring Sarah Bernhardt. By the 1930s Paramount's publicity department in Hollywood had twenty publicists plus secretaries. It was still possible for staff members to start at the bottom and work their way up. Photographer John Engstead began as a messenger and office boy in the Paramount publicity department, making $18 a week. Eventually Engstead handled the fan magazines and then headed the still department.

While Terry DeLappe headed Paramount's publicity department on the west coast, Bob Gilliam was his counterpart in New York. In Los Angeles publicists worked at desks in a circle, sometimes collaborating. Later in the decade, when Hal Wallis moved his production unit to Paramount, Wallis employed his own publicity team, headed by Walter Seltzer. Seltzer arranged a campaign for each of Wallis's films. Lizabeth Scott recalled that when she was discovered, Seltzer wrote letters to editors and film critics all over the United States, telling them that she was Elizabeth without the "E."

Harry Brand, the publicity head at 20th Century-Fox, was highly respected by those who worked with him, and Brand returned that respect with loyalty. Decades after he left the studio, he refused to talk about the stars he represented at Fox, maintaining that to do so would be a breach of confidence. Brand was a private person who appeared happy in his work; he was ably assisted by Sonia Wolfson. As at other major studios, the 20th Century-Fox publicity department seldom missed a chance to have its stars' names in print. It mailed out sanitized biographies of every star on the lot to newspapers from coast to coast and covered in detail the making of each of the studio's pictures, down to the lowliest B production.

At Columbia, a smaller studio, life in the publicity department was more casual. During the 1930s William C. Thomas, later co-producer with William H. Pine of several adventure films, was Columbia's publicity director. He was succeeded during World War II by Bob Taplinger. In 1936 Gail Gifford, a recent college graduate, started at the studio in the stenographic pool, although she could barely type. She was sent to the publicity department to relieve an employee going on vacation, and she stayed for nine years. Soon she became an assistant to Maggie Maskel, who was in charge of both fan and national magazines. Gifford realized quickly that success at the studio called for versatility, since the staff there was small.

All the publicists at Columbia worked with Rita Hayworth, Janet Blair, and the studio's other stars, but every picture was assigned a unit person. In the mid-1930s Columbia's publicity staff consisted of one person who did advertising copy, a planter or two, and a bull pen of three or four writers. Over the next decade the department expanded, and its procedure became more systematized. Writers turned copy in to an editor, who sometimes infuriated unit publicists and reporters with extensive changes.

The publicity department at Columbia had two addresses—the official one at 1438 North Gower and an unofficial one across the street at Brewer's Cafe, where publicists went after work to drink and mingle with actors and wranglers. Living up to his reputation as a monster, studio boss Harry Cohn could be as abusive to publicists as he was to his writers and performers. Lou Smith, who survived two terms as publicity director under Cohn, developed an effective technique for dealing with the studio boss. Smith had a hearing problem and wore a hearing aid. Anytime Cohn became particularly unpleasant, Smith would pretend not to hear. He'd just smile ingratiatingly at his boss and turn his hearing aid down.

In 1951, after leaving Columbia, Gail Gifford went to work for the publicity department at Universal. Well organized under the direction of David Lipton, the Universal office was bigger and more departmentalized than Columbia's. Universal had unit publicists and a large bull pen of writers. They had a person in charge of fan magazines and another in charge of national magazines. The art department was much bigger than Columbia's, and they had a telephone planter, as well as a downtown planter. At Universal publicity was housed in its own building, where stars frequently dropped in for a visit. "They used to come up there and sit around and talk and laugh," Gifford recalled. "They'd bring their dogs up and leave them in my office. It was very family-ish."

At Christmas the various publicity departments had office parties, and stars would be invited. Publicists from Columbia or 20th Century-Fox might go over to Warner Bros. or Paramount to celebrate with the people there, particularly if they knew workers in the other departments. Most publicists had little contact with publicity people in rival studios, except when one of their studio's stars was on loan to another lot, in which case the two departments shared publicity.

Delighted to have everything arranged for them, most stars cooperated with the publicity office. Whenever stars were interviewed, someone from the publicity department usually accompanied them, and newcomers were carefully instructed on what to say. Studio publicists loved puffery. "If you had a farm, it was an estate," said Celeste Holm. "If you had a field, it had to be full of horses." Warner Bros. claimed young contract player Nanette

Fabray's father was a concert pianist; he was a fireman on a train. Warner publicists said her mother had been queen of the Mardi Gras, which wasn't true. They said that Fabray had been in the *Our Gang* comedies, when she had only been an extra in a crowd scene. On one of Joan Leslie's birthdays the young actress was asked to step outside and have her picture taken with Jack Warner as he presented her with keys to a brand new car, which Warner supposedly was giving her as a birthday gift. "I was so trained on how to pose for pictures at that point," Leslie remembered. "I'd just look a little bit off vision, because your eyes photograph better that way." They took the pictures and then collected the keys and drove the car away. "I certainly never received a car from Mr. Warner," Leslie said. "It was a business of falseness."

The big studios carefully orchestrated publicity, building stars' images to catch and hold the public's interest without overexposing the stars so that fans became bored. Metro had a policy about not quoting money, feeling that to do so placed commerce ahead of glamour. MGM's publicity staff was never to mention what the stars earned, nor was anyone supposed to know. The public, however, liked to know personal things about screen actors and actresses, so publicity offices fed selected information to the press. Still, thousands of letters poured in each week from fans demanding more details.

Publicity work could become discouraging. "There were many nights when I'd come home and was going to quit," Emily Torchia confessed. "Not *everyone* was helpful, and not all the stars were that crazy about me." Torchia worked well with Katharine Hepburn and Spencer Tracy and for a time was assigned to Clark Gable. She routinely ate in the MGM commissary because that's where she heard many of her stories. Esmé Chandlee worked with Robert Taylor and Greer Garson and for a long time handled MGM's "baby ballerinas": Pier Angeli, Leslie Caron, and Debbie Reynolds. Cyd Charisse was assigned to Ann Straus. Although publicists focused on the stars, featured players were represented, too, so that no one under contract was overlooked.

"The major studio publicity departments did such a superlative job of blowing up the stars into larger-than-life figures," director Edward Dmytryk wrote, "that everyone sooner or later succumbed to the snow job—the public, the stars themselves, the front office, the studio workers, and finally even those who had started the whole thing, the publicity departments. All this gave the stars great power within the studios." (Dmytryk, *It's a Hell of a Life*)

Some stars hated publicity, and a few even refused to cooperate. Jean Arthur, after she became established, did well without it. Greta Garbo was never enchanted with the picture business and spoke her mind in interviews, often criticizing Hollywood. "I insisted that a member of the publicity department be present at every interview," Howard Dietz recalled. "Garbo said if she weren't allowed to speak freely, she wouldn't have any interviews, and she didn't have any interviews for the next 48 years." (Dietz, *Dancing in the Dark*) Deanna Durbin not only hated Hollywood, she detested stardom and despised publicity. Hailed during the late 1930s as America's Sweetheart, Durbin remained private and aloof in her personal life. When she and her producer-husband, Felix Jackson, traveled to New York on the train, they didn't get off until everyone else had left. Then they exited the station by a freight elevator with a police escort. Durbin resented fans crushing in on her, preferring to stay in her hotel room during publicity tours. Still, she was adored by millions. One fan sent her a gift of two thousand tulip bulbs; another worked ten years on a sweater for her. Yet the invasion of privacy was constant. "I remember when we brought our baby home," Felix Jackson said. "A publicity man assigned to our unit called me and said Louella Parsons had called to see if it was true that now that the baby was born we were getting a divorce!" Durbin found it an impossible way to live and retired soon afterwards. "My job was to guide her career," Jackson said, "but if you have a star who doesn't want to be a star, it's very difficult."

Most stars, however, were ambitious and realized the advantages of skillful publicity. Prudent performers made themselves available whenever something newsworthy occurred. When a parking garage was excavated under Pershing Square, across from the Biltmore Hotel in Los Angeles, a time capsule was embedded in a corner of the park. Warner Bros.' Virginia Mayo was on hand to officiate as the capsule was placed in cement, and an autographed photograph of the actress, inscribed "To Whom It May Concern," was placed on top of its contents. The studio's Janis Paige christened McNary Dam on the Columbia River and became Miss Dam Site, and Doris Day was pictured advertising National Printers' Week. "I never refused any part or any request to do publicity," Warners' Patricia Neal wrote, "no matter how silly. I marched with the Junior Rose Bowl Queen and posed at Pacific Ocean Park with a rifle in one hand and a Kewpie doll in the other. I even drove a bulldozer in ground-breaking ceremonies for a Salvation Army boys' club. I accepted all the arranged-for citations. I was honorary nurse at the Huntington Memorial Hospital and Burbank's lady mayor for a day." (Neal, *As I Am*) Loretta Young, always a perfectionist, studied fashion magazines to see how professional models posed, imitating them in her publicity shots.

Many publicity gimmicks occurred serendipitously. Veronica Lake was posing for a photographic session when her blonde hair fell down over one eye. Her producer and director saw that look and used it in Lake's next picture. "It was just chance," noted publicist Teet Carle. "We capitalized on the girl that became famous by hiding one eye." (Carle int., AFI) Lana Turner made an impression as a "Sweater Girl" walking down the street in her first picture at Warner Bros., *They Won't Forget* (1937). Emily Torchia heightened that image at Metro during a portrait sitting with photographer Clarence Bull. Initially Turner wore a blouse under her sweater. Torchia suggested she take off the blouse and her brassiere, and Lana's Sweater Girl reputation was created.

Young stars needed publicity and were constantly in their studio's publicity office, eager to see their names in print. Jeannette Morrison Reames arrived at MGM before she became Janet Leigh, spending hours with the Metro publicists. "Their job," Leigh said, "was to make a 'Who?' into a 'Wow!'" Starlets generally worked with newcomers in the publicity department, graduating to the senior publicists once they became more important. Youngsters were involved in anything that would attract the press's interest, not just national news. "You couldn't ask Bette Davis to go open a supermarket," Bill Hendricks said, "but you could ask one of the younger stars." Hometown stories were concocted for every player.

Part of the publicity department's job was to create an image for each actor and actress. Once the image had been determined, it was important for that would-be star to dress accordingly at all times. Barbara Rush must always look like a lady; Anita Ekberg should look sexy, although her blouse mustn't be cut too low. Even when young stars went to the commissary, they needed to maintain their image.

Walter Seltzer was at Columbia when the studio signed two young men— Glenn Ford and William Holden—and three young ladies—Evelyn Keyes, Janet Blair, and a gorgeous Spanish-American girl named Margarita Carmen Cansino, who became Rita Hayworth. Columbia's genial publicity director, Lou Smith, called his department of fourteen together and said, "Look, there's not too much activity going on. If we want to keep the department intact, we have to do something spectacular." The Columbia department collectively went to work on these five youngsters and developed them into stars. (Seltzer int.)

During the 1940s the press eagerly reported the behavior and indiscretions of anyone whose face appeared on a movie screen. If studio publicists didn't provide lurid enough stories, jounalists made them up. Columnists Louella Parsons and Hedda Hopper, Hollywood's duennas of gossip, gained unparalleled power. They could be kind, but they could also be vicious, marring personalities, even destroying careers. Parsons had the Hearst press behind her and an intelligence network that included hotel busboys, beauty-parlor assistants, telegraph operators, and doctors' and dentists' receptionists. She operated out of her Beverly Hills home, talking on the telephone for hours, sometimes writing her column on trains, in airplanes, at the Santa Anita racetrack, even in the ladies' room at Ciro's. Married to Dr. Harry Martin, a Hollywood urologist, Parsons had access to various medical laboratories around town and often knew that an actress was pregnant before the woman herself knew. Hedda Hopper, Louella's celebrated rival, had originally been an actress. "Louella Parsons," Hopper said, "is a newspaper-woman trying to be a ham. I'm a ham trying to be a columnist." (Eells, *Hedda and Louella*) Together they claimed a readership of nearly seventy-five million.

Hedda and Louella terrified most newcomers, and even veterans were on their best behavior when they talked to them, revealing as little as possible. "They could crush you in one sentence," Lizabeth Scott said. Jayne Meadows, admittedly a defensive, insecure young actress when she arrived in Hollywood, walked off the set of *Song of the Thin Man* (1947) in a huff one day, convinced her part wasn't good enough and that movies were not to be taken seriously. "When I walked off that picture," Meadows recalled later, "Louella Parsons knew it in five seconds. I don't know whether the studio called her or what, but she wrote about it in her column the next day: 'My advice is Jayne better walk right back on the picture if she wants to have a career.' Louella could really fix you, so I walked back and fast."

Parsons was a tough lady who drank a great deal and often seemed not to be paying much attention, when in fact she missed very little. "She was just like a tape recorder," Bill Hendricks commented. "She had a fantastic memory." Jayne Meadows remembered going to Parsons's house for an interview and meeting Montgomery Clift coming in as she was on her way out. Clift was being evasive and rude to her, kidding her about her drinking. Clift kept saying, "You aren't going to remember any of this in the morning. Why don't you write it down?" But Louella seemed to retain everything.

Hedda Hopper, a long-legged woman with flashing eyes and brown hair, seemed almost a caricature. "The first time I walked into her office," recalled actress Jeff Donnell, "Hedda was behind her desk with her feet propped up on it. She had this huge hat on and a run in her stocking." But Hopper was not to be taken lightly. She became disenchanted with Joan Bennett at one point and wrote several sniping items about the actress in her column. Hopper implied that a friend of Bennett's had appeared at a party drunk, which the actress claimed wasn't true. Harry Crocker, another columnist, took exception to Hedda's inferences and wrote a harsh article about responsible reporting, without mentioning Hopper by name. Bennett took out a full-page ad in *The Hollywood Reporter* and *Variety*, placing the two columns side by side, and since it was close to Valentine's Day, had them outlined in hearts. The page was headlined: "CAN THIS BE YOU, HEDDA?" To add to the insult, the actress ordered a live, deodorized skunk delivered to Hopper's door; Hedda named it Joan. (Bennett and Kibbee, *Bennett Playbill*)

Parsons claimed "a story wasn't a story unless I got it first." (Parsons, *Tell It to Louella*) Generally an item appeared in either her column or Hopper's, but not both. Since Parsons's column ran in five or six hundred papers daily and hers was the largest circulation, publicists favored her. During his first week as publicity director at Warner Bros., Bill Hendricks nearly lost his job because he sent a big story to Hopper rather than her rival. Parsons promptly called Jack Warner and said, "Jack, you've got to fire that boy." Warner summoned Hendricks to his office and said, "You'd better make peace with that old broad." For a week Parsons refused to speak to Hendricks, even after he sent her flowers. Through her nephew, who happened to be a close friend, the publicist finally arranged to have breakfast with her. "We can get along," Parsons said. "All you have to do is just give me all the stories." Hendricks had to admit she was cooperative from that point on.

When handled carefully, both Hedda and Louella were great supporters of the industry, ballyhooing the latest productions. Parsons's zealous announcement of January 28, 1952, was typical: "Certainly the most unexpected and hottest teaming of 1952 (so far) will be Lana Turner and Kirk Douglas pitching woo in 'Tribute to a Bad Man' at MGM." (The title of the picture was later changed to *The Bad and the Beautiful*.)

Studio executives noted carefully how much fan mail a young star received. An increase in fan mail for an actor normally insured a raise at the next option period. Sacks of mail, bundles for each star, were delivered daily to every studio. The fan mail department was separate from publicity. Sometimes a star's secretary answered fan letters; other times fans were sent a form letter reply, or pictures were sent for a nominal fee. Some stars insisted on autographing photographs mailed by the fan mail department, while others used forgers who signed for them. Joan Crawford employed a secretary who handled her fan mail, but if she liked a letter someone had written, she'd write a note back herself.

Until World War II most unit publicists were men; this was considered appropriate at the time since the job required traveling with the company while the production was being shot. Unit work involved covering the set, profiling the players for a press book, making sure publicity and group stills were taken for the press kit, and seeing that magazines and newspapers received sufficient information. Howard Strickling required a certain number of stories every week from his unit people at Metro. Announcements of new cast members, reports of an accident or any interesting development on the set, and descriptions of parties at the end of production and special plans for a premiere were all sent to likely sources. Reporters were brought to the set almost daily, sometimes from around the world, and stars might be interviewed two or three times a day between takes. A still photographer took pictures of every scene, many of which appeared in fan magazines, sometimes even before the film was finished.

Emily Torchia's first experience working on location came with a Greer Garson picture called *Desire Me* (1947), when the company was sent to northern California, near Monterey. Torchia's department told her that no matter what happened she was to send a telegram. They said they were too busy to answer the phone. Garson and actor Richard Hart were out in a boat in the ocean when a wave came along and swept the actress into the sea. A fisherman caught her and pulled her out, although the cameraman had actually gotten to her first. "We wanted the fisherman to save her," Torchia said. "I telegraphed the department, but I couldn't get anyone to believe me!" The publicist finally called Ralph Jordou, Metro's head planter, and said, "I can't get anyone to believe me, Ralph, and Miss Garson is in the hospital. She's just cut to pieces." So Jordou hired an airplane, put the press on it, and flew them to San Francisco. The episode made headlines all over the world. "We had to get into the hospital to get Miss Garson made up," Torchia recalled. "I remember photographer Virgil Apger's pushing me up to the hospital window, so I could climb in. The doctor didn't want the press in taking pictures, but that's how the publicity department worked."

After a film was shot, the unit publicist, aided by the rest of the department, worked on the advertising campaign for the movie. Fifty or a hundred ideas might be suggested, but only four or five would generally be used as the basis for the sales campaign once the picture opened. Main title billing was worked out before the film was shipped, so that the advertising department knew the order and exact size of each name listed, all in accordance with the various contracts. Actors and actresses knew they had become stars when their names appeared above the film's title. Esmé Chandlee recalled that one of the greatest dilemmas in her experience with billing occurred on *That Forsyte Woman* (1949), which starred Metro's Greer Garson and Walter Pidgeon, along with Errol Flynn, whom the studio borrowed from Warner Bros. Garson always received first billing, but when Flynn's agreement came through, Warners insisted he get first billing. "So we had a real problem," Chandlee recalled. "We all sat down and tried to figure out how we were going to get around it. We were all doodling, and I finally saw this curve." Chandlee suggested the names be put on a curve with Flynn's first, then Garson's at the top, and Pidgeon's to the right. The solution worked.

While the advertising campaign was being prepared, the press from all over the world was shown bits of the film being readied for distribution. Full-page ads were designed for fan magazines, and newspaper strategy was carefully coordinated. The challenge was to come up with something new and striking for each film. During the big studio era billboards were widely used—twenty-four sheets, six sheets, down to three sheets. Paramount had 360 twenty-four-sheet locations in New York City alone. Publicists went in for lurid types of material, Eugene Zukor said, "taking elements from the film that would lend themselves to a romantic struggle, or a rough night on the sea, or the hero galloping through flames. Those lithographs were colorful, and the artists had a field day." Billboard advertising was an important factor in film campaigns.

Most executives and producers wanted their advertising to be truthful but exciting. During the hard times of the 1930s, when film biographies were popular, two billboards were joined in a V shape at a prominent Warner Bros. location; one sign read: "George Arliss in *Alexander Hamilton*" and the other: "WHAT DID GEORGE WASHINGTON DO WHEN AMERICA'S PROSPERITY WASN'T WORTH TEN CENTS ON THE DOLLAR? He sent for ALEXANDER HAMILTON." (Roddick, *New Deal in Entertainment*) Eugene Zukor described Paramount's first telegraph pole advertising, instigated by John C. Flinn for *The Miracle Man* (1919), starring Lon Chaney. Flinn had cards imprinted with a blue circle and the statement "The Miracle Man is Coming." There was no Paramount trademark nor anything else on the card. He sent out runners to every town in the country to tack up his signs. "People didn't know whether this was a religious meeting, or a Chautauqua, or what it was," Zukor remembered. "Just that this mysterious thing was coming, the Miracle Man. That's what sold the picture."

Not all advertising was tasteful. In *Beyond the Forest* (1949), her last film for Warner Bros., Bette Davis played Rosa Moline, a popularized version of Emma Bovary, a role Davis hated. "Bette Davis as a twelve-o'clock girl in a nine-o'clock town" hailed the picture's ads. When Joan Leslie, who heretofore had been Miss Homespun, arrived in New York to publicize *Repeat Performance* (1947), she was horrified to discover her billboard image atop a Broadway building in a nightgown she wore in the picture, now inexplicably torn off her shoulder and dramatically hanging to one side.

Studio publicists in charge of special events arranged for a picture's premiere and parties and banquets to launch it. The opening of Howard Hughes's *Hell's Angels* (1930) featured, among other excesses, a three-day cocktail party for the press at the Astor Hotel in New York. Later, stunt pilots buzzed Grauman's Chinese Theater in Hollywood. Bill Hendricks said the premiere Warners staged at Grauman's for *Giant* (1956) was the biggest event of his years with the studio. Grauman's stood next to a large parking lot, so Warner Bros. built bleachers there and filled them with fans. A battery of twenty-five or thirty searchlights was placed at the back of the lot, its beams over the heads of onlookers like a curtain. The studio laid red carpet on the sidewalk up to the Roosevelt Hotel, about a block away, and stars walked on the carpet to the theater. The famous area of footprints in Grauman's courtyard was surrounded by radio and television cubicles. "It was spectacular and netted a tremendous amount of publicity for the picture," Hendricks said.

If a picture did well at the box office, the producer and studio executives usually claimed it succeeded because it was a fine example of Hollywood craftsmanship. If the picture failed, studio heads and everyone involved attributed the failure to a poor advertising campaign. Actors and directors who seemed embarrassed by advertising, or who considered publicity buildups demeaning, soon realized how out of step their values were. "When I was a kid and fascinated with the theater, only cheap vaudeville acts advertised," said director George Cukor. "But in Hollywood, everybody advertised."

After a premiere, a studio usually sent some of the picture's stars on a national tour. Decked out in new wardrobes, which the studio generally provided, and accompanied by a member of the publicity department, stars visited New York and other major cities, appearing on stage several times a day. Schedules were tight, but the work was usually fun. Studios had a field man in all the principal cities, while the publicist looked after the stars and handled the press. If necessary the publicist in charge groomed the players for the interviews. Bill Hendricks was on the road most of his first year at Warner Bros. His first assignment was in Gallup, New Mexico, on a picture called *Pursued.* To publicize *The First Time* (1953), Columbia sent Jeff Donnell and her friend Barbara Hale on a tour of thirty-six cities in thirty-five days. "We had a trunk full of laughs," Donnell said years later, "such happy memories. That tour was made without a hairdresser, without a script, without anything!"

Performers who could dance or sing had an easier time with personal appearances than straight actors, since they could entertain audiences. Jane

Powell and Doris Day sang; Virginia Mayo and her husband Michael O'Shea did a vaudeville routine; Burt Lancaster and his former partner, Nick Cravat, performed their circus act during a publicity outing for *The Flame and the Arrow* (1950). "Nick would balance a thirty-foot pole on his head," Bill Hendricks explained, "and Burt would climb this pole and do the acrobatics they had done in the circus. Getting that thirty-foot pole on trains and planes was no easy job." In other cases the studio's drama coach devised a short comedy sketch or a speech from one of their films for performers to use to entertain live audiences.

Often tours involved a parade, with the star riding in an open car, waving to the crowd. When *Gone With the Wind* premiered in Atlanta, a parade traveled from the airport through the city, down famed Peachtree Street. Rather than marching bands, there were musicians and choirs stationed in strategic locations along the route. Fifty flower-bedecked automobiles whisked Clark Gable, Vivien Leigh, Olivia de Havilland, Evelyn Keyes, Ann Rutherford, Ona Munson, Laura Hope Crewes, and other cast members through a sea of confetti and music. The population of Atlanta in 1939 was close to three hundred thousand, but an estimated million and a half people lined the streets that day.

Even Lassie, Metro's prize collie (a male), was in demand for personal appearance tours. Actually Lassie's trainer, Rudd Weatherwax, took four dogs on tour and would substitute one of the other three if he felt Lassie had had too strenuous a day. Lassie was treated like a star; at Metro he even had his own dressing room.

After the stars returned to Hollywood, the theaters across the country continued to publicize pictures that lent themselves to special promotion. Jungle pictures occasioned stuffed apes and fake palm trees, or box offices fashioned like tree huts. Horror films brought forth piercing screams and maniacal laughter from loudspeakers over the theater's entrance. Arthur Mayer, who ran the Rialto Theater on Broadway, became known as the Merchant of Menace. Mayer achieved a sort of immortality when a gossip columnist reported one day that Sam Goldwyn had visited the Rialto and left saying, "When I see the pictures they play in that theater, it makes the hair stand on the edge of my seat." (Mayer, *Merely Colossal*) Between more spectacular offerings, theaters contrived an atmosphere conducive to family entertainment, offering a touch of elegance to the average American. During the months just before the bombing of Pearl Harbor, amid the "Bundles for Britain" campaign, Karl Hoblitzelle of Interstate Theaters in Texas had his cashiers knit socks when they were not selling tickets.

Between making movies, stars spent a great deal of time in the studio's portrait gallery doing fashion layouts and special art work. At MGM contracts specified that one day at the end of each picture must be devoted to portraits and fashion stills. Scores of pictures were taken, although most stars hated these sessions and considered them a bore. Ann Straus became the fashion editor at Metro; it was her job to dress the stars for magazine covers and photo layouts for such magazines as *Ladies Home Journal, McCall's,* and *Good Housekeeping.* The studio's portrait gallery was located next to the dance bungalow, in front of a machine shop. Straus usually ordered a limousine to bring the star to the gallery, worked with the still photographer on the background and props, and scheduled the makeup artists, wardrobe women, hairdressers, and crew. The crew was paid whether the star arrived or not, and the cost was charged against the publicity department. If the star came, and she decided she was tired and wanted to go home, the publicist cajoled her. Ann Straus put flowers in the star's dressing room, ordered tea and coffee, tried everything to keep her happy. "The portrait gallery was a very ugly, uninspiring place," the publicist noted.

Since Straus lived in Beverly Hills, she often stopped by the shops and designers' studios on her way to work to pick up clothes for that day's sittings—much to the dismay of the teamsters union, which insisted she use a studio car and driver. "Time was very precious to me in those days," Straus said, "because we had so darned many sittings, and I didn't have too much time to prepare. I'd come flying in like seventy devils were chasing me, with my car full of clothes. We were always doing winter clothes in the summer and summer clothes in the winter, because you had to work that far ahead to meet the publication dates. So it was always a hassle getting the clothes. And, of course, I had to know what each person's likes and dislikes were—if they didn't like green or purple or hated prints—and know what their measurements were."

One day Straus suggested a green negligee for a sitting to the dignified Greer Garson. Garson put the negligee on and, posing in the doorway of the portable dressing room, said grandly, "What do you think of this?" Straus exclaimed, "Oh, Miss G., you look divine!" Garson turned, looked down at the publicist, and said, "Aren't you dishing it just a bit?" For a moment Straus forgot she was speaking to the queen of the Metro lot and replied, "I'm not dishing it, I'm shoveling it!" Garson convulsed with laughter, and the two became lasting friends. "From that moment on," Straus said, "I could tell her to put on a lampshade and she'd do it."

Straus believed in using real jewelry and beautiful clothes, since MGM's hallmark was glamour and luxury. Her arrangement with the shops and designers was that they would receive credit for any fashions used, and that she would return clothing in the same condition it left the store. Should anything be damaged, the studio would pay for it. "That always made me nervous," Straus confessed. "I'd say to the actresses, 'Now watch your glands. Don't perspire.'"

Makeup people worked from the chin up; body makeup people worked from the chin down. "Nobody was allowed on anybody else's territory," Ann Straus said, "because of union regulations. I wasn't supposed to pick up a powder puff, I wasn't supposed to pick up a comb, and I really wasn't supposed to pick up pins." Stars working in the portrait gallery normally ate lunch there, since time was limited and no one wanted to return the next day if they didn't have to. When proofs came in from the lab, the studio's fashion editor would "kill" the unflattering shots and order any necessary retouching. Some still photographers—Clarence Bull at Metro, John Engstead at Paramount, George Hurrell at Warners—became famous in their own right, known for their glamorous images.

Holiday art provided another publicity dimension—color photographs of celebrities dressed in Thanksgiving or Christmas attire, for instance, used in the rotogravure sections of Sunday newspapers. Alfred Hitchcock was once pictured at Warner Bros. with three laughing Santa Clauses, all holding their stomachs in Hitchcock's familiar pose. Gail Gifford remembered Universal actress Julie Adams in a pilgrim costume, with a turkey and a hatchet.

The big studios squelched information as well as disseminating it. If a sex symbol became pregnant, her studio tried to keep her condition out of the news. The studio's head of security, who was not above using bribery to keep an indiscretion secret, dealt with drunk driving charges, illicit romances, and drug offenses. If studios objected to articles in *Photoplay* or *Modern Screen* or some other fan magazine, they withdrew advertising. Studios thereby exercised considerable control, since the major companies all advertised heavily during the 1930s, 1940s, and 1950s.

Stars normally cooperated in building a positive image. Mickey Rooney at thirteen or fourteen, however, was irrepressible, causing his Metro publicist, Les Peterson, to live in fear of what his charge might do. During the Depression, when MGM celebrities were not supposed to flaunt ostentatious living, young Rooney bought a new Cadillac and hired a driver. "Here was this pint-sized kid riding around in a new Cadillac with a chauffeur," Walter Seltzer recalled, "and he just loved to talk about it. Les Peterson, present at an interview, would be biting his nails."

But although studios usually wanted only positive press coverage, they also used negative press for their own purposes. A recalcitrant actor would read in the gossip columns that he was misbehaving, or that his fans were becoming annoyed, perhaps even that his wife was thinking of a divorce. If the pressure was great enough, the actor usually capitulated. When Joan Fontaine, under contract to David Selznick, refused to make the Technicolor spectacle *Frenchman's Creek* (1944) on loan-out to Paramount, Selznick barraged her with telegrams. "Hourly telephone calls from lawyers and agents would harass me," Fontaine said. "Columnists were coerced into printing that I was difficult, ungrateful, temperamental, uncooperative, swell-headed. The pressure became intolerable." (Fontaine, *No Bed of Roses*) Eventually she played the part.

Hollywood marriages frequently suffered from too much attention in the press. When Douglas Fairbanks and Mary Pickford began having marital troubles, their problems were magnified, as Douglas Fairbanks, Jr., put it, by "too much attention paid and the fierce glare of the arc lights on this world-famous romance and marriage." That kind of spotlight centered on all the Hollywood stars, and the persistent publicity, largely artificial and manufactured by the studios, was in many cases damaging to human relationships and personal growth. Situations were magnified, if not actually invented, and falsehoods were churned out in such volume that celebrities lost touch with reality. "I have looked recently at interviews I allegedly gave," Douglas Fairbanks, Jr., said of his own career. "I know perfectly well those interviews never took place. I didn't even know who the writers were, and they quoted me as saying things I never would have dreamed of saying."

Young Fairbanks and Loretta Young made six pictures together, and the Warner Bros. publicity department hinted they were having a romance. "It couldn't have been less true," the actor said. "We didn't even like each other. She couldn't stand me, and I thought she was a snippy young twerp, getting far too big for her britches." Not until later did they become friends. Western star Dale Evans, on the other hand, was twenty-eight when she arrived in Hollywood, had been married, and was already the mother of a twelve-year-old son. "That was like one foot in the grave," Evans later declared. "My agent just about had a spasm." Studio publicists insisted the actress claim her son was a younger brother.

It was the publicity department's job to create stars and maintain the public's interest in them. Publicists manipulated news, enhancing or exploiting events in ways that were both beneficial and cruel, and they controlled the information the public received, suppressing some, rationing the rest. In the process they built careers, turning studio contract players into household names. From the publicists' standpoint, working at a Hollywood studio meant a good life. "I got to meet royalty, great writers, religious leaders, and fine musicians," Emily Torchia remembered in retirement. "I was able to travel, with studio cars at my disposal. That was a part of life I could never have afforded without a studio behind me. I think that enriched me; it made a different life for me entirely."

9

Writers and the Story Department

Ohio exhibitor J. Real Neth on a tour through the Story Department at MGM. Courtesy of the Academy of Motion Picture Arts and Sciences.

Although it is generally conceded in Hollywood that making a great picture out of a bad script is impossible, writers within the big studio system were vastly underrated by top executives and comparatively underpaid. "The writer was very important in Europe," director Fritz Lang observed, "but here he is transformed into a mechanic. In any major studio there are ten writers working on a single script." (Lang int., Rosenberg and Silverstein) Screenwriters persistently felt exploited, earning far less than stars, directors, and producers, yet they considered themselves the creative backbone of the industry. At a time when many stars were paid $10,000 or more a week, major writers were earning no more than $1,500. The average salary for writers was closer to $800 a week. Studio executives considered writers a special breed—necessary but troublesome, often independent and irreverent toward company authority. "Their ideas were frequently considered radical," Warner Bros. screenwriter Howard Koch said, "a potential threat to any corporate structure committed to the status quo. If scripts could have been turned out by a machine, I think most studio heads would have preferred it as a more tractable alternative." (Koch, As Time Goes By)

Writers resented being pressured to adhere to studio policy, having scenes rewritten by somebody else or eliminated, being treated as inferior by insensitive executives who sometimes expected them to punch a time clock. It was easy for a writer to become typed, assigned mainly war pictures, westerns, whatever his or her biggest success had been. The advantages of a studio contract were steady employment with a substantial weekly paycheck, the conviviality of stimulating colleagues, and the excitement of working in a glamorous setting on projects that were assured a mass distribution. Yet in a medium largely shaped by directors, the position of Hollywood writers was seldom easy, requiring a great deal of confidence to succeed.

With the advent of sound the old scenario writers and gag men of the silent days were edged out as film producers began looking to the theater for writers who could create dialogue for talkies. Playwrights from New York and London were rushed in, along with an influx of literati from the publishing world. The results were not satisfactory, since playwrights and novelists were unaccustomed to the rapid pace necessary to keep scenes in motion pictures from becoming dull. Often in early sound features, actors simply faced one another and talked, making synthetic and lifeless sequences. In many instances the scenario writers were reinstated as assistants to dialogue writers, helping them fashion scripts that would play cinematically.

During the 1930s and 1940s most major studios employed a junior writers' program, bringing in promising youngsters, mainly college graduates who had attracted literary attention in some way, such as publishing an exceptional short story or journalistic piece. Several came from George Pierce Baker's esteemed playwriting workshop at Yale. A studio might bring together a group of ten or twelve apprentices, assigning each to an established screenwriter already under contract, allowing the youngsters to ask questions and work on scripts without credit. By the late 1940s junior writers sometimes earned as much as $250 a week, but they frequently lasted only a few months, when they were replaced by others who seemed more promising. Since experienced writers were not eager either to answer novices' questions or to read beginners' rewrites, the program disintegrated, although by 1950 some of the young people involved had worked their way up through the studio ranks, often starting with B pictures and eventually graduating to better assignments.

Many Hollywood writers never aspired to anything more than formula treatment of routine A or B pictures. "They knew their craft and enjoyed the comfortable life it afforded," Howard Koch explained, "taking no risks with more demanding or controversial material." (Koch, *As Time Goes By*) Others developed into real craftsmen, adept at script construction, aware that writing for the screen required different skills than writing for the stage. Not only must dialogue be pared down, but scenes needed to be structured to hold an audience's attention visually, advancing the story at the same time. "I try to make the story move," screenwriter Walter Bernstein said. "I try to make it have a kinetic quality so it's never static." Motion picture writers cannot be pure wordsmiths; they must always keep the physical aspect of picture making in mind. In films visual imagery is essential; words have to be turned into images. "You can be a damned good screenwriter and not write exceptional prose," Academy Award–winning writer Robert Pirosh noted. "You can be a wonderful novelist or short story writer and not be a good screenwriter. The dialogue is much different." While some screenwriters' strength lay in dialogue, others were better at story continuity—one reason for having several writers assigned to the same project. Since most movies in the Golden Era were adaptations, the screenwriter was called upon to transform somebody else's stage play, novel, or short story into a film script. Hollywood writer Jo Swerling once compared the process to fine cabinetmaking, a craft demanding taste and skill, but not the pure creation of an artist. "I'm a writer, but not an author," 20th Century-Fox scriptwriter Philip Dunne said. "I'm a moviemaker, and I made a few good ones. That's enough for me."

During the big studio era it was almost impossible for an outsider to sell an original screenplay. Most stories filmed were either adaptations or scripts written by contract personnel. The major studios had a reading department, a subsidiary of the story department, in which staff readers (or screen analysts, as they were later called) made a synopsis of every book, play, and short story submitted. Readers were expected to cover a story a day (unless the book was unusually long), which involved preparing a ten-page summary and a one-paragraph opinion. Studios were inundated with books and stories, often in galley proofs, each of which had to be evaluated to avoid possible lawsuits, unless the envelope containing it was returned unopened. "The trash that was sent to those studios was unbelievable," said Marjorie Fowler, who for a time worked as a reader at 20th Century-Fox.

The head of the reading department at Fox was Julian Johnson, who received the recommendations of his staff and issued a weekly bulletin to the studio's producers, including a synopsis of the recommended material. If a producer liked a story description, he would ask to see the material in its entirety. Studios always had a vast backlog of material, since every week they received another batch from publishers and agents.

Most readers were privately working on original screenplays of their own, hoping the studio would ultimately buy one; several of Hollywood's top writers began that way. Philip Dunne, the son of Finley Peter Dunne, the creator of "Mr. Dooley," started in films as a reader at the old Fox studio. Later he wrote *How Green Was My Valley* (1941), *The Ghost and Mrs. Muir* (1947), and *The Robe* (1953) for Darryl Zanuck. Jesse Lasky, Jr., also began as a Fox reader, earning twenty-five dollars a week. Lasky became a ghostwriter, then a screenwriter, and eventually was a writer on *Union Pacific* (1939), *Samson and Delilah* (1949), and *The Ten Commandments* (1956) for Cecil B. DeMille. Lester Cole, in 1947 one of the Hollywood Ten called before the House Un-American Activities Committee, took a job at $50 a week during the early 1930s as a reader at Metro-Goldwyn-Mayer, where Samuel Marx and later Edith Farrell headed the script department. Cole was assigned a desk in a large office with a dozen or more other readers, among whom was young Lillian Hellman. In his youth Budd Schulberg, author of the novel *What Makes Sammy Run?*, served as a reader in the story department for David O. Selznick.

During the late 1930s Selznick's Eastern story editor, Kay Brown, sent her boss a copy of a new novel she had read in manuscript. The book was entitled *Gone With the Wind*, but since Selznick didn't have time to read it, he passed the copy on to his Hollywood story editor, Val Lewton, and an aspiring young writer named Ring Lardner, Jr., another of the Hollywood Ten, who was working for Selznick at the time. Both Lewton and Lardner argued that the novel should not be made into a movie, finding it reactionary. But Lardner's future wife, Silvia, also read the book and joined Kay Brown in recommending *Gone With the Wind* to Selznick, who was prevailed upon to read a long synopsis. Lardner maintained that Selznick never actually read the book, though he decided to make the film.

At most studios writers were shepherded by the head of the story department, although they were an individualistic lot—intelligent, humorous, most of them dedicated to turning out quality work. If writers came up with an idea for a picture studio heads thought was good, they were given the freedom to pursue it. "At that time I couldn't have afforded to do the kind of research I did," Metro screenwriter Robert Pirosh said. "I wouldn't have had the entrée into places. But the studio made arrangements for me to go wherever I wanted to go."

The big studios maintained research departments with extensive libraries geared toward motion picture production. Major studios kept four or five researchers on their payroll, who with the assistance of the university and public libraries around town could locate nearly anything a writer needed within twenty-four hours. "The next day your desk would be stacked high with material," said Academy Award–winning scriptwriter Edmund North, "pictures and text as well." Twentieth Century-Fox writer George Seaton was such a thorough researcher it was frequently quipped that if Seaton had been assigned Dante's *Inferno*, he would have gone to hell to find out everything he could about the place.

Rarely did serious Hollywood writers settle for shortcuts. When Virginia Kellogg was preparing the script for Warner Bros.' *Caged* (1950), an authentic story about prison life, she had herself incarcerated in a women's prison in California. Kellogg said later that if the inmates had known she was a "plant," her life would have been in jeopardy. While the picture was being filmed, Warners hired a matron from a women's penitentiary in Illinois in an advisory capacity to insure a semidocumentary flavor. (Garde int.) As concerned as studio writers were with entertainment value, the serious ones also hoped to make a social comment. Upon completing *Along Came Jones* (1945) with Gary Cooper, screenwriter Nunnally Johnson wrote critic Walter Kerr: "I was again using the cold, unflinching eye of the camera to probe a sick Society.

Never for one second did I think of Cooper as a tramp cowhand; to me he was Western Man, eternally gallant, eternally defeated, and the picture itself one long bitter laugh at life." (Johnson and Leventhal, *Letters*)

Yet entertainment always came first, and screenwriting in the big studios generally was a team effort. At 20th Century-Fox, Darryl Zanuck considered the final script sacred, even though it had been created by an assembly line. Since Zanuck was short of stars and had been a writer himself, he looked upon the script as his real star, and he depended primarily on scripts written under his supervision. Zanuck was fortunate in having many talented writers under contract: erudite, civilized Philip Dunne; witty, gracious, intelligent Nunnally Johnson; dignified, gentlemanly Lamar Trotti. Both Johnson and Trotti were southerners, and all three had Zanuck's confidence. Trotti was responsible for *Drums Along the Mohawk* (1939), *The Ox-Bow Incident* (1943), and *Wilson* (1944), Zanuck's favorite film. Nunnally Johnson wrote the adaptation of John Steinbeck's *The Grapes of Wrath* (1940), as well as *The Keys of the Kingdom* (1945) and *The Gunfighter* (1950). Steinbeck respected and trusted Johnson so implicitly that when the scriptwriter was adapting a second Steinbeck story, *The Moon Is Down* (1943), the author wrote him a note that ended, "Tamper with it." (Bowdon int.)

A somber man privately, Johnson had an acid humor, quite different at the dinner table or in letters from what reached the screen. Discussing an unruly actor he once commented, "Well now, some say he's not so bad, unless you insist on a human being." On another occasion Johnson and Philip Dunne attended a Stanford-Alabama football game together, during which the Alabama team ran like deer, including a ninety-eight-yard drive, finally winning the game by a substantial margin. Johnson dourly watched the Stanford team lose. Finally he asked, "Whatever became of all those Southern boys I used to know that had hookworm?" (P. Dunne int.)

Both Johnson and Lamar Trotti had suites in the 20th Century-Fox executive building, but Dunne refused to move from the old writers' building even after a new building had been constructed, with space for a secretary outside each writer's office. Although Colonel Jason Joy, a large, booming-voiced, courtly gentleman, was the administrative head of writers on the lot, Darryl Zanuck himself oversaw the major scripts. Zanuck made a practice of sending scripts completed by one contract writer to another for cross-fertilization. Dunne would receive Trotti's scripts, and Johnson would receive Dunne's. "We weren't supposed to tell the other fellow, but of course that only lasted until the script arrived," Dunne said. "Then I'd be on the phone right away: 'I just got your script.'" Often the second writer could make useful suggestions.

Zanuck's story conferences became legendary, with the production head pacing about the room swinging his polo mallet, exuding nervous energy. He could be dictatorial, yet seldom forced writers to accept mandates unless policy was involved. "Zanuck was a man who would give you twenty ideas," remarked Edmund North, "four of which would be sensational, and the rest you'd have to deal with as best you could." If Zanuck's taste on scenes was not infallible, his dialogue was sometimes downright bad. Still, he understood the process of scriptwriting and had a keen eye for cinematic construction. "You could argue with him," North continued, "but you'd better have a good case."

Generally Zanuck held a preliminary conference, so that everybody involved knew what kind of picture they were making; then there would be a second conference after the writer had finished his story treatment. Once the first draft of the screenplay was completed, a third conference would be called, followed by another after the second draft, and so on through the final revisions. At each session Zanuck made extensive comments, often cutting out motivation to speed up the action. "It was the movement that seemed to be important," Ring Lardner, Jr., observed; Zanuck firmly believed that movies should be fast paced.

Present at each story conference was Molly Mandaville, who had a facility for putting everybody's remarks into acceptable English, then typing them up and seeing that everyone concerned received a copy the following morning. In addition Zanuck sent out memos clarifying his suggestions. "Zanuck was the only studio head I knew who would commit himself on paper," Edmund North said. "His ego was strong enough, and he wasn't afraid to do that." Sometimes his memos were long, underlining decisions that had been reached, detailing specific changes, rearranging the writer's material, adding some of his own. "Okay, boys," Zanuck might conclude, "if you follow this and don't deviate from this outline, I think we'll come out with a good picture."

After Zanuck approved a script, it was set. A director could ask for minor changes, but nothing was altered without approval. Since Zanuck watched all the dailies, directors dared not take liberties with scripts the studio head had helped shape. John Ford might improvise, but most contract directors knew they had to film the approved script. The results were uncommonly high quality for commercial entertainment. *How Green Was My Valley*, *The Ox-Bow Incident*, *The Song of Bernadette*, *Wilson*, *The Razor's Edge*, *Gentleman's Agreement*, *The Snake Pit*, *A Letter to Three Wives*, *Twelve O'Clock High*, and *All About Eve* were all nominated for Academy Awards within a ten-year period, 1941–1950. Philip Dunne fashioned *David and*

Bathsheba (1951) into a literate script, half of which was written in blank verse, rather than a DeMille epic. Actor Gregory Peck considered Nunnally Johnson's *The Gunfighter* "letter perfect." As Peck explained, "A script like that is really a gift from the writer. I don't think we changed a syllable as we shot it. All we had to do was try to give him a hundred percent of what he wrote."

At Warner Bros. there was more regimentation, with Jack Warner insisting writers be in their offices by nine o'clock and leave no earlier than five-thirty in the afternoon, producing a minimum number of pages per week. The writers, however, slyly avoided Warner's rules and developed a strong sense of fraternity among themselves. "The official climate at Warner Bros. was chilly," Edmund North recalled. "People who went across the street to the drugstore were chided for it the next day. But the climate among the writers was very pleasant. There was a nice group there."

Warner himself knew the value of scripts, but he never got involved in script preparation the way Zanuck did. Rarely did he enter the writers' building, although he liked to know the people who worked for him and frequently conferred with them. He irritated writers; on the surface he seemed little more than a clown with his unsophisticated approach. He once advised production head Hal Wallis to assign a team of writers to create biographical stories for Paul Muni, during the time of Muni's success with *The Story of Louis Pasteur* (1936), *The Life of Emile Zola* (1937), and *Juarez* (1939). "Anything but Beethoven," Warner insisted. "Nobody wants to see a movie about a blind composer!" (Lawrence, *Actor*)

For writers, daily life at Warner Bros. was full of cliques, internal rivalry, and camaraderie. There was the usual clamoring for prime assignments and friction over credits, as there was at every studio. Around 1940, Warners employed between thirty and thirty-five writers, housed on the west side of the lot. Robert Lord and Norman Reilly Raine, both old-timers, were still there, as were the Epstein twins, whose specialty was comedy. Howard Koch, who had won attention with the Mercury Theater broadcast of "War of the Worlds" before collaborating on *Sergeant York* (1941) and *Casablanca* (1942), was at Warners. So were Robert Buckner, who wrote *Yankee Doodle Dandy* (1942) and several of the Errol Flynn pictures; John Huston, who was a major screenwriter before becoming a director on *The Maltese Falcon* (1941); and later Melville Shavelson, Jack Rose, Ranald MacDougall, and I. A. L. Diamond.

Robert Buckner was assigned a series of westerns because he had started out with a picture for Errol Flynn called *Dodge City*, which had been highly successful. "Somehow I just inherited one western after another," the writer said. "I got a little bored after a while." Buckner came to Warner Bros. in 1937 and stayed for fifteen years. A southern gentleman, he was pragmatic, willing to compromise literary values or historical accuracy when necessary. "You could take dramatic license with a western that you couldn't with a straight biography," he explained. *Dodge City*, *Virginia City*, and *San Antonio* were based on facts, but he was primarily interested in an entertaining story. He wrote several pictures directed by Michael Curtiz, whom Buckner judged a cinematic genius. "Mike could make a picture when he didn't know what it was about," the writer claimed, referring to the Hungarian director's success with filmed Americana. Buckner had been interested in Virginia City ever since he read Mark Twain's *Roughin' It*. When he wrote *Virginia City* in 1940, he wanted to capture some of the Twain flavor, but found he and Curtiz were at odds. "Mike was more interested in big brawling scenes, where bars were broken up and people were shooting each other," the screenwriter said. "So you couldn't get very much finesse into a picture like that. Mike didn't understand anything subtle; it went right over his head."

Howard Koch, who also worked on *Virginia City*, arrived at Warner Bros. three years after Buckner; the studio considered him a Martian expert because of his success with "War of the Worlds" on radio. Koch sat around for six weeks without an assignment. He visited all the sound stages and explored the back lot, then began to grow restless. He pleaded with Walter MacEwen, head of the Warners story department, for a project, but still sat idle. Then producer Henry Blanke, on John Huston's recommendation, sent Koch a thirty-page treatment of *The Sea Hawk* (1940), planned as Errol Flynn's next picture. The treatment Koch received was typical Flynn fare—much physical action but little in the way of nuances or characterization. The writer informed the producer that he'd like to do some research, suspecting that what actually happened with privateers of the period might be more exciting than another formula adventure story. When the studio's research department found some interesting data on Sir Francis Drake, Koch decided to use Drake as his model for the Flynn role, taking a few liberties with history, particularly regarding the navigator's love affairs. The picture was a financial success, and Koch soon began drawing choice assignments, earning salary increments at every option period.

Julius Epstein remained at Warner Bros. for fourteen years, collaborating with his twin brother, Philip. They turned out three or four pictures a year. "We decided early we were not making art, we were making a living," said Julius Epstein. "It was not called the motion picture *industry* for nothing. It was like working at belts in a factory; they turned out a picture a week."

Completed scripts were normally put into production right away, so that writers quickly saw how their scenes played.

At Metro-Goldwyn-Mayer screenwriters were shown more respect than at Warner Bros., and Metro had some distinguished writers under contract. While MGM's scriptwriters enjoyed little control over their material, neither was there the pressure to complete scripts quickly to show they had earned their weekly paycheck. Since Metro regarded itself as the industry's elite studio, writers there were expected to dress like professionals and assume a dignified attitude. During the early 1930s MGM had approximately forty writers on its staff, half of whom were women. A few old-time scenario writers were still around from the silent days, although they were rapidly phased out by incoming playwrights—what Hollywood veterans called "Eastern scribblers." Already the studio's story department received twenty thousand submissions a year from literary agents and publishers around the world.

Metro had a writers' building, even though the higher-paid writers occupied quarters in the Thalberg Building, the studio's main office building. By the 1940s there were between seventy and eighty writers on the lot, the overflow housed in one-room bungalows. While Metro employed some 125 secretaries, not counting the typists' pool, only the most esteemed writers merited their own. If a "rush" script came through, secretaries across the lot discovered that before they could punch out for the evening they were expected to type three pages of script. Those pages would be rushed to the mimeograph room so that copies could be distributed on the set the next morning. It was not unusual in the midst of production to have pink pages with changes coming down to the sound stage each morning.

At Metro junior writers accompanied their mentors to story conferences, which after Irving Thalberg's death were usually chaired by individual producers. By the late 1930s most of a writer's dealings, once assigned to a project, were with the producer in charge, although the studio's story editor might make suggestions. Louis B. Mayer himself read few scripts; he preferred wholesome, family stories and hated unhappy endings. He was convinced that if a star died in a film, his or her box office draw would suffer in the next picture. Mayer didn't realize until too late that Long John Silver is finally apprehended and sent to prison in *Treasure Island* (1934), and he argued vehemently against allowing Camille to die in the Garbo film.

In its heyday Metro basked in luxuries that later would not be financially possible. For both *A Night at the Opera* (1935) and *A Day at the Races* (1937), the studio sent the Marx Brothers on tour after the rough draft of the script was finished. Like a road company, they performed hour-long versions of their upcoming films between features in big movie houses in Duluth,

Chicago, Minneapolis, and San Francisco, with the writers in attendance to modify the jokes for bigger laughs. The comedians had not fared well in previous pictures at Paramount, and Thalberg's theory was that Marx Brothers material needed to be tried out on the stage first, as their successful *Animal Crackers* (1930) had been on Broadway. *A Night at the Opera* and *A Day at the Races* both became hits.

MGM, like other studios, also relied on sequels and formulas. If RKO made *Flying Down to Rio* with Fred Astaire, MGM was likely to follow with *Flying Down to Buenos Aires* starring another male dancer who could perform similar numbers. Metro's *Andy Hardy* series, inexpensive to make and popular with audiences, was widely copied by other studios. When MGM's *Father of the Bride* (1950) proved a comedic gem, the studio quickly concocted *Father's Little Dividend* (1951) for the same director and cast. "It wasn't so good," producer Pandro Berman admitted, "but it was a money-maker." *Anchors Aweigh* (1945) had been a hit with Gene Kelly and Frank Sinatra, and Metro was eager to team them again, but had trouble finding a story. When Kelly was handed a script in 1948 he thought was bad, he complained he could write a better story himself. He called his assistant, Stanley Donen, and the two of them sat up all night working on a treatment about a couple of baseball players who performed in vaudeville during the winter. Kelly and Donen took their outline to the studio the next day and MGM bought it, assigning Harry Tugend, a professional screenwriter, to complete the script. *Take Me Out to the Ball Game* was born, Gene Kelly said, "in complete self-defense."

Sometimes Metro-Goldwyn-Mayer hired distinguished writers at big salaries for the prestige of having them on the lot. Producer Arthur Freed brought in novelist Robert Nathan, author of *Portrait of Jennie*, but didn't like what he did. "I wrote picture after picture for Freed," Nathan said, "and he would tell me it was one of the finest scripts he'd ever seen. But that would be the end of it. For six or seven years I kept writing scripts that were never made." Eventually the novelist wrote an Esther Williams movie filmed as *Pagan Love Song* (1950), the title of an Arthur Freed song the producer was hoping to revitalize on the ASCAP list. "None of the characters I wrote were in the picture," Nathan said in disgust. "That was the last thing I did for Freed."

At Paramount, writers enjoyed more camaraderie than at any other studio. Although Paramount made serious films, its specialty during the 1940s was comedy with songs. The comedy writers were housed on the fourth floor of the writers' building, where they gathered around the coffeepot by the switchboard to exchange ideas. Simone, the telephone operator, served as den mother to the group, and the exchange between offices was consistently stimulating and pleasant. If a writer struck a snag, everybody would pitch in and try to help out.

After work Paramount writers frequently gathered across the street at Oblath's, a favorite bar, and sometimes met for lunch at Lucey's. "Then, too, we had a very inept writers' baseball team," said Charles Marquis Warren, who wrote and directed a number of westerns for the studio. The team played 20th Century-Fox, Metro, and Warners, managing to hold their own against players who were equally bad. "I played third base and was quite proud," Warren said.

Writer Melville Shavelson's first producer at Paramount was Paul Jones, who had worked with Preston Sturges and made some of the *Road* pictures with Crosby and Hope. "His great contribution was his ability to laugh," Shavelson said. "If you could make him fall out of his chair, you knew you had a good scene." When Jones didn't laugh, he'd often say, "Come on, let's go to the racetrack on the studio." He and his writers would take the rest of the day off and head for the track, with the writers feeling guilty, determined to try harder once they got back to their desks.

Billy Wilder and Charles Brackett were both brilliant screenwriters on the Paramount lot, turning out dramatic films such as *Double Indemnity* (1944), *The Lost Weekend* (1945), and *Sunset Boulevard* (1950), all classic examples of film noir. Wilder was a master at creating images, repeating visually an idea expressed in dialogue earlier. Brackett functioned as a counterbalance, keeping the creative Wilder within productive limits. Their collaboration resulted in some of the more subtle screenplays old Hollywood produced. Brackett and Wilder's characters are complex, deeply psychological, with shades of darkness even in moral types.

At nearby Columbia, conditions were not nearly as relaxed. Harry Cohn himself read all the scripts, judging them from the viewpoint of the average moviegoer. Writers were required to report to the studio no later than nine and expected to stay in the writers' building and write. Cohn even occasionally came around to check on their progress, but contract veterans learned how to escape his prying eyes. "Harry didn't realize what was happening," said former screenwriter Roy Huggins, "but writers would come in and play gin." Basically they wrote when they felt like writing. Most were serious about their careers, even if they weren't serious about keeping Harry Cohn's time schedule.

Jo Swerling, a former newspaper and magazine writer, was among Columbia's finest scriptwriters, along with Robert Riskin, director Frank Capra's most frequent collaborator. Virginia Van Upp, eventually executive producer at Columbia and a confidante to the studio head, shaped several of the important Rita Hayworth films, including *Cover Girl* (1944), *Gilda* (1946), and *Affair in Trinidad* (1952). But most of the pictures Columbia turned out were little more than service pictures, B programmers sold to the theaters on the strength of the Capra films and the Hayworth musicals.

Universal assigned its writers to one of several small bungalows on the lot. Henry Koster wrote the stories for his early Deanna Durbin movies there, with the assistance of experienced Hollywood screenwriters since the German-born Koster couldn't write in English at that time. Later Felix Jackson, another German refugee, wrote the Durbin scripts. When screenwriter Walter Bernstein worked at Universal during the late 1940s, he too was installed in a studio bungalow; he remembered walking around the lot looking at the sets for *Phantom of the Opera* (first filmed in 1925) and other classics made there. Bernstein said that writers then were normally on fifteen-week projects and had the security of knowing they would be working on a particular script until it was finished.

Universal specialized in modest-budget pictures that brought modest returns at the box office. Among the studio's popular features were the Abbott and Costello movies and the *Francis* and *Ma and Pa Kettle* series. Bud Abbott and Lou Costello had been burlesque comedians, and much of their material was crude, based on gags they had performed for years on the stage. The studio would assign a regular contract writer to create their scripts, but later John Grant, Abbott and Costello's gag man, would inject the comedians' own routines, which they'd refer to on the set simply as "Spin the Dragon" or "Slowly I Turn," much to the confusion of the director.

Samuel Goldwyn employed some of the most illustrious writers ever to work in Hollywood. Goldwyn respected writers, but at the same time he resented their intellectual superiority. His story department was headed by a succession of talents, among whom Marguerite Courtney was the only woman, although for a time Goldwyn's story editor on the east coast was Beatrice Kaufman, wife of the distinguished Broadway director-playwright George S. Kaufman. Sam operated on hunches. "He was the kind of guy who didn't know anything about art but knew what he liked," Robert Pirosh said.

Literary giants generally found adjusting to the rigid studio system difficult. F. Scott Fitzgerald arrived in Hollywood late in his life, when he was emotionally unstable and drinking heavily, referring to himself as a Philistine because he was working for the studios. Fitzgerald experienced little success in films; his only real achievement in Hollywood was in gathering material for his uncompleted novel *The Last Tycoon*. Despite frustrations,

William Faulkner fared better, collaborating on several scripts for Warner Bros., notably *To Have and Have Not* (1945) and *The Big Sleep* (1946). Whenever his daughter Jill needed dental work, Faulkner would write his friend Robert Buckner and ask if he had anything he could do for a few months. Buckner would find him an assignment, and Faulkner would board a train for California.

Dorothy Parker, the notorious wit of the Algonquin Round Table in New York, worked steadily in Hollywood, with considerable success. When Parker was initially brought to MGM and asked her preference in offices, she replied, "All I need is room to lay my hat and a few friends." (S. Marx, *Gaudy Spree*) Robert Benchley, Moss Hart, John Howard Lawson, George Kaufman, John O'Hara, Christopher Isherwood, P. G. Wodehouse, James Hilton, Bertolt Brecht, and Aldous Huxley all worked for the big studios, often proclaiming their talent was being wasted. Others, like Somerset Maugham, tasted Hollywood and fled with their integrity intact. Howard Lindsay, who enjoyed many triumphs on Broadway, ranging from *Life With Father* to *The Sound of Music*, hated the studio system and never did anything important for pictures. Those who stayed fretted about compromising their standards and frequently claimed Hollywood's shoddy methods were corrupting their artistry. "Let's face it, darling," actress Anna Lee admonished her husband, novelist Robert Nathan, "you sold out to the flesh pots. You stopped being a good writer when you left New York. You sold out just as I did." (Nathan int.) Playwright S. N. Behrman felt the same, but rationalized with wry humor: "It's slave labor," said Behrman, "and what do you get for it? A lousy fortune!" (S. Marx, *Gaudy Spree*)

Yet with such esteemed screenwriters as Dudley Nichols, responsible for *The Informer* (1935) and *Stagecoach* (1939), and Herman J. Mankiewicz, who with Orson Welles created *Citizen Kane* (1941), the Hollywood film industry reached distinguished heights. Both Nichols and Mankiewicz came from the New York newspaper world, although Mankiewicz had been a critic and a member of the Algonquin Round Table. He arrived in Hollywood in 1926, three years before Nichols, shortly becoming head of Paramount's story department. Brilliant and witty, Mankiewicz established himself as a Hollywood intellectual. He had a serious drinking problem, but once quipped, "Isn't Herman Mankiewicz drunk better than almost anybody else sober?" (Jackson int.) With *Citizen Kane* he turned mass entertainment into cinematic art.

Part of Hollywood's bad reputation among serious writers stemmed from the collaborative process most studios employed. "What I don't like about Hollywood films," Orson Welles said in 1939 before a symposium composed mainly of New York academics and intellectuals, "is the 'gang' movie and I don't mean the Dead End Kids. I mean the assembly line method of manufacturing entertainment developed in the last fifteen years or so . . . . When too many cooks get together they find, usually, the least common denominator of dramatic interest." (Naremore, *Magic World of Orson Welles*) Writers who functioned well within the big studio system accepted motion pictures as a composite art, admitting that a second or third mind could often improve what they had written. Some even found the collaborative process beneficial for other reasons. "I've had good collaborators and bad ones," Robert Pirosh said. "With George Seaton it was stimulating. If I were in an unhappy mood, he'd be cheerful; if he came in a little depressed, I'd be cheerful."

Occasionally a producer would take over the writing himself. Pandro Berman remembered having to go to New York on business during the shooting of *Roberta* (1935) at RKO. Since the script needed revision, Berman took screenwriter Allan Scott with him on the Santa Fe Chief. All the way to New York the producer and his writer discussed the *Roberta* script, with Scott making changes, adding humor to the dialogue. They wired the changes back to the studio so those scenes could be shot the next day.

A notorious example of studio collaboration occurred at Warner Bros. on *Casablanca* (1942). Howard Koch and the Epstein twins were assigned to the script, although nearly every writer on the lot contributed to it before the film was completed. "I know at least six writers who worked on *Casablanca*," Robert Buckner claimed. Buckner himself turned the job down, telling production head Hal Wallis that he didn't believe in either the story or the characters. "This guy Rick is two-parts Hemingway, one-part Scott Fitzgerald, and a dash of cafe Christ," Buckner wrote Wallis. (Behlmer, *Inside Warner Bros.*) Koch and the Epstein brothers wrote well together, until the Epsteins were called to Washington to work on Frank Capra's *Why We Fight* series. Howard Koch continued alone on *Casablanca*, under pressure to meet deadlines, the studio sending mimeographed pages to the various departments before the script was even near its final form. Suddenly revised pages began coming out of the mimeograph office from sources unknown to Koch. He eventually discovered that the Epsteins, having returned from Washington, had rejoined the project. Since the members of the team had no joint conferences, they followed radically different paths. The Epstein twins saw the film in comic terms, whereas Koch attempted to produce a serious melodrama. The final weeks were hectic, especially since the writers were not always in agreement with director Michael Curtiz. When the revised script was finished, not even the writers knew what kind of film would emerge from their composite efforts. Even after the picture opened in January 1943, Howard Koch wondered why the public and critics were so excited. A year later *Casablanca* won the Academy Award for best picture, and the film remains a classic.

Writers were expected to fashion scripts for a particular star, tailoring the character to enhance that actor or actress's image. When MGM was preparing *Marriage Is a Private Affair* (1944), story editor Edwin Knopf told the initial screenwriters, "We have to cheapen this script for Lana Turner." Knopf gave them the choice of either staying on the project or letting another team make the required changes. The original writers, who included Ring Lardner, Jr., decided not to go ahead with the Turner vehicle. (Lardner int.) Often a script would be completed, a star assigned to it, and then further changes would be considered necessary. Robert Lord and Robert Nathan spent almost a year on a screenplay based on John Galsworthy's *The Forsyte Saga* and were satisfied with their work. Metro executive producer Sam Katz, however, didn't like their sophisticated script. The studio decided to make the picture if Greer Garson would play the part of Irene, but Garson said she would only if the story were rewritten so that she remained the pivotal character throughout. Six pairs of writers worked on the script before MGM finally approved *That Forsyte Woman* (1949). (Nathan int.)

Seasoned screenwriters found it advantageous to know the actor they were writing for, his or her limitations and capabilities, so they could write knowing their script would fit the personality and talent of the person involved. Powerful stars might demand approval, and if the results didn't suit them, they occasionally wrote scenes themselves, as Errol Flynn and Miriam Hopkins did on *Virginia City* (1940).

Sometimes studios stole ideas from each other; *The Black Shield of Falworth* (1954), for example, was Universal's version of Metro's *Knights of the Round Table* (also 1954). In certain instances writers were asked to produce sequences according to the shooting schedule rather than the script's continuity. If the weather suddenly turned bad and exterior scenes couldn't be shot on a given day, the writers were sometimes expected to prepare interior scenes overnight, so the cameras could roll the next day.

The studio's sales department determined film titles, with the result that writers were often dismayed when their work reached theaters with names that had little or no relationship to their script. The picture made from James Dunn's book *Deadlier Than the Male* was called *Born to Kill* (1947), since RKO found the original title too "high-toned." *The Shamrock Touch* at 20th Century-Fox became *The Luck of the Irish* (1948); Universal's *Live Today for Tomorrow*, a title that made little sense, was used because production head

William Goetz had had an earlier success with a picture using "Tomorrow" in its title.

In the big studio days, according to Edmund North, if the company planned to make four pictures with a given star, they bought A, B, C, and D properties; writers worked on them, and the pictures were made. "That makes it sound like a sausage factory," North said, "but as an enjoyable way of life, the studio system was truly great, although we complained bitterly. There was a sense of participation in a process that you don't get doing independent productions."

With rare exceptions writers were not welcome on the set, unless they were called upon to solve a specific problem. Melville Shavelson remembered dropping by the filming of a Doris Day musical he had written at Warner Bros. and finding director Roy Del Ruth with the script on the floor. "He would turn the pages with his foot," Shavelson said, "because he didn't want to get any closer to the words than that." When the writer suggested that Doris Day and Gordon MacRae couldn't logically sing a duet together, since at that point in the story they were supposed to be 250 miles apart, Del Ruth ordered him off the set.

The director often shortened dialogue, and sometimes actors made up lines as they went along. Dale Robertson claimed RKO started *Son of Sinbad* (1955) with four pages of script. "Nobody knew what the story was going to be about," Robertson said. "At the end of the day you'd get a couple of pages for the next day. We'd shoot and laugh and have water fights; there was no need in being serious. I had more fun doing that picture than any I ever made." Even a classic like *Gunga Din* (1939), its script by Charles MacArthur and Ben Hecht, required on-the-set rewrites and spontaneously added gags. "We'd sit around all morning and rewrite scenes," Douglas Fairbanks, Jr., claimed. "The script went out the window."

Usually a solid script by an established writer makes the actors' work a pleasure. Actress Virginia Mayo found making *The Best Years of Our Lives* (1946) for Samuel Goldwyn the easiest role she ever did. "Doing that picture was like doing a play," said Mayo, "because any word that Robert Sherwood wrote didn't need any changing. The character of Marie was so well delineated that I didn't have to use my imagination much. I could see that girl completely. Some writing is sketchy and actors have to fill in what they think, but everything in *The Best Years of Our Lives* was there. We didn't want to change anything."

On lesser movies, particularly B pictures, many of the scripts were ludicrous. It was up to the actors to make their lines as believable as possible. "We used to call them rainbow productions," 20th Century-Fox actress

Lynn Bari said. "They'd have different colors for each rewrite. At the end of the picture my script would include all the hues from lavender to red. Sometimes we'd shoot twenty pages or so of dialogue a day." Writers disliked B assignments as much as actors did, although for younger ones, seeing their name on the screen was encouragement. Smart beginners realized that although they were not writing films of consequence, they were receiving good training. "You knew you weren't working on something that was going into the archives," Edmund North said, "but at the same time it was a challenge to make a B picture as good as it could be made. We didn't slough anything deliberately."

Since a studio's writers, stars, and directors all had to be kept working, a salvage unit would sometimes be created: a producer would be assigned certain unemployed personnel, and a team of writers would be expected to devise a script for them. By cutting corners the studio could piece together an acceptable low-budget picture that would help fill their quota in the theaters and balance the time sheets of contract personnel. The ten good pictures turned out by major studios each year would sell the other forty because of the block booking arrangement. "We tried to keep B pictures from being looked upon as slumming," Eugene Zukor said. "They were an opportunity for writers and technicians to show their resourcefulness, to make something that looked like an important picture."

If the writers couldn't think up anything original, they would delve into the studio's storehouse of previously filmed scripts. Warner Bros.' *The Unfaithful* (1947) with Ann Sheridan was a modernization of Somerset Maugham's *The Letter*, which the studio had made in 1940 with Bette Davis. MGM remade the Clark Gable–Norma Shearer picture *A Free Soul* (1931) as *The Girl Who Had Everything* (1953) with Elizabeth Taylor, and Columbia's *The Boogie Man Will Get You* (1944) was a takeoff on *Arsenic and Old Lace* (also 1944). Bryan Foy, head of the B picture unit at Warner Bros., kept a stack of scripts on his desk. Whenever a writer was stuck for an idea, Foy would reach under the pile and pull one off the bottom. "This is about horse racing," he told Ring Lardner, Jr. "Maybe we could make it about automobile racing."

Westerns were recycled most often, sometimes every two or three years. Paramount in the early years had an agreement with novelist Zane Grey that the studio could rework his stories as many times as they liked, so long as they paid the author $12,500 every time they did. The first picture director Henry Hathaway made was *Heritage of the Desert*, which Victor Fleming had filmed before. Hathaway recalled that he took the old script and a writer to a cabin and came back with a new script in two weeks. Charles Starrett, who played the Durango Kid at Columbia, alternated decent scripts with bad ones in the series, until the actor went through his paces like an automaton. Finally Starrett wondered, "If we don't change the story, why can't we at least change the horses?"

Censorship was another problem that plagued writers during Hollywood's Golden Age; avoiding the taboos of the production code became a challenge for them. Will H. Hays, the censorship czar and a former lawyer, had been postmaster general under Warren G. Harding, but in 1922 he was appointed head of a self-regulatory organization created by the studio moguls in an effort to improve Hollywood's image after a string of embarrassing scandals. In 1930 the Hays office issued the Motion Picture Production Code, which went into effect in 1934 with Joseph Breen as director. Terrified of politically appointed local censorship groups, Hays issued twelve commandments for filmmaking, which ran the gamut from how long a kiss could be held to a ban on the explicit presentation of crime. As comedian Frank Fay commented, "Just like Hollywood, two more than even God felt the world needed." (O'Brien, *Wind at My Back*)

Every script had to be submitted to the Breen office for approval before it could be filmed. Normally Breen returned the copy with extensive comments and a list of items his office found objectionable. A conference would be held with representatives from the Breen office and the picture's writers, director, and producer, during which both sides usually had to compromise. Some filmmakers later argued that the production code forced writers to be more creative, insisting that films made during the censorship era were often erotic, yet in subtle ways. Others claimed the code made serious screen drama impossible.

Metro's *Tea and Sympathy* (1956), for instance, had to include a prologue and an epilogue to show that the wife who had sex with a boy accused of homosexuality was sorry and had paid for her sins. "We solved the problem by adding a prologue," producer Pandro Berman recalled, "which was damn foolishness. I hated it." With *The Macomber Affair* (1947) the filmmakers involved thought they had one of the finest Hemingway pictures ever made— until the last eight minutes, at which point they were never able to solve the censorship problem fully. Since the Hays office wouldn't allow them to use the author's ending, in which the woman who shot her husband went unpunished, director Zoltan Korda, screenwriter Casey Robinson, and actor Gregory Peck tried to elicit from Ernest Hemingway some suggestion that would satisfy the censorship office without ruining his story. They sent letters, telegrams, even tried to reach the author in Cuba on the telephone, but received no response. The least compromising ending the filmmakers could come up with was to have Mrs. Macomber sent to prison for involuntary manslaughter, hinting that the white hunter might be waiting for her in Africa when she got out of prison. "Unfortunately it was such a wrench in style that what we did at the end seemed tacked on," Peck said.

The Children's Hour, with its lesbian theme, could not be filmed in 1936 as Lillian Hellman had written it for the stage. Hellman explained to director William Wyler that the story was not about lesbianism; it was about the power of a lie. The film was made as *These Three* and was far more successful than Wyler's 1962 remake, after censorship had ended, which dealt openly with lesbianism and used Hellman's original title.

More frequently the changes Breen demanded produced less satisfying results. Instead of only the virgins on campus wearing college beanies, RKO's version of *Too Many Girls* (1940) altered the script so that all the girls who had never been kissed wore beanies. In *Arsenic and Old Lace,* when the lunatic aunts reveal to the Cary Grant character that he's illegitimate and therefore not a candidate for hereditary insanity, Grant couldn't say to the girl he wants to marry, "Darling, I'm a bastard!" Instead he turned to the girl and said, "Darling, I'm a son of a sea cook!" And in Zion, Illinois, the picture based on the Broadway musical *Damn Yankees* (1958) was even billed on a local theater marquee as *Darn Yankees* in order not to offend local townspeople.

Writers appeased themselves by complaining about the system, finding solace in their own club. Since the number of screenwriters was relatively small, they knew almost everybody working at the other studios. Musso-Frank's Restaurant on Hollywood Boulevard became the writers' informal social club during the cocktail and dinner hours. "Without prearrangement," Howard Koch wrote, "many of us gathered at tables of four or six in the rear of the wood-paneled taproom. Dry martinis, doubled without asking, washed away whatever problems we brought with us. Conversation flowed easily—the films we were working on, the front office foibles—and, of course, sex reared its beguiling head." (Koch, *As Time Goes By*)

The Screen Writers Guild was organized in 1933, later renamed the Writers Guild of America. "Going to a guild meeting was like going to the Elks," Julius Epstein commented. "Most of the work in Hollywood was done by about two hundred people." The Writers Guild arbitrated disputes over screen credit and was officially nonpolitical, although the orientation was always liberal. The guild succeeded in gaining better working conditions for members, but it did not win them control over their own material. Many screenwriters complained that the only way they could protect their scripts was by becoming the producer or director. Nunnally Johnson hated what John Ford did with his script of *Tobacco Road* (1941) and vowed to become a director himself, which he eventually did. The Epstein brothers were elevated to producers at Warner Bros. after their success with *Casablanca*. Metro gave Robert Pirosh an opportunity to direct his next picture after he won an Academy Award for writing *Battleground* (1949).

Screenwriting was hard, lonely work, yet it could also be satisfying, even with the restrictions imposed by the big studios' regimentation. Most successful scriptwriters were craftsmen rather than artists, pragmatic in attitude and approach. Although executives might belittle writers' contribution, most actors and directors recognized the importance of talented writers who could work productively within a commercial atmosphere. "I feel nostalgic for the camaraderie, the fun at the writers' table in the commissary, and things like that," Julius Epstein said, "but for the method of operation I don't feel any nostalgia at all."

10

Musicals and the Music Department

Scene from Warner Bros. musical Wonder Bar *(1934),*
choreographed by Busby Berkeley. Courtesy of the Mary McCord/
Edyth Renshaw Collection on the Performing Arts, Hamon Arts
Library, Southern Methodist University.

Each of the major studios maintained a music department that included a permanent stable of staff composers, arrangers, orchestrators, orchestra and chorus, conductors, choreographers, and vocal coaches. Making a musical required approximately twice the work of a dramatic picture or straight comedy, necessitating weeks of rehearsal and many meetings to plan song and dance routines, decide on sets, coordinate costumes, and determine budgets for each sequence. Composers and arrangers worked diligently to meet deadlines. At MGM musicians were housed on the lot in bungalows. "Those little bungalows were always lit up at night," said film composer David Raksin. "Somebody was always in them at four A.M." Like scriptwriters, studio musicians established their own rapport and found satisfaction in the Hollywood of the Golden Era. "It was like a happy family," songwriter Harry Warren recalled.

Screen musicals arrived with sound; Al Jolson sang more than he spoke in *The Jazz Singer* (1927), Warner Bros.' revolutionary talkie. But the era of creative film musicals really began at Warners in 1933 with *42nd Street*, a cinematic milestone created from the genius of former Broadway dance director Busby Berkeley. Eccentric and egotistical, Berkeley was not so much interested in dance steps as he was in camera movement. "The camera was the audience," Berkeley said, "and that's why I made a great study of it and am noted for my camera work." (Berkeley int., Steen) If Berkeley's show girls knew basic steps and could do elementary ballroom dancing, that was enough for him. Above all, they had to be beautiful and sexy. Berkeley picked his girls himself, fell in love with many of them, and used the same ones in numerous films. "If you hired Busby Berkeley, you got his girls with him," said screenwriter Robert Buckner. "They had done the same routines so many times that they didn't have to rehearse. He'd just say to them, 'All right, girls, 13-A,' and they'd go right into it. Buzz knew exactly how to put a musical together. He was a real product of Hollywood and spoke a kind of argot that was peculiarly his. There was no nonsense about him, no waste of time. He was a businessman, and that's why he was so successful."

Berkeley meticulously planned his numbers, originating the top shot, in which the camera looked down on a number. Sometimes this technique meant ripping a hole in a sound stage roof to get the camera far enough back to take in an entire scene. "I did my cutting with the camera," Berkeley said. "That is, the cutter had to put it together the way I shot it. I set every shot in every film I ever made. I got behind the camera and showed the operator exactly what I wanted." (Berkeley int., Steen) His numbers often lasted for ten or fifteen choruses. "He never knew when to stop," declared Harry Warren, who worked with Berkeley repeatedly. "That was his one fault. But he was clever with the camera. He was the daddy of that."

Since Berkeley was a perfectionist, his pictures were expensive, frequently involving a hundred girls on stage at once. His films launched a new vogue in screen musicals. After *42nd Street* came *Gold Diggers of 1933*, *Footlight Parade* (also 1933), and *Dames* (1934), all featuring songs by Harry Warren and Al Dubin. "Half the time the scripts were horrible," Warren admitted. "There was nothing there to write about; we had to concoct songs that could be produced. They weren't book songs; nothing came out of the dialogue or story. They were all big production numbers that could be shot by Busby Berkeley."

Harry Warren had started in movies with Vitagraph in New York during the silent days, playing the piano for ballroom scenes to give dancers a rhythm. Although he never took a music lesson, he began making up songs, long before he had any notion of becoming a songwriter. By the time Warner Bros. bought Remick Music Corporation, Harms Music, and M. Witmark and Sons (three music publishing firms), Warren was working for Remick as a staff composer. He was shortly sent to California to write numbers for an upcoming movie, which turned out to be *42nd Street*. He delivered the score within twelve weeks, earning a salary plus royalties on his songs, and remained at Warners for six years. He was constantly busy, accepting any assignments and rarely taking a vacation even though his contract stipulated a twelve-week layoff.

Normally Warren preferred to work in his office at the studio, which he found more convenient and quieter than working at home. Jack Warner didn't insist on his songwriters' working at the studio, although like screenwriters they did not command great respect. Richard Whiting and Johnny Mercer were also on the Warner Bros. lot at the time, so that newcomers not under contract had little chance of having their songs heard. "They weren't about to let new people in," composer Saul Chaplin said.

Melody was Harry Warren's strength, and his reputation was established in 1935 when "Lullaby of Broadway," a number he wrote for Berkeley's *Gold Diggers of Broadway*, won the second Academy Award for songwriting.

"I really loved to write," Warren recalled shortly before his death. "I used to get a big boot out of hearing that sixty-piece orchestra play my music. When you hear a big orchestra playing your songs for the first time, it's really something."

In making the early screen musicals the orchestra was on the sound stage with the performers; singing and the accompaniment were recorded simultaneously, as routines were filmed. Later, music was prerecorded, which insured better sound, with performers mouthing lyrics for the camera or dancing to a playback. Berkeley was shooting *Gold Diggers of 1933* when Los Angeles suffered one of its worst earthquakes. "At the time it hit I had fifty beautiful girls up thirty-five feet high on platforms," Berkeley recalled. "I shrieked and hollered to them not to jump. They would have killed themselves. I yelled, 'Sit down! Sit down!' No one was hurt, but the walls of the stage were shaking." (Berkeley int., Steen)

Chorus members on early Warner Bros. musicals earned $10 a day for rehearsing, $15 a day for recording, and $15 a day as an extra if they appeared on camera. Choristers worked an eight-hour day, unless the director decided to go into overtime. Choral directors, like Dudley Chambers at Warners, were experienced professionals, and it was great training for youngsters to work under them.

After Warner Bros.' success with *42nd Street*, RKO quickly entered the field, creating a sensation by pairing dancers Fred Astaire and Ginger Rogers in *Flying Down to Rio* (1933). The film's popularity not only rescued the financially troubled studio, but established Rogers and Astaire as perhaps the most remarkable dance team in Hollywood history. A shy, private man, Fred Astaire was a perfectionist who rehearsed for weeks, sometimes months. "You have to rehearse like hell for long periods to get your material working," the dancer said. "If you're working with a girl, you have to rehearse with her. I was never really aware of what I did. I just sort of did it. I'd look at it in the mirror when I was rehearsing, and if it didn't look good to me, I changed it. I never really analyzed it."

According to Astaire's longtime choreographer, Hermes Pan, the dancer's style was unique. "I think his magic was his feeling," Pan said. "You were almost hypnotized by his movement, even though it was something so simple that he didn't really seem to be doing anything." Astaire had a sense of humor and liked working with props—dancing with a hat rack, revolving around barrels in a fun house, playing golf, or igniting firecrackers. He devised many of his own routines, but his ideas on dance meshed with those of his choreographer. "The first time I saw Fred dance," said Pan, who bore a close physical resemblance to Astaire, "he did all the things that I felt. It was uncanny that his rhythms and mine were almost exactly alike. We even had a language of our own. I knew exactly what he meant, and he would know exactly what I meant. It was amazing how much we thought alike on our dancing."

Together Astaire and Pan introduced a new dimension to screen dance. "Before Ginger would come into the rehearsal hall," Hermes Pan recalled, "Fred and I would work together, and I would dance Ginger's part on various things. We'd work out certain moves and ideas, and then Ginger would come in, and I would show her. It got to be more or less like a collaboration. Ginger was not what you would call a great dancer, but she was very complementary to Fred. Of all the partners that Fred had, even though the others might have been much better, I don't think any of them approached the magic that Fred Astaire and Ginger Rogers projected together. Their chemistry was just right."

Rogers had the ability to "sell" a number with her personality, while Astaire exhibited exceptional rhythm. With original scores by George Gershwin, Cole Porter, Irving Berlin, Jerome Kern, and Vincent Youmans, the Astaire-Rogers pictures brought together a rare combination of talent. "We worked with two formulas," producer Pandro Berman stated. "We didn't want all those pictures to be alike." Berman first produced *The Gay Divorcee* with Astaire and Rogers alone. Then he produced *Roberta* and decided on having Irene Dunne and Randolph Scott in the picture, too, so that it was a four-person picture, with comedians backing up the stars. Next Berman decided to do *Top Hat* with Rogers and Astaire alone. Then they went back to the second formula in *Follow the Fleet*, in which the producer used Harriet Hilliard and Randolph Scott to vary the story. The strategy worked for nine pictures, although *Top Hat* (1935) probably stands out as their best.

Astaire and Pan wanted musical numbers to grow out of the dialogue. "Fred and I recognized that we had to get into a number so that you wouldn't be stunned by suddenly going from dialogue to music," Pan said, "going from reality to fantasy. So we used to plan ways of getting in and out of songs." Since Astaire came from the Broadway stage, he was wary at first about how his dances were going to look on the screen. Gradually he learned that motion pictures could enhance dance, showing the performer from different angles, moving the camera without making the audience conscious of it. "You get more action if your camera stays still and pans across something," Astaire said. "High shots are effective, but we didn't want to do what Busby Berkeley did. His was mostly ensemble work; ours came from an entirely different concept. We were careful to see that our musical numbers fit into the plot, which wasn't always easy."

Astaire and Rogers were not close friends off camera, and he often would have preferred not to use her. On *The Gay Divorcee* (1934) Astaire felt Rogers was incapable of playing an English lady. "You're absolutely right," producer Pandro Berman told him. "But the American public won't know the difference." Berman later explained, "The great unwashed crowd that went to the movies at that time didn't give a damn whether it was an English girl or not. I had a terrible struggle with Fred about Ginger on every picture. Finally I had to promise him that if he'd use her in a certain picture, I'd give him another girl in the next one." Having been linked with his sister Adele on the stage, Astaire was reluctant to become identified again with any one partner. For *A Damsel in Distress* (1937) the studio teamed him with Joan Fontaine, who was no dancer, but who was under contract to RKO at the time. The picture was far less successful than Astaire's predecessors had been, and he was immediately paired again with Ginger Rogers in two more RKO films.

Universal found its key to financial solvency in Deanna Durbin, a teenage singer with classical training, whom producer Joseph Pasternak featured in a series of popular musicals, beginning with *Three Smart Girls* (1936). Durbin's vehicles had the largest budgets of any pictures made on the lot, but also showed the highest grosses. Charles Previn, the uncle of André Previn, headed the music department at Universal during Durbin's early reign. Musical selections for her pictures ranged from standard operatic arias to an original score by Jerome Kern.

The rest of Universal's musicals, however, were much less sophisticated and involved far less budget. Following Metro's success with Judy Garland and Mickey Rooney in the early 1940s, Universal teamed Donald O'Connor and Peggy Ryan, two talented youngsters from vaudeville, in a succession of light films made quickly and inexpensively, with Louis DaPron responsible for most of the choreography. "They used to make those Universal musicals in thirty minutes," dancer Dan Dailey remarked. The studio featured Dailey and O'Connor in a picture with the Andrews Sisters called *Give Out Sisters* (1942), which the cast laughingly referred to as "Give Up Sisters." "At Universal everybody was treated like second-class citizens," Maxene Andrews said. "Our pictures were made in ten days, and they decided they weren't going to make any special wardrobe for us. We just went to the wardrobe department and picked out three of something they could fix up for my sisters and me. We had to do most of our own choreography, which we made up as we went along. You never had any feeling of class; everything at Universal was done cheaply." While Deanna Durbin had her own bungalow across the way, the Andrews Sisters were assigned a dressing area in a bungalow attached to another performer's.

At Paramount the Bing Crosby musicals and the Mae West pictures were the big money-makers during the 1930s. Originally for radio, Crosby's singing style established a new vogue, and although he couldn't dance well and didn't read music, the easygoing crooner became one of the most popular personalities in films. "Crosby was wonderful," said Leo Robin, who wrote lyrics for several Bing Crosby pictures, including "Blue Hawaii" for the movie *Waikiki Wedding* (1937). "Any song you'd give him, he'd say, 'All right, I'll sing it.'"

Although during the Depression Paramount didn't have the money to lavish on its pictures that Metro did, Paramount still emphasized musicals, especially under Buddy DeSylva, since the studio head had been a songwriter himself. In its heyday Paramount had twenty-two composers under contract (including Irving Berlin for a while in the 1930s) to supply material for its musicals. Leo Robin and Ralph Rainger emerged as a successful team, writing "Please," "Love in Bloom," and "June in January" and winning an Academy Award for "Thanks for the Memory," Bob Hope's theme song. Later James Van Heusen, Johnny Burke, Jay Livingston, and Ray Evans were under contract to Paramount; the studio eventually added its own music publishing company and record label.

Metro-Goldwyn-Mayer produced the most opulent and innovative screen musicals. By the 1940s three major producers at Metro specialized in musicals—Arthur Freed, Joseph Pasternak, and Jack Cummings. Arthur Freed, a gifted songwriter himself, wrote "Singin' in the Rain," among countless other songs, and produced the most sophisticated Hollywood musicals. He used the finest talent on the MGM lot and brought in the giants from Broadway and Tin Pan Alley. Freed had grown up in vaudeville, understood music, and knew the leading figures in the popular music field. He made actors of Judy Garland, Gene Kelly, and Cyd Charisse; he brought June Allyson, Van Johnson, Nancy Walker, and choreographer Robert Alton from Broadway; he gave art director Vincente Minnelli and choreographer Charles Walters a chance to direct; and he filled his staff with brilliance—Lennie Hayton, Kay Thompson, Ralph Blane, Hugh Martin, John Green, and Roger Edens. Freed's method was to select the right person for a job and let that person work to the best of his or her ability. "We had all the time in the world when we were writing," said songwriter Ralph Blane, who with Hugh Martin composed the score for the Freed hit *Meet Me in St. Louis* (1944). "We had a free hand to come and go as we pleased—write at the studio, write at home, it didn't matter. Freed just wanted it to be great." Mayer gave the producer complete authority, so that the Freed unit became virtually autonomous. Freed justified Mayer's

confidence in him by turning out a string of MGM's top-grossing films—
Easter Parade (1948), *The Barkleys of Broadway* (1949), *An American in
Paris* (1951).

Joe Pasternak's pictures had an old world charm and sweetness. He loved
fables and fairy tales and believed in what studio workers called "Pasternak
Land." "There were rules in Pasternak Land," explained director George Sidney: "no terrible people, and everything came out okay in the end. The
mother and father were respected, the country was respected, the flag was respected—things that today are considered corny." Pasternak preferred soprano ingenues capable of singing the familiar classics—Kathryn Grayson, Jane
Powell, Ann Blyth—yet he liked José Iturbi playing the piano and helden-
tenor Lauritz Melchior's voice. Pasternak produced the Mario Lanza pictures, including *The Great Caruso* (1951). "Of course the picture is simplistic," musical director John Green said, "but it's a lovely picture. The public
was given a chance to pat themselves on the back and say, 'Oh, I know that
aria!'"

Jack Cummings's approach was somewhere between those of Freed and
Pasternak. Although Cummings loved musicals and was musically sensitive,
for technical knowledge he relied on his associate producer, Saul Chaplin, a
versatile musician. Cummings's most important musical was *Seven Brides for
Seven Brothers* (1954), which contained energetic choreography by Michael
Kidd. Dancers rehearsed for five weeks before filming began, since there
were seven major dance numbers in the show. Director Stanley Donen recalled that his problem was to convince the studio to cast the picture with
dancers, even bringing in Jacques D'Amboise from ballet. "They kept saying
to me in words of one syllable, 'You can't have a bunch of guys jumping
around; the audience will think you've got some pansy backwoodsmen!'"
But the picture proved a lasting success, appealing to generation after generation. Cummings was also responsible for several of the Esther Williams hits,
as well as such reworked Broadway fare as *Lovely to Look At* (1952) and *Kiss
Me, Kate* (1953).

Remarkable talent was brought together in Metro's music department.
Herbert Stothart, who with Rudolf Friml had written the operetta *Rose
Marie*, initially was the studio's chief conductor, a position taken later by
John Green, who headed the department from 1949 to 1958, bringing it to
its height. Roger Edens, the department's musical factotum, was a man of
taste, artistic integrity, and seemingly endless creativity. "Roger would sit
down at the piano and sound like he had four pairs of hands," songwriter
Ralph Blane said. In addition there were arrangers Connie Salinger and
teenage André Previn, orchestrator Adolph Deutsch, Austrian-born dance di-

rector Albertina Rasch, vocal coach Arthur Rosenstein, the department's troubleshooter Lela Simone, and a lengthy roster of other talent who worked well together. Actress Marsha Hunt viewed the operation as an outsider. "The moment I knew I had fifteen minutes between camera setups," Hunt recalled, "I would repair to the music bungalows, where all those brilliant composers and arrangers were noodling songs, and I'd hear snatches of music they were working on."

When John Green was offered the job of heading the studio's music department, Louis B. Mayer told him he wanted the greatest music division that had ever existed in the entertainment field. Green took Mayer seriously and fired orchestra members near retirement he thought had grown sloppy in their work. "There were fifty people in the contract orchestra," Green said, "and my first explosive official act was to fire twenty-eight of them the first week. Then I became the pirate of all time; I robbed people from every great orchestra in the country. At that point I was able to pay unheard-of salaries." Besides the contract orchestra of fifty players, Green had an additional forty-five on first-call, so that the MGM Symphony Orchestra often consisted of ninety-five members. The studio's choral director was Bobby Tucker, who built a chorus of almost three hundred voices at full strength. "That was a gorgeous chorus," recalled opera star Nan Merriman, who sang at Metro in her youth.

John Green was a tough but able administrator, sending memos to everyone. He conducted the studio orchestra and continued his own composing. He called department meetings once a week, during which members shared ideas. If another assignment came up and a member of the department was pulled off a project, there was always a volunteer replacement to take over and finish the number. The environment not only was mutually helpful, it was stimulating. Musicians were free to experiment, since they were getting paid a weekly salary no matter what they wrote.

Until he became too busy, Roger Edens rehearsed stars on their songs during the leisurely preproduction period, playing the piano for Judy Garland, Lena Horne, Betty Garrett, or June Allyson. Later, Ralph Blane and Hugh Martin took over some of these duties. If a voice was to be dubbed in a picture, dozens of singers would be auditioned to match the *speaking* voice of the star involved, since audiences would never hear that star *sing* in the movie. Once a song was prerecorded, a half dozen people watched while the number was being filmed to make sure lips were synchronized with the recording. Big production numbers were painstakingly rehearsed. Nancy Walker remembered working six weeks on "Milkman, Keep Those Bottles Quiet" for *Broadway Rhythm* (1944), even though the number lasted only three and a half minutes on the screen and performers had learned it within two weeks. "That's boring and energy wasting," Walker complained. "We used to show up and play cards."

Metro's choreographers were the finest the New York stage had to offer. Robert Alton, who had enjoyed several successes on Broadway before entering movies, was adept at precision work, handling crowds of dancers with a minimum of commotion, standing on a ladder and pointing out positions with his foot. A handsome, flamboyant man, Alton had a delightful sense of humor and was pleasant to work with. Dancers appreciated his enthusiasm and respect for their talent, although he worked them hard. Charles Walters, also from the Broadway theater, made the transition from stage choreography to film even more successfully. He proved immensely gifted at intimate numbers, sliding his dancers smoothly in and out of dance routines without interrupting the plot. In special cases ballet choreographers (among them Eugene Loring and David Lichine) were brought in by Metro producers.

If an MGM actor or actress needed to dance in a picture, instructors were on the lot to teach them the polka, the mazurka, whatever was required. Metro kept sixteen boys and sixteen girls under contract for dance ensembles, and costumes for each picture would be made expressly for them. During rehearsals an arranger was usually present to add a few bars of music here, change the rhythm there, whatever the choreographer thought necessary. Since MGM had money to spend during those years, everything could be done on a grand scale, with no detail too small for careful attention. Time rarely seemed a factor; numbers that had taken weeks to prepare might be eliminated after they had been shot if the finished picture proved too long.

Dancer Gene Kelly was responsible for many of Metro's most innovative musicals, perfecting screen choreography. A compulsive worker, Kelly came to Hollywood following his triumph in Rodgers and Hart's *Pal Joey* on Broadway. He spent months studying the Hollywood system, visiting the major studios. Fred Astaire was working on his "Firecracker" number for *Holiday Inn* (1942) when Kelly arrived at Paramount. Astaire was in the middle of kicking firecrackers around and noticed Kelly standing on the side of the rehearsal hall. Later they would work together.

Athletic and intellectual, Kelly set about finding ways to make movement interesting on the screen. He realized quickly that what he could do for five or six minutes on the stage and hold an audience's attention didn't work on the screen. It had to be cut by almost two-thirds, since dance on the screen looked flat. "I didn't jump out of the screen the way I did on the stage," Kelly said. "It took me a while to realize that dance had to be photographed in a special way." On a two-dimensional screen, audiences sensed no danger in a dancer's leap or pirouette. The tension of whether the performer would achieve a movement or fall was almost completely lost. Kelly learned that screen dances worked best if they were photographed outdoors, where there was a semblance of a third dimension—as illustrated by his "Singin' in the Rain" number. In film, audiences had to be constantly surprised, and Kelly's athletic approach added a dynamic quality. Movement across the screen against a background image proved most effective; movement away from or toward the camera lost the feeling of third dimension. Panning with a camera produced a sense of greater speed, giving movement strength. "I was wallowing in all kinds of aesthetic luxuries at MGM," Kelly said later. "I had the chance to experiment. In those days if you did have a flop, nobody really cared. They could put it in the theaters as a second picture, a program picture."

Much of Kelly's innovation started with *Cover Girl* (1944), a Rita Hayworth picture he made on loan to Columbia. Since Harry Cohn had begun shooting the movie without a leading man, the Columbia boss was eager to borrow Kelly from MGM, although Kelly agreed to make the picture only after Cohn promised he could devise his own choreography. Kelly danced down long streets and around corners in *Cover Girl*, enhancing the feeling of space and movement. "I remember I had a big fight with Cohn about opening the walls of the two big studio sets and bringing the street down through both sound stages," the dancer recalled. "That had never been done at Columbia. Cohn said, 'That'll cost too much money. Just shoot it here.' But I told him, 'No, I have to have that space.'" Cohn finally agreed, recognizing that Kelly was trying to move beyond dances that looked like stage dance photographed. "I thought the ideal would be an endless expanse of plain," Kelly said, "so the dancer could just keep coming into the camera with all his kinetic forces going full speed, with plenty of room to move. The camera with its one eye has no peripheral vision. It takes a dancer and isolates him against a particular sector of the background. To get him over to another section, you have to pan him there. You have to do tricks to get him there. Neither do you have the personality of the dancer, no living presence."

The most remarkable number in *Cover Girl* is Kelly's "Alter Ego" number, in which he talks to and dances with his own image. "We panned and dollied a camera in double exposure," the dancer explained, "which had never been done before. That was a real technical feat. I'm still very proud of doing that as a young man. I showed cameraman Rudy Maté the ways that we could move the camera to pan and dolly and do double exposure, letting the sound track be a rhythmic guide." Director Charles Vidor claimed the idea would never work, but Kelly and his assistant, Stanley Donen, planned the details with the help of their experienced cinematographer. "In fact it did work," Donen said later. "We did it crudely, because there were no computers in those days, and the camera movements had to be exactly duplicated. It was very cumbersome. We did it all with bits of chalk and tape and counts on the sound track and me counting, 'One, two, three, four—pan; two, three, four—tilt,' and so on."

Back at Metro, by then Kelly's home studio, producers realized what the talented dancer could do and gave him the freedom he needed. In *Anchors Aweigh* (1945) for Joe Pasternak he danced with Jerry, MGM's animated cartoon mouse, overcoming major technical problems. The sequence took a year to prepare, involving the cartoon department and the optical department. Originally Kelly was supposed to dance with Mickey Mouse, but plans had to be changed when Walt Disney refused to let Mickey appear in an MGM picture.

On the Town (1949), which Kelly co-directed with Donen, broke further ground. Freer in form and choreographed with a burst of energy, its dances were performed on actual locations, giving the picture a realistic look. *An American in Paris* (1951) closed with a seventeen-minute ballet to Gershwin's music, which cost the studio a half million dollars. Irene Sharaff designed approximately five hundred costumes for that sequence alone. Composer Irving Berlin heard what Kelly, Arthur Freed, and director Vincente Minnelli were planning and warned, "I hope you boys know what you're doing." (Freed int., Kobal) Mass audiences were unaccustomed to such an artistic approach, and even in *On the Town* most of the original Leonard Bernstein score had been replaced by one intended to be less sophisticated for general listeners, written by Roger Edens.

But MGM did attempt to integrate songs into their scripts, much as the Broadway theater had done in *Oklahoma!* and *Carousel*. Martin and Blane's score for *Meet Me in St. Louis* had a close relationship to the story, and their songs helped develop characterization as well. Large sections of *Summer Holiday* (1948), based on Eugene O'Neill's *Ah, Wilderness*, were set to music. "The words seemed to sing themselves," composer Harry Warren declared. "I think that's one of the best musicals I ever did. All those songs were written right for the book." In a scene where Mickey Rooney goes into a barroom and ends up getting drunk with Marilyn Maxwell, the sequence, dialogue and all, was music, and an introductory scene found Walter Huston, Agnes Moorehead, Mickey Rooney, Gloria

DeHaven, and the entire family of characters singing in harmony. "*Summer Holiday* is a classic," lyricist Ralph Blane agreed, "really delicious material. One of the songs, 'Spring Isn't Everything,' is right out of O'Neill's text."

When Gene Kelly broke his ankle shortly before filming began on *Easter Parade* (1948), Fred Astaire was coaxed out of retirement to take his place, winning acclaim for his work opposite Judy Garland. Astaire remained at Metro for a reunion with Ginger Rogers in *The Barkleys of Broadway* (1949) and for *Royal Wedding* (1951), *The Band Wagon* (1953), and *Silk Stockings* (1957). In *Royal Wedding* he did a trick number inside a contraption that made a room spin, so that it looked as if he were dancing on the walls and ceiling. "The whole room had to go around," the dancer explained, "and I'd have to meet it just as it reached a certain point. A man with an electric switch controlled the movement of it. For a while I had to carry a little diagram to show people how I did it. But we shot that all in one piece." Perhaps Astaire's finest number, one of his longest, was the "Girl Hunt" ballet from *The Band Wagon*, which he performed exquisitely with Cyd Charisse. "I absolutely loved that," he said later. "Cyd Charisse was a dream dancer."

In addition to the songwriters under contract, Metro acquired the services of top Broadway tunesmiths and lyricists to create original material for their films—Cole Porter, Irving Berlin, Harold Arlen, Ira Gershwin, Alan Jay Lerner, Frederick Loewe. But no matter who wrote the songs or who the stars were, an MGM musical bore the stamp of its producer. An Arthur Freed picture, especially, exhibited a unity of style, tone, and structure pervading every aspect of the film from sets and costumes to choreography and story texture. "Those were exciting times," reflected George Sidney, who directed musicals for Freed, Pasternak, *and* Cummings. "People now talk about the classic days of the MGM musical, but at the time we were too busy making them to think we were making a classic anything."

With MGM's reputation in the field of musicals established during the 1940s, every studio began trying to imitate Metro's success. Although tiny Columbia couldn't match Metro in most areas, it did compete with a limited offering of musicals, usually featuring Rita Hayworth. Two of the Hayworth pictures, *You'll Never Get Rich* (1941) and *You Were Never Lovelier* (1942), paired her with Fred Astaire, who had known her father, Eduardo Cansino, in vaudeville. "I was a good deal older than Rita," said Astaire, "but we had lots of fun together. She was a beautiful dancer, a different type than Ginger or any of the others that I've worked with. She had a great Spanish touch. Whenever I worked with somebody, I tried to do something that suited both of us. Rita had a certain style, and I tried to make proper use of it." Jack Cole, who specialized in sexy numbers, choreographed many of Hayworth's pictures, including *Gilda* (1945) and *Down to Earth* (1947); Jerome Kern and Ira Gershwin collaborated on an original score for *Cover Girl*, which contained the classic "Long Ago and Far Away."

Harry Cohn wanted the Hayworth pictures to exude sophistication, as well as his personal imprint. Yet Cohn's taste was often overshadowed by his ambition. Saul Chaplin, who did the vocal arrangements for *Tonight and Every Night* (1945), recalled playing one of his choral renditions for Cohn and having the studio head insist he wanted it to sound like the Broadway cast recording of "Oklahoma!" When Chaplin tried to explain that they weren't doing *Oklahoma!*, Cohn fumed, "If you can't make it sound like that, I'll get somebody else!" Whereupon the arranger went back to his office, distorted the song a bit, and created a pulse exactly the same as "Oklahoma!" He was sure Cohn would throw it out once he heard the results. "He loved it!" Chaplin exclaimed. "That's the way it is in the film."

In addition to Rita Hayworth, Columbia had Janet Blair, Larry Parks (who starred in two highly successful Jolson pictures), and Ann Miller for a series of budget musicals, until Miller left for MGM. Morris Stoloff headed the music department at Columbia, and Fred Karger was the studio's vocal coach. Columbia's music department consisted of four people. As Saul Chaplin explained, "I arranged. I conducted choruses. I wrote songs. I wrote incidental music. I wrote lyrics. I taught people. I recorded on the piano. I did everything you could think of except conduct the orchestra." Although expensive pictures were intensively prepared (Larry Parks studied Al Jolson's movements and practiced imitating them to a playback for weeks before the Jolson films were shot), B musicals were put together quickly. "I'd get a call and someone would say, 'Listen, we're sending over a script. I need a song on page fifty-four,'" Saul Chaplin remembered. "I wouldn't have time to read the whole script, so I'd read from page forty to page sixty and write the song; the girl would be in the next day to learn it."

Twentieth Century-Fox also had success with musicals—first with Alice Faye and then with Betty Grable. Faye was the better singer of the two, starring in such pictures as *Alexander's Ragtime Band* (1938), *Tin Pan Alley* (1940), *Hello, Frisco, Hello,* and *The Gang's All Here* (both 1943). Grable was primarily a dancer, known for her beautiful legs; she was fun to be around and possessed a bawdy sense of humor. She was never a great singer, only a fair dancer, and not a dramatic actress. But Grable projected magnetism on the screen. "Betty was a hard worker," co-star Cesar Romero said, "and she was a hell of a good little performer. She knew how to use what she had." Grable hated to rehearse, preferring to spend her day at the racetrack, but worked hard when she had to. Her pictures included *Down Argentine Way* (1940), *Springtime in the Rockies* (1942), *Coney Island* (1943), and *Diamond Horseshoe* (1945), all of which contributed to her being selected the number one "pinup girl" by servicemen during World War II.

Grable later worked mostly opposite Dan Dailey, a talented entertainer who starred with her in *Mother Wore Tights* (1947), *When My Baby Smiles at Me* (1948), *My Blue Heaven* (1950), and *Call Me Mister* (1951). Dailey developed a drinking problem and was often hung over on the set, yet he remained a gifted performer, and his pictures with Grable were consistently popular. At 20th Century-Fox, where budgets were generally under a million dollars, there was not the money for innovation that Gene Kelly enjoyed at Metro; but according to Dailey, he was never really competing with Kelly, or with Fred Astaire. "Gene was a serious dancer and leading man type," Dailey said. "Fred Astaire was a light, breezy, white tie and tails sophisticate. I was the happy hoofer, doing burlesque and vaudeville material."

Besides Betty Grable and Alice Faye, Fox had June Haver, Vivian Blaine, and the "Brazilian Bombshell" Carmen Miranda under contract for musicals; Dick Haymes, Tony Martin, and Perry Como were Fox's equivalents of Bing Crosby. Seymour Felix, an old-fashioned dance director, did much of 20th Century-Fox's choreography, while Alfred Newman was head of the music department and Ken Darby the studio's principal vocal coach. Harry Warren, a contract songwriter on the lot from 1939 through 1945, claimed that much of his work consisted of providing the choreographer with something interesting to do. On *Diamond Horseshoe*, for instance, Warren wrote two ballads with Mack Gordon, "The More I See You" and "I Wish I Knew." He also composed "In Acapulco" for Grable, a specialty number that led into a colorful dance routine.

Warner Bros. reentered musicals during the 1940s with a succession of modestly priced pictures featuring Doris Day, Gordon MacRae, Gene Nelson, Patrice Wymore, and occasionally Virginia Mayo and the veteran James Cagney. "All Warner Bros. was doing was following MGM's suit and kind of reaping the residual profits from their exploration," dancer Gene Nelson said. "Warners, like Fox and Columbia, had an obligation to make a certain number of musicals a year. So therefore they had to have so many actors under contract who could sing and dance—not too many, just enough to keep up. That was the attitude. So consequently there was nothing progressive about what they did. Metro had the great experimenters. I had to fight tooth and toenail at Warners to get some of my numbers in, and then they had to be done within a limited schedule." While MGM might spend two and a half million dollars on a musical, Warner Bros. was stretching to spend half that much, mostly recycling songs the studio already owned.

Doris Day was introduced with Warners' *Romance on the High Seas* (1948), then went on to popularity in *Tea for Two* (1950), *On Moonlight Bay* (1951), and *Calamity Jane* (1953). A former band vocalist with dance training, Day had a unique singing style and worked well with Ray Heindorf, head of the Warner Bros. music department. "Ray knew when I wanted to breathe, when I wanted to stop, when I wanted to slow down or pick up," she said. "He just knew what to do." On their recording session for "Secret Love," the hit ballad from *Calamity Jane*, Day and the orchestra began recording at one o'clock. "I was on my way home at a quarter after one," the singer said. "Of course, the musicians had been there earlier to rehearse the arrangement, but I came at one and we just did it. Ray was brilliant."

Day remembered working on dance routines in an empty rehearsal hall with Gene Nelson's wife Miriam and a pianist. Comedian Billy DeWolfe had given Day the nickname of Clara Bixby, and everyone closely associated with her at Warners called her that. Whenever Day claimed she couldn't do a step or that a combination was too difficult, Miriam Nelson would yell encouragement: "Come on, Clara, you can do it!" The singer said, "They'd push me out there, and I'd do it. It was fun making those corny musicals."

Most of the choreography at Warner Bros. by that time was supervised by LeRoy Prinz, a feisty little man who always had a cigarette dangling from his lips and who looked more like a bartender than a choreographer. Simplicity was Prinz's keynote, and like Busby Berkeley, he was good with camera angles. "I used to stage with the camera in mind," Prinz explained. "I knew before I had a number finished every angle, where I was going to be at a particular point. I was born with a natural instinct." Prinz had a flair for period pieces and devoted hours to research. But for the choreography itself, he would lay out his ideas and then leave for a meeting, letting his brother Eddie or another assistant worry about the steps involved. "He seemed to be more of a political figure than a choreographer," one of his former chorus boys recalled.

Republic, which made mostly westerns, had no music department when Gene Autry arrived in 1935. The singing cowboy mostly selected his own songs. By the time Dale Evans reached the studio in 1943, to star opposite

Roy Rogers, Republic had added a music division under Martin Scott. Operating on limited budgets, Republic frequently used songs in the public domain, but occasionally bought new material if a song seemed essential to one of its bigger productions. Songwriter Jule Styne, under contract to the studio in the early 1940s, found Herbert Yates, the studio's head, almost laughably ignorant in musical matters. But Cy Feuer, who briefly headed Republic's music department, later of Broadway's Feuer and Martin, gave Styne his first opportunity to compose for films, as he had lyricist Sammy Cahn, Styne's partner at the time. A minor studio, Republic was happy to reap what it could from promising young talent who would work for low wages. "Gene Autry and those cowboys lived in another world," Styne said. "Republic was good training because I learned to deal with the toughest kind of people."

By the mid-1950s the era of the original screen musical was approaching an end. Production costs had soared, and a more sophisticated public had tired of simplistic song-and-dance extravaganzas. When Betty Garrett made a publicity tour for Columbia's *My Sister Eileen* in 1955, distributors whispered in her ear before press interviews, "Don't say it's a musical." The market was nearly dead. *There's No Business Like Show Business* (1954) was 20th Century-Fox's last big original musical until *Star!* (1968), a notorious disaster. *Seven Brides for Seven Brothers, High Society* (1956), *Les Girls* (1957), and *Gigi* (1958) represented Metro's final giant efforts. "Those little pictures that Paramount made with me could have been quite successful five years earlier," singer Rosemary Clooney pointed out, "those tiny things like *The Stars Are Singing*, which I did with Anna Maria Alberghetti. Even the typical Bob Hope musical, such as *Here Come the Girls*, with a built-in audience, would have been a lot more successful five years before. This was the tail end of musicals in pictures." What remained were mainly screen adaptations of big Broadway shows—*Oklahoma!* (1955), *Carousel* (1956), *The Pajama Game* (1957), *The Music Man* (1962), *My Fair Lady* (1964), *The Sound of Music* (1965), *Oliver!* (1968), *Cabaret* (1972).

Another area of film music, however, was just gaining public attention, in part because of the introduction of long-playing records and the sudden popularity of sound track albums. In the early years of talking pictures film scoring for dramatic pictures did not command much respect, at least not outside a limited circle. Max Steiner had shown the impact a dramatic composer could make on a film with his score for RKO's *King Kong* (1933), but the importance of music in creating atmosphere and enhancing screen action was slow to gain recognition. Astute film composers accepted the fact that their function was to underscore what the eye sees and that music had to serve the drama, often influencing audiences subliminally. They knew they had to write forceful music without overpowering the dialogue, and even had to be willing to opt for silence to protect the dialogue.

In most instances dramatic scores are written after a picture is finished, since the composer needs to know the film's texture intimately. During the post-production period, the composer works closely with the producer and director, who frequently want less music or more obviously commercial music than the composer does. "The composer can ruin what you're trying to achieve," veteran director King Vidor commented, "if he goes against what you had in mind. One idea must underscore the whole picture." On the other hand, music can magnify the dramatic impact of a picture if the union is effective. Rock Hudson remembered watching while the score was added to *Magnificent Obsession* (1954), for which Universal employed the Roger Wagner Chorale and a full orchestra for parts of Beethoven's Ninth Symphony. "I just became a blithering idiot," the actor said. "There I was up on the screen, and with that wonderful music I couldn't stop crying. Ross Hunter, the producer, was crying, as was everybody else. Even the conductor was crying."

From the mid-1930s until the late 1940s Warner Bros. lavished more care on dramatic scores for its pictures than any other studio in Hollywood. Leo F. Forbstein, who headed the music department at Warners before Ray Heindorf, was a superb conductor and orchestrator. The studio also had under contract two of the top composers in the industry—Max Steiner and Erich Wolfgang Korngold, both child prodigies and both from Vienna.

Steiner, who had studied with Gustav Mahler and become a professional conductor at age sixteen, came to Hollywood by way of Broadway, eventually writing over 250 scores. He worked first at RKO, then at Warner Bros., remaining at Warners for almost thirty years. More than any other composer, Steiner pioneered the use of original music as background scoring for films. He never wrote music from a script, since characters often appeared different to him in the picture than they had on paper. He would view a film once, at most twice, then tell his editor where he wanted the music, writing everything on cue sheets. "I can help a scene that may be too slow," Steiner said, "or I can help a scene that's too fast. You might change the tempo of your music, and if it's an uninteresting scene, the music can make it so exciting that you think they really said something, even though they didn't say a damn thing." (Steiner int., Rosenberg and Silverstein) He was also excellent at creating bridges between scenes. A droll little man who wore thick glasses and had a penchant for second-rate puns, Steiner composed the scores for eighteen Bette Davis pictures. "Max understood more about drama than any of us," the actress claimed years later. (T. Thomas, *Music for Movies*)

Since Jack Warner wanted volume in the music for his movies, he insisted that Steiner put as much brass into each score as possible. But for *Gone With the Wind* (1939), which he worked on during a loan-out to David Selznick, Steiner was able to achieve a lusher sound; this score proved to be his masterpiece, although he was nominated for eighteen Academy Awards. Only 30 minutes of *Gone With the Wind*'s 222 minutes are without music. "I did it in three months," Steiner recalled, "and I wrote a picture called *Intermezzo* at the same time. I would start to work at nine o'clock in the morning, lie down about midnight for a few hours; I'd be awakened by my butler at five A.M., and then I'd go back to work. The doctor would give me an injection to keep me awake, and I'd keep going." (Steiner int., Rosenberg and Silverstein) In addition to composing, Steiner organized the Screen Composers Association in 1942, which became one of Hollywood's principal guilds.

Erich Korngold was already an established composer in Europe before coming to Warner Bros. in 1935; he remained a classicist. The first composer of international stature signed to a studio contract, Korngold worked at Warners for twelve years, becoming the highest paid composer of his time. He produced only eighteen scores, but won Oscars for *Anthony Adverse* (1936) and *The Adventures of Robin Hood* (1938). Korngold composed lusty, martial music for Robin and his lads of Sherwood Forest, but after he had been at Warners ten years, producer Henry Blanke said to him, "Erich, you don't write as beautiful music as in the beginning." Korngold, who always looked upon Hollywood with disdain, replied: "I tell you why. Now I understand the English dialogue." (Blanke int., AFI)

Alfred Newman, head of the music department at 20th Century-Fox from 1940 until 1961, proved not only a fine administrator and an outstanding conductor, but one of the most prolific composers in Hollywood, winning nine Academy Awards. Since Darryl Zanuck trusted him implicitly, Newman had an authority at Fox few musicians enjoyed, as well as a huge annual budget. Although he demanded a great deal from his staff, he also protected them from insensitive producers and executives and made sure his people were well paid. Newman's own scores include those for *How Green Was My Valley* (1941), *The Song of Bernadette* (1943), *Captain from Castile* (1947), and *All About Eve* (1950), although he often was assisted by other members of his department. Since he was known to be a master, all the producers on the lot wanted Newman to write music for their films. Frequently he'd hand his thematic material to a staff composer to finish. "Sometimes schedules were so crazy that a group of us worked on a film," 20th Century-Fox composer David Raksin said. "We had to do scores which should have taken two months in three weeks, sometimes less."

Raksin's break came with the film *Laura* (1944), a great success; its theme became a popular song. "Usually a composer gets a dozen or half a dozen letters, *if* he gets any at all," Raksin pointed out. "With *Laura* I had over 1,700 when I stopped counting them." His next assignment was one of the worst pictures he was ever associated with. "My employers at the studio wanted to make sure I didn't get any ideas about being somebody," the composer later remarked. Raksin felt film music gave him a chance to experiment and to have his music heard in ways it would not be in the concert hall. Having studied with Arnold Schoenberg, Raksin was interested in writing modern music rather than pretty melodies. "Ordinary people who go to concerts would never in a million years have accepted the dissonances which we routinely used in films," the composer declared. "But if they saw a film which was full of violence and the music *worked* with that, they automatically accepted the music as valid. There's a reciprocal thing that goes on between the visual and the aural."

During the scoring of a picture composers frequently worked twenty hours a day. In most instances the momentum of the project, the esprit de corps, and the excitement of meeting an impossible deadline sustained the musicians. Historical films usually required research to acquire a feel for the music of that period. The challenge was to write effective film music that would also stand on its own merit. "I liked to take thematic material and make it metamorphose and change and develop during the course of a picture," David Raksin said. The disappointment came when a composer's music was reduced almost to inaudibility in the dubbing room to protect inconsequential dialogue, a regular enough occurrence that Aaron Copland referred to the dubbing and mixing process as the "composer's purgatory." (Green int.)

By the early 1950s MGM had seven or eight dramatic composers under contract, in addition to four or five orchestrators and ten or more copyists and assistants. Miklos Rozsa, one of the giants in film music, specialized at Metro in "epic" scores: *Quo Vadis* (1951), *Ivanhoe* (1952), *Julius Caesar* (1953), *Knights of the Round Table* (1954). Rozsa attempted to be authentic in his work, using medieval musical sources for *Ivanhoe*, relying on Malory and Tennyson for inspiration on *Knights of the Round Table*, and creating music appropriate for a Shakespearean drama rather than "Roman" music for *Julius Caesar*. Bronislaw Kaper, who had come to Metro earlier from Germany, remained at the studio for thirty years, composing scores for *Keeper of the*

Flame (1942), *Gaslight* (1944), *Green Dolphin Street* (1947), and *The Swan* (1956). Kaper possessed a keen sense of humor, yet was a serious musician, convinced that the film composer could add an important dimension to films.

John Green, the eventual department head, also wrote dramatic scores at MGM, adapting Copland's "El Salon Mexico" as a piano rhapsody for *Fiesta* (1947) and scoring the entire picture. For *Raintree County* (1957), Green produced one of his richest scores, yet wisely decided to let his music disappear during a long scene in which Elizabeth Taylor's character relates what happened the night a mysterious fire occurred. "I must have run that scene alone in my projection room fifty times if I ran it once," Green said. "Every time I ran it the tears started running down my face. All I could figure out is that if I were four Beethovens and six Berliozes and three Wagners, all my music could possibly do to that scene was wreck it. It was *there*—the way Eddie Dmytryk directed it, the way it had been written, the way it was shot, the way it was lit, and the incredible way in which it was performed. It was complete!"

David Raksin, after his success at 20th Century-Fox, came to Metro in 1949 but stayed only three years. During that time he wrote a remarkable score for *The Bad and the Beautiful* (1952), but with his penchant for modernism, Raksin found the atmosphere at Metro unconducive to creativity. "MGM was a place where the primary purpose was dedicated to preventing art," Raksin claimed. "They saw to it that you were so exhausted from battling the system that you had no time to make waves or do anything that would startle anybody. Their immersion in a kind of corrupt pragmatism was such that when anybody demurred or was unwilling to be compliant, he was considered a nut. But if they thought you were talented and they needed you, they kept you on. Eventually you could achieve a certain degree of autonomy. John Green understood this and protected the guys in his department."

Dimitri Tiomkin, who won acclaim at Columbia with his score for *Lost Horizon* (1937) before going elsewhere to write music for *Duel in the Sun* (1947), *High Noon* (1952), *The High and the Mighty* (1954), and *Giant* (1956), was not nostalgic about his early years in Hollywood. Salaries were initially poor for musicians, and working conditions were worse. Tiomkin claimed film composers of that era were almost prisoners, given a projection booth, a tiny screen, and a piano of uncertain quality. "A great deal of music was needed," Tiomkin explained, "and anyone who could think of a tune and put a few notes together might be hired to concoct motion picture scores. Stray violinists and piccolo players were placed under contract as composers." Although musicians with high standards later appeared, Tiomkin wrote of Hollywood, "Many of the studio orchestra conductors were routine second-class musicians who had picked up the trick of timing music to film footage and neither knew nor cared how to get the best out of an orchestral score." (Tiomkin and Buranelli, *Please Don't Hate Me*)

Franz Waxman, however, who wrote music for *Rebecca* (1940), *The Paradine Case* (1948), *Sunset Boulevard* (1950), and *A Place in the Sun* (1951), was a real craftsman. A pianist since early childhood, the German-born Waxman could be difficult, yet his sense of drama in scoring was unsurpassed. He worked for Selznick and Warner Bros., then took up residence at Paramount, winning two Academy Awards there. Victor Young, who did most of his work at Paramount, either wrote, arranged, or supervised approximately 350 films. Young's music was distinguished by warmth of melody, evident in *For Whom the Bell Tolls* (1943), *Golden Earrings* (1947), *Samson and Delilah* (1949), and *Around the World in Eighty Days* (1956). "Young wrote music from his heart," Paramount staff member Bill Stinson said. "He may have been the best melody writer Hollywood ever had. But Victor had one failing; he always did too much." Young frequently wrote scores for eight or nine pictures a year at Paramount alone, employing an orchestrator who knew his style.

By contrast Bernard Herrmann never used an orchestrator throughout his long career. "I write for the orchestra," insisted Herrmann, who had the advantage of being fast and brilliantly gifted. (Green int.) He came to Hollywood with Orson Welles to write the score for *Citizen Kane* (1941), immediately showing himself an uncompromising perfectionist. "He was not the easiest man in the world to get along with," *Kane*'s editor, Robert Wise, declared. "But what a brilliant composer!" Herrmann later worked extensively for Alfred Hitchcock, on such films as *Vertigo* (1958), *North by Northwest* (1959), and *Psycho* (1960).

Alex North, who began his Hollywood career by writing the jazz-oriented score for *A Streetcar Named Desire* (1951), has sometimes been credited with bringing film music into the twentieth century, although that claim is overstated. George Antheil, Virgil Thomson, and Aaron Copland, among the more renowned American composers, all had earlier written film music using the modern idiom.

By the time North arrived, every producer in Hollywood wanted a title song as a means of publicizing his pictures, one that could be recorded and perhaps make its way onto the popular record charts. Most of the major studios already owned music publishing companies, but during the late 1940s they began acquiring record companies as well. Most serious film composers found the emphasis on theme songs distressing, especially when their lyrics had little or no connection with the picture itself. Those who survived recognized that they must adapt to a changing industry. As the big studio system entered its final phase, new techniques of marketing had to be found; a hit record was part of the mix.

11

Design, Hairstyling, and Makeup

Hand-painting fabric in the Wardrobe Department at MGM.
Courtesy of the Academy of Motion Picture Arts and Sciences.

All of the big studios excelled in the quality of their craftsmen, paying salaries that assured the best in every field. Nothing seemed impossible for Hollywood set designers, costumers, prop men, hairstylists, or makeup artists, and each studio projected an individual look. Metro-Goldwyn-Mayer emphasized sumptuousness; Warner Bros. preferred stark black-and-white realism; 20th Century-Fox leaned toward the vivid and garish. In various ways the old studios transported audiences into a world of fantasy and, in most cases, one of voluptuous luxury. Romance and glamour reigned throughout Hollywood's Golden Era; the big studios, including Warner Bros., seldom depicted squalor except in unique or foreign circumstances.

At MGM, Cedric Gibbons served as supervising art director for thirty years; he insisted that every inch of his sets be lighted fully, so that an overpowering whiteness became his trademark. Born in New York and educated in Europe, Gibbons was witty, effete, and intelligent. He sought perfection both in his designs and in the women he collected, including his wife, Dolores Del Rio. An architect by training, Gibbons could be severe and demanding. He chose his staff of nearly two hundred wisely; most stayed with the company many years. More than at other studios, each of the various departments at Metro formed a self-contained principality with its own hierarchy. Gibbons was among the most autonomous of the department heads. More executive than functioning artist, he had a clause in his contract giving him sole credit for every picture MGM produced in the United States, despite the fact that he mainly administered the work of others.

Gibbons's visit to the 1925 Exposition in Paris shaped his style for the next twenty years. He returned to Hollywood and imposed a spacious Art Deco design on the Joan Crawford vehicle *Our Dancing Daughters* (1928); his striking sets became the first to fully exploit the Modernist decor. Other studios quickly began using Art Deco in their pictures, which had a major impact on American design, creating a vogue for glass brick and blonde wood as symbols of wealth and fashion. Gibbons loathed wallpaper but loved decorative plaster, a preference that helped establish Metro's glossy look. He created a new standard of taste in films, emerging at the forefront of Hollywood art directors. Not only did Gibbons himself design the Oscar, but he collected eleven Academy Awards for best art direction.

Besides art directors, the MGM art department included draftsmen, scenic and matte painters, a model shop, and property and construction departments. As administrative head, Gibbons assigned each film to an art director, approved all budgets and sketches, and represented his staff in meetings with studio executives and other department heads. Preston Ames went to work for Gibbons in 1936, labored at the drawing board for a couple of years, then was assigned a picture to design, remaining in the industry for thirty-five years, mainly at MGM. Although an art director must serve as the architect of the set, Ames insisted that an ability to get along with people was equally important. "You have to have the patience of Job," he said. "And you have to have strength in your convictions to be able to tell a director yes or no. You don't fight the director, you work with him." It became essential for art directors to know color, landscaping, marine architecture, anything the script might call for. "I've probably traveled with [Metro] a quarter of a million miles just to see the world and prepare new pictures," Ames declared. (Ames int., Steen)

The designer's work begins with the script. Once he or she has absorbed the script's requirements, a production meeting is called with the producer, director, writers, set decorator, costume designer, and in the case of musicals, the composer and choreographer. Colors must be coordinated and everyone involved must understand the texture of the proposed film. Some veteran directors readily accepted the art director's designs; others, like Vincente Minnelli, a former designer himself, could be difficult, demanding repeated changes. After his initial conference with the producer and director, Preston Ames figured the cost on each scene to be played in a given area. Then the designer prepared a detailed budget to present to studio executives. "If the writer hasn't had diarrhea of the pencil," Ames remarked, "you can keep it down to a fairly tight amount of sets. Sometimes you'll find you have a hundred and fifty sets, because the writer feels his story should move from one set to another." (Ames int., Steen) Once the budget was approved, Ames proceeded with sketches and layouts, sometimes making scale models, and eventually turning those over to the department's draftsmen, who made a series of architectural drawings. An estimator reviewed finished drawings, scrutinizing every detail—paint, plumbing, landscaping. If the cost exceeded the approved budget, adjustments had to be made, perhaps even paring down the original concept. After construction began on a set, a cost verification was issued every week, which the set director analyzed to make sure the project wasn't going over budget.

In an effort to cut expenses, sometimes sets made for an earlier picture were reused. One of the *Picture of Dorian Gray* sets (1945), for example, was repainted and redressed at Metro for the "Limehouse Blues" number in *Ziegfeld Follies* (1946). More often new sets were specially built for big pictures. Vincente Minnelli insisted that the street in *Meet Me in St. Louis* (1944) be constructed expressly for that picture, with the houses absolutely authentic. For *The Clock* (1945) Penn Station was re-created, accurate down to small details.

Although an art director might be working on a picture long before the actors and cameramen were engaged, he or she had to design the sets so they could be lighted and so the director would have ease of operation. Historical pictures usually required research, but designers often preferred to give the illusion of authenticity rather than copy historic styles. "In motion picture making," Preston Ames noted, "copy is rather a dirty word." Capturing the spirit of the period generally was more important. For instance, when Metro made *Marie Antoinette* (1938) with Norma Shearer, several architects visited the studio and congratulated Cedric Gibbons on his authentic re-creation of French Renaissance architecture, particularly the design of Versailles. Gibbons accepted their praise but explained that had they copied the architecture of Versailles, the delicate molding and decoration would never have come across on the screen. "Consequently, we had to redesign the entire thing so it would photograph properly," he said. (Ames int., Steen)

At the height of the big studio era the art director most prominently listed in the screen credits was the department head, who served as a supervisor. Since no individual could perform all tasks with equal polish, responsibility was delegated. Most studios had specialists for interiors, exteriors, castles, ships, nightclubs, and so on. If the director or choreographer had special needs, he or she talked with a scenic designer with expertise in a particular area. In preparing the appearance of Vera-Ellen as Miss Turnstiles in *On the Town* (1949), Gene Kelly told Jack Martin Smith, "I need to see that girl as if she came out of the poster when we bring her to life." Smith pondered the matter and came up with a concave cyclorama, a device that provided the necessary illusion. "Things like that came out at MGM through the cooperative talents of many people," said Kelly.

Art directors worked as a team with the set decorator. The decorator and the art director conferred frequently with the costume designer, the director, and the cameraman, coordinating colors and deciding which would photograph best. "When color film came in, there were two schools of thought," Preston Ames said. "One was: Don't load your set and decor with color because it interferes with the action and story, it distracts. The other school was to put color everywhere you could think of. There was quite a hassle for a while, but the first school won out." Many continued to feel that well-photographed black-and-white pictures were the most aesthetically appealing. When filming in black-and-white, colors on the set still had to be selected and controlled. No two colors that would photograph tonally alike could be used together, since they would offer no contrast. "We used to go into warm grays and cold grays deliberately so that we knew exactly how it would come out on the screen," said Ames. But with the introduction of the three-strip Technicolor process, the situation became more complicated, since colors seldom came out exactly right. "We had to go to these highly experienced technical men and have them design a color that would actually photograph to look like the color we wanted on the screen," Ames explained. "Reds were difficult. Blues were an impossible situation. The palest blue would turn out to be a brilliant blue. Everything had to be brought down to the laboratory with color swatches and color schemes." (Ames int., Steen)

Like the art director, the set decorator began a project by studying the screenplay. "The script is like a Bible," said Arthur Krams, who went to work for Metro in 1945, "and is followed as the law whether it be a contemporary comedy or a drama, a Western, a musical, a period picture, or a larger-than-life spectacle." While the set decorator functioned in tandem with the art director, he or she also had conferences with the producer and the director during the preshooting period. "The decorator starts with a bare set," Krams explained, "three walls and a floor, provided by the art director. Seldom is there a ceiling or partial ceiling since the set must be lighted from this area by the huge lamps on the high catwalks which surround the set. The fourth wall is open for the camera, but later on may be mounted in place for reverse shooting." (Krams int., Steen)

During his training period with Metro, Krams worked on sets for several pictures at once. His first complete film assignment was *Holiday in Mexico* (1946); later he won an Academy Award for Hal Wallis's production of *The Rose Tattoo* (1955). The decoration of a set, Krams said, is inspired by the characterization of the actors and the screenplay itself. Eventually the set decorator would need to consult the wardrobe people to make sure the costumes of the star and the colors of the set complement each other.

Research was essential, not only for period pictures, but on any assignment in which a special kind of decor might be required. Sets had to be synchronized with the action. On days when filming on a new set was scheduled, the decorator and his crew might come into the studio as early as seven o'clock so that shooting could start by nine.

Not only did Metro own vast reserves of stock scenery, standing or stored, the company also had warehouses filled with props of every conceivable kind. Ed Wallis, who headed the studio's prop department, spent thirty years collecting objects from all over the world, from the finest furniture and French antiques to Victorian chamber pots, vast collections of silver, and thousands of perfume atomizers from all periods. "The prop department at MGM was like a museum," contract player Jane Powell remembered, "a glorious collection of real and unreal, of every period, every style you could imagine, and some you could *never* imagine. If they didn't have something, they found it. If they couldn't find it, they made it, and made it accurate in every detail." (Powell, *Girl Next Door*)

Lyle Wheeler, who for eighteen years was supervisory art director at 20th Century-Fox, earlier had served as a unit art director for Metro. He later worked on pictures for David Selznick, including *Gone With the Wind* (1939), which took nearly three years to prepare. Wheeler had studied architecture at the University of Southern California, but entered the motion picture business during the early 1930s when construction was depressed. Ultimately he supervised almost five hundred pictures, initialing every sketch the artists under him produced. At 20th Century-Fox fifteen to twenty unit art directors worked under Wheeler at a time, as well as thirty to thirty-five draftsmen. If a problem arose on the set between a unit art director and the producer or director, it was Wheeler's job as department head to settle it.

Wheeler favored an eclectic approach; he felt that every design must fit the story. Characters should never appear in a background foreign to where they belonged. "Otto Preminger opens *Laura* with the camera panning around the set," Donna Wheeler, Lyle's wife, pointed out. "Very rarely does an art director get that, but that established Laura's personality. It told you something about this woman whom you didn't know and were going to learn about strictly from voice-over." (Wheeler int.) Wheeler insisted on $500 in the budget of each picture for library acquisitions, building a substantial research collection during his years at 20th Century-Fox. He also supervised putting in a huge garden, with all kinds of plants that could be used to dress a variety of sets.

Wheeler was a master at taking a permanent set, using segments of it in different ways, and having it look fresh and newly created. But if he needed a new set, the money for it was usually available. Although much of his time was taken up with administrative paperwork, Wheeler checked on each set in use on the lot every day. He talked to the cameramen, consulted with the special effects department, and viewed tests made by costume designers; he was awarded screen credit on every picture he supervised.

There was always a stable of art directors under Wheeler so that should one quit, there would be continuity. During coffee breaks scenic designers talked about art and problems they were struggling with, such as the best material for a particular set, thus borrowing from the genius of others. If a decorator needed a prop, he or she went to one of the studio's warehouses, some of which were blocks long. Inside, suspended from the ceiling, was every kind of chair conceivable, in sets of six, ten, or twenty-four, all carefully organized. Once a year 20th Century-Fox had buyers travel to different parts of the country and purchase furniture at auctions and estate sales.

While 20th Century-Fox's set designers exhibited solidity and a sense of volume during the late 1930s and 1940s, Paramount at the time favored a curvilinear airiness. Hans Dreier, in charge of Paramount's art department, had served his apprenticeship in Germany and preferred a continental sophistication. Under Dreier Paramount films possessed a pictorial elegance and a shadow-veiled atmosphere in sharp contrast to the clean, brightly lighted sets of Metro. As director Ernst Lubitsch once remarked: "There is Paramount Paris and Metro Paris, and of course the real Paris. Paramount's is the most Parisian of all." (Webb, *Hollywood*)

Harry Horner, who won an Academy Award for his black-and-white design of *The Heiress* (1949) at Paramount, was a protégé of William Cameron Menzies, perhaps the greatest of all the Hollywood production designers. Horner was also influenced by the legendary German director Max Reinhardt in his subtle combination of realism and expressionism. For *The Heiress* Horner began by reading the script and studying the characters and their relationships. Then he researched the period, compiling volumes of notes on the dress, furniture, manners, and mores of the 1860s. "I find that when I study the life of the people in a period, I can create an entirely different dimension," said Horner. (Horner int., McBride II) Most of his research file consisted of text, not pictures. Then he went to New York, the setting for *The Heiress*, to experience the feel of Washington Square. Next Horner made rough sketches to convey the moods of his sets to director William Wyler. Most of the film takes place in a house on Washington Square, and the designer took great pains in his conception of the staircase. Toward the end when Catherine Sloper climbs the stairs in defeat, realizing that her young suitor is not coming to marry her, her climb needed to be poignantly dramatic. Horner thought the designer could enhance this moment by creating an exceptionally steep staircase. He felt much of the designer's work was primarily emotional rather than visual.

The supervising art director at Warner Bros. for twenty years was Anton

Grot, born and educated in Poland and strongly influenced by German impressionism. Grot designed eighty pictures at Warners between 1927 and 1948, turning out thousands of pencil or charcoal sketches and building sets on limited budgets. Warner Bros.' New York street was recycled endlessly, although the studio rarely economized on research and by the 1950s its prop department had a reputation for quality. When director Morton Da Costa went to Warners to re-create *Auntie Mame* (1958) for the screen, prop man Roby Cooper came to him and said, "Mr. Da Costa, I've read the script, and I see that you have flaming drinks in it. Could you sketch the vessel that those are in?" The next day Cooper returned with a wooden model of the vessel described. After Da Costa assured him it was exactly what he had in mind, Cooper replied, "All right, now we'll cast it in silver." The director started to walk away when Cooper asked, "Just a minute, Mr. Da Costa, wouldn't you like a monogram?" Coming from Broadway, Da Costa wasn't used to such precision. (Da Costa int.)

Van Nest Polglase headed the art department at RKO, where designs were often restricted by tight budgets. Art directors and set decorators at all the studios learned to economize on B pictures, using a few tables to suggest a restaurant or making a stack of bleachers with three hundred extras look like a packed stadium. While Hollywood always admired those who thought big, scenic designers were craftsmen and customarily made do with the budget assigned them. Art directors accepted B pictures as an essential part of their studio's overall financial package, although few wanted to linger on budget productions longer than necessary.

Broadway designers, like Joseph Urban (Ziegfeld's celebrated set designer) or Howard Bay (who later created a brilliant set for *Man of La Mancha* in New York), occasionally worked in Hollywood, but for the most part studio art directors were a unique category. William Cameron Menzies emerged as the most universally respected, the first Hollywood designer to think in terms of participating in the creation of drama rather than simply contributing background decoration. A short, stocky man with a warm, gentle personality, Menzies was a raconteur and a genius in his work, making drawings for every shot in his pictures on huge pieces of cardboard. He knew how to economize without lowering his standards, refusing to build an entire porch when only a door was needed. "Menzies could take two walls and make them look like a great ballroom," claimed actress Arlene Dahl, who starred in *The Black Book* (1949), which Menzies designed for budget-minded Eagle-Lion. David O. Selznick insisted on having Menzies work jointly with Lyle Wheeler on *Gone With the Wind*, realizing that a man of Menzies's talent would be essential on such a lavish production. The producer wanted to have *Gone With the Wind* prepared down to the last camera angle before shooting began; on such an expensive film, thorough preparation could save hundreds of thousands of dollars.

Selznick was no less concerned about the thousands of costumes for his Civil War epic. Gowns used for several days on any set had to be made in duplicate, in case one should tear or become soiled. There were twenty-seven copies of Scarlett's battered calico dress, which Vivien Leigh wore through the burning of Atlanta and the perilous journey back to Tara, a pair for each stage of disintegration. Walter Plunkett, a Hollywood designer who specialized in historical costumes, working first at RKO and then at Metro-Goldwyn-Mayer, created Vivien Leigh's wardrobe for the picture, influencing fashion across the country. "The nation is now burdened with snoods for women's hair as a result of the Scarlett and Melanie snoods," David Selznick wrote his publicist, Russell Birdwell. "I assume you know that the costumes of *Gone With the Wind* are the basis of at least fifty per cent of fashions at this present moment, and I am sure the whole business of the return to corsets is due to *Wind*. All of this is trivial and laughable in a world that is shaken by war, but women being what they are, I think it could make for excellent publicity. Because of my feelings about its trivial nature, though, I would suggest that the publicity be confined to the women's magazines, fan magazines, etc." (Behlmer, *Memo*)

Although directors had costume approval during the big studio era, an aggressive producer like David O. Selznick involved himself with every aspect of production. When Broadway costume designer Lucinda Ballard worked for Selznick on *Portrait of Jennie* (1949), she received endless telegrams about details of various garments from the producer, one of which was five pages long.

During Hollywood's Golden Age few A pictures were made for which studio designers did not create clothes. Practically nothing was bought off the rack. "In those days it was a fabulous, fabulous empire," designer Edith Head remarked. "The pictures we were making were glamorous, the stars were glamorous, clothes were fun, designing was at its peak." Hundreds of thousands of women went to the movies each week just to see the clothes. But costume designing for films, even in the Golden Era, was no mere style show. The requirements of the drama itself—the story's mood, its situations, its personalities, its time period—dictated costuming. "Designing costumes for motion pictures *begins* where fashion designing ends," Michael Woulfe wrote. (Woulfe, "Costuming a Film") Even when creating costumes for a film with an ultra-modern setting, the designer had to keep in mind that the picture might not be shown in the theaters for a year or more. It was therefore necessary to modify existing fashion trends so they wouldn't be dated, no matter when the film was released.

Metro as usual led the way in the creation of beautiful costumes, splendid in every detail. From 1928 until 1941 MGM's principal costume designer was Adrian, a tall, lean man who carried himself with an air of distinction. Adrian began his career in Paris, where Irving Berlin saw his work and brought him to New York to design the costumes for his next shows on Broadway. Later Adrian came to Hollywood, where he worked with Valentino, designed the costumes for DeMille's *The King of Kings* (1927), and finally became the head designer at Metro-Goldwyn-Mayer, where he worked with almost unlimited budgets. At Metro he gowned Garbo, Norma Shearer, Joan Crawford, Myrna Loy, Jeanette MacDonald, and Katharine Hepburn. He was an ideal counterpart to production designer Cedric Gibbons. Adrian had a passion for sequins and glitter, preferred straight lines, liked bows, loved black against white, and avoided middle grays. He was responsible for Joan Crawford's broadly padded shoulders, yet realized that Myrna Loy's strength came from underplaying and that too many details in her attire distracted from the subtlety of her performance. With Garbo's *Camille* (1936) Adrian tried to interpret the story through clothes, while in *The Wizard of Oz* (1939) he utilized his flair for the fantastic. Both versatile and productive—for *Marie Antoinette* (1938) he designed four thousand costumes—he soon was earning $75,000 a year, a salary equal to that of the president of the United States.

Adrian refused to equate drabness with utility. "For my money," he said, "a good dress has to be becoming, useful, and beautiful." (Markel, "Adrian Talks") For the screen, clothes, above all, must be photogenic. Comfort and practicality were of little concern. Many of his gowns were too tight for actresses to sit in, requiring them to recline on "leaning boards" between takes. Yet he pampered his stars with luxurious touches the camera would never reveal, like real diamonds and emeralds. Gradually Adrian began to lose his autonomy at the studio and was sometimes forced to make several sketches before he created one that satisfied everybody involved. He left MGM in 1941, opening his own salon in Beverly Hills the following year.

His place was taken by Irene Sharaff, who as Metro's executive designer shared her predecessor's enthusiasm for glamour. A tall beauty herself with glorious hair, Irene loved clothes. "Her color sense was impeccable," singer Jane Powell remembered. "She created a wonderful greenish-blue color. Technicolor green, I called it. . . . Even her fingernail polish was hers alone. It was pink with a bit of blue in it." (Powell, *Girl Next Door*) During fittings Irene would have actresses move, paying attention to such things as neckline detail, making sure it wouldn't detract from the face in close-ups. "Her evening clothes had great movement," actress Mary Astor noted, "material that would flow around the thighs, revealing and releasing. She paid attention to accessories, dragging out a whole box full of jewelry, pawing through for just the right pin or a clasp for a bag, earrings that would shape the ear and not waggle and detract from the face." (Astor, *Life on Film*)

Later Walter Plunkett and Helen Rose joined the MGM wardrobe staff, and Dolly Tree was hired by the studio to design for young players. Specializing in period and ethnic costumes, Plunkett arrived in 1945. He spent months preparing the many gowns Lana Turner, Donna Reed, and Gladys Cooper wore in the epic *Green Dolphin Street* (1947) and rummaging through Salvation Army stores to find old quilts he could turn into authentic dresses for the girls in *Seven Brides for Seven Brothers* (1954), rather than using material that simply *looked* like the patchwork called for in the script. Sometimes, however, beauty took priority over authenticity. When stage director Elia Kazan joined Metro in 1946 and Plunkett showed him sketches for Katharine Hepburn's costumes for *The Sea of Grass*, the costumes were handsome but far from the homespun clothing the director had wanted for Hepburn's character. "This story is supposed to take place, is it not, in the backcountry?" Kazan asked. Plunkett looked surprised. "Actually," the designer replied, "this picture takes place in Metro-Goldwyn-Mayer land." Kazan accepted his fate philosophically. "The Metromill was grinding," he wrote later, "and I was between its teeth." (Kazan, *A Life*)

There was an MGM mode to which designers were expected to conform. Studio legend had it that Helen Rose, among Hollywood's most admired designers during the 1940s and 1950s, arrived at Metro wearing a nondescript black dress, with her slip showing and her hair pulled up in an unbecoming knot. One day she was introduced to Louis B. Mayer, who ordered Rose to fix herself up or be fired. Her designs, however, fit into the fantasy world MGM endeavored to create. Rose specialized in contemporary female fashions, with even the undergarments elegantly made, frequently of lace—all part of the glamorous image touted by the studio's publicity department. Metro-Goldwyn-Mayer continued to emphasize glamour into the 1950s, when other studios were turning toward realism.

A Hollywood costume designer began by reading the script to determine the number and kind of costumes a film required. He or she then discussed the wardrobe with the stars, the producer, the director, the color expert, the set designer and decorator, making sure all agreed on the idea behind each scene. Once the colors and look had been agreed upon, the designer made a sketch for each costume, which had to be approved by the stars, the director, and the producer. Period pictures especially required research to assure authenticity in styles, fabric, colors, and accessories. At MGM each player had a body form in the wardrobe department to minimize fittings. Since costumes normally were made in advance of shooting, stars might be involved with another picture or available for only brief periods. Costumes for historical films were sometimes tailored abroad; those for *Quo Vadis* (1951), for instance, were made in Florence.

The Metro wardrobe department was organized under a general manager who supervised the sewing and checked every detail. "Such care was taken in those days," Jane Powell commented. "Even our petticoats were of the finest materials. Ribbons and lace, silk or chiffon, crystal pleats." (Powell, *Girl Next Door*) Embroidery was generally done by hand, even on the hems of dresses where it wouldn't show. Sixteen Mexican women worked on the department's second floor. "They brought their lunch in brown bags and never went out on the lot," recalled publicist Ann Straus. "They just sat and beaded and embroidered all these beautiful costumes that were designed by the MGM designers. They never knew the people for whom they were doing the work. They just knew it was production number 1420 and that was all." A dye lady operated from a little cubicle that resembled a modernistic painting. "There were dyes sprayed all over the walls, Ann Straus said. "If they had a certain piece of material and they wanted it to be plaid in certain shades, she would mask off stripes and spray on the colors. They also had tailors who were marvelous. If you wanted a plaid material, they could stitch the plaid out of thread on the fabric."

During the shooting of a film a night wardrobe woman freshened up costumes that had been worn during the day. If a problem arose on the set, the designer had to be prepared to solve it quickly, so the unit wouldn't fall behind schedule. Musical numbers could prove especially troublesome, with complications arising during final stages of production. Censorship frequently forced last-minute changes. After 1934 the Hays office mandated an anticleavage rule; and while it was all right to show a man's navel, women's navels were taboo. The wardrobe department was often called upon to supply a diamond belt or pearls to hide a dancing girl's navel before shooting could resume.

Contract player Ruth Hussey, who played a supporting role in *Marie Antoinette*, remembered having five gorgeous gowns made especially for her, and wigs being fashioned for each cast member. As individual stars became more important, they reported either to Adrian's or Irene's private quarters. "Mr. Adrian's fitting room was nirvana," actress Ann Rutherford said. "He had white art deco sofas and marvelous sketches on the walls, rugs deep as the middle of the ocean, and he served tea and things. There was a maid in uniform to help you in and out of your clothes." For Rutherford the wardrobe department became the center of the Metro lot. "It wasn't just floor after floor of clothes," she said. "It was ringed with a cast-iron balcony so that you could stand there and look at these marvelous racks brimming with costumes. You could see pirates. You could see Marie Antoinette. You could see every costume and style under the sun." (Rutherford int., Wagner)

Until the emergence of Metro in the mid-1920s, Paramount was the richest and most stylish of the studios, operating both in Hollywood and New York. In 1923 Howard Greer, a New York couturier and musical comedy costumer, was invited to Paramount to work on Cecil B. DeMille's silent version of *The Ten Commandments*; he quickly became the studio's head designer. The following year another top designer, Travis Banton, was added. Banton shared assignments with Greer until 1927, when Greer resigned to open a salon of his own. Banton, who had earlier been apprenticed to New York society dressmakers and designed costumes for the *Ziegfeld Follies* on Broadway, not only became Paramount's chief designer but strengthened his contract in 1929 to include absolute authority over the wardrobe actresses wore in Paramount pictures. He dressed Carole Lombard, Claudette Colbert, and Marlene Dietrich, accentuating the body, using the costliest fabrics, and proving a master of texture and trim. When Banton left Paramount in March, 1938, Edith Head succeeded him as the studio's principal designer.

Head started out as a Spanish teacher in La Jolla, working as a studio sketch artist during summers. Eventually she was hired as an assistant under Greer and Banton; she worked mainly on B pictures and westerns. When she wasn't racing between projects, Head spent every possible moment watching Banton, studying his style. "After all," she said, "he was the best designer, bar none, in the world. And he taught me everything I knew about designing." (Head and Calistro, *Edith Head's Hollywood*) Since Paramount was then making approximately fifty pictures a year, Banton and Greer designed for the stars, letting Head dress the rest of the cast, especially the men, sometimes out of stock wardrobe. "I did the aunts, the grandmothers, the men, and what were called the background people," she said. "It was tremendous experience; we tried to keep the whole look of the picture together."

Eventually Head became one of the most creative designers in Hollywood, bridging the gap between the era of glamour and the later emphasis on authenticity. She won eight Academy Awards (two in 1950), received thirty-four nominations, and remained Paramount's chief designer until 1967. She contributed to more than 750 films. Her clothes were classic—elegant and timeless. Often there were fifty or sixty people working under her. The caste system initiated by Greer and Banton continued: Edith Head designed for the stars, occasionally accompanying them on loan-outs to other studios, while her assistants took care of the second leads, the character people, the bit players, and the extras.

Head created the Veronica Lake look, designed Dorothy Lamour's sarong for *The Jungle Princess* (1936), and started a Latin American vogue with Barbara Stanwyck in *The Lady Eve* (1941). But rather than consciously setting fashion trends, she attempted to transform the stars she worked with into the characters they portrayed on the screen. "Most people labor under the delusion that a designer has a field day and designs what they want," Head said. "But when you get a script, the script tells you exactly the kind of person you're creating for—their status, their social background, their locale. The script tells you what to do, what is correct for that particular character." If an actor or actress was known for a certain kind of clothes, Head would often veer in a different direction, unless the person was an established personality like Mae West. "We're not designing for the pure sake of creativity," she said. "We're asking the public to believe that the same star is a different person in every picture. It's a transformation, it's magic, it's camouflage. A good designer will design subtly enough to get the feeling of today without dating a film."

While Head had critics, no one could deny she was a hard worker; she normally arrived at the studio at daybreak and remained long after almost everyone else had gone. Her method was to break a script down into a wardrobe plot, defining every role in terms of clothes. Then, taking a sketch pad and pencil, she would meet with the producer and director. She rarely settled for one design, presenting instead two or three options. Then she consulted the star, the cinematographer, the art director, and the set decorator. She found that certain patterns tended to be distracting on the screen; checks and plaids could be problematic, making a costume test necessary in some cases before a final decision could be reached. "We have the power through the medium of clothes of translating people into anything we want," Head claimed.

In Paramount's *The Heiress*, Catherine Sloper needed to appear clumsy and awkward. "I had to get across how uncomfortable she was with herself in whatever image she projected," Edith Head explained. "I could not do it by giving her inexpensive or ugly clothes, because her father was a wealthy man and everything she wore was of the finest quality." So rather than giving Olivia de Havilland, who played the part, a perfect fit, the designer made her clothes purposely gap apart or wrinkle in the wrong places. "I would cut a collar too high, or a sleeve a bit too short," declared Head. "If *The Heiress* had been a silent picture, anyone could have told the story simply by watching the transition of costumes." (Head and Calistro, *Edith Head's Hollywood*)

If a script called for high fashion, Head could produce some of the most lavish creations ever devised for Hollywood films. In *Lady in the Dark* (1944) Ginger Rogers played a sophisticated woman with a taste for glamorous clothes. Edith Head created a mink skirt, split up the front and lined with ruby glitter. "Ginger's famous legs were encased in long sheer tights," the designer said. "Under a mink jacket was a blaze of the same ruby glitter. The mink alone cost fifteen thousand dollars." (Head int., Steen) At the height of the Golden Era budgets posed few restrictions; if luxury was called for, designers on big pictures were allowed to be extravagant. Yet if a star was required to wear an old, ragged apron, a studio designer made it.

Head found historical films easiest, since she was simply re-creating something. The secret was enough research to understand the period, and then to interpret it in terms of the project, the personality of the star, and the requirements of the script. During Hollywood's early years historical accuracy didn't matter. "If the director was shooting a picture in 1885 or 1887 and didn't like a bustle," Head recalled, "we just didn't use a bustle. It didn't matter, because the public wasn't particularly astute. The idea was to look pretty and to hell with reality." While Cecil B. DeMille sought historical correctness, he also wanted glamour and sumptuousness. For instance, he demanded that his women wear high heels, so in *Northwest Mounted Police* (1940) Paulette Goddard wore wedgies concealed in her moccasins. "I want clothes that will make people gasp when they see them," DeMille told the Paramount wardrobe department. "Don't design anything anybody could possibly buy in a store." (Chierichetti, *Hollywood Costume Design*)

Seven designers were assigned to DeMille's spectacular *Samson and Delilah* (1949). Edith Head created the wardrobe for Hedy Lamarr, who played Delilah, while Dorothy Jeakins supervised the men's clothes. There was even a designer to take care of the horses. For the sequence in which Samson brings down the Philistine temple, Lamarr wore a gown with peacock feathers cascading down a long cape. Edith Head voiced concern over finding enough feathers, but DeMille informed her that he had several peacocks on his own ranch and there was no problem. "We waited for peacock molting season," the designer recalled, "and then my wardrobe people trudged up to his ranch to collect almost two thousand feathers. The cape, which attached at the shoulders and formed a train several yards long, was covered with the iridescent feathers, each hand-sewn or glued in place." (Head and Calistro, *Edith Head's Hollywood*)

Head thought working in color was easier than black and white, since much more attention had to be given to line and detail in black and white. In Technicolor even the simplest dress could be effective because of the beauty of the color alone, although Alfred Hitchcock gave the designer sound advice: "It's really very simple, Edith," Hitchcock told her. "Keep the colors quiet unless we need some dramatic impact." (Chierichetti, *Hollywood Costume Design*)

Like Metro, Paramount had a fabric store that included laces and satins and chiffons from Switzerland, Paris, Rome, all over the world. "When we got ready to design clothes," Head remembered, "we'd go into our little store and say, 'Ah, there's that beautiful velvet; I'll do this cape in it.' When you've worked as long as I have, you instinctively know what fabrics to use. We had fabulous brocades, cut velvets, hammered satin." Women would sit at looms and weave cloth especially for a star's costume, sometimes taking six or eight months to complete enough for a dress. "It was another world," said Head.

Since each studio constituted a family, costumers knew most of the actors and actresses they'd be designing for and could plan ahead. They worked with the same cinematographers year after year, the same set decorators, the same art directors. People in one department looked forward to working again with friends in another, spending at least three months together preparing a picture, or in the case of a DeMille epic, fifteen months or longer.

Among her other talents, Head was a diplomat, well schooled in studio politics and aware of the value of publicity. "Edith Head was one of the most talented politicians I've ever encountered," observed fashion designer Oleg Cassini, who otherwise dismissed her as "an uninspired but competent designer" whose "answer to every fashion question was the shirtwaist." On the other hand, Cassini acknowledged, "She was a fabulous talker. She spoke a curious language, a cross between Hollywoodese and fashionese." (Cassini, *My Own Fashion*) Head could sell her ideas to producers, joking with them, appearing modest and never critical, making her players feel comfortable and at home, easing their insecurities. "She loved maneuvering and manipulating people," Hollywood designer Yvonne Wood said. "She was defensive and very clever about handling her competition. I admired her survivorship. She had excellent taste, but Edith's clothes were always the safest."

At Warner Bros., Orry-Kelly was the major costume designer from 1932 to 1943. Australian-born, Orry-Kelly proved not only one of the best Hollywood designers, but one of the wittiest. His style emphasized simplicity, relying on middle grays and dull-finished wools, chiffons, and velvets. He achieved high fashion with few frills. Orry-Kelly created the costumes for Busby Berkeley's musicals and most of the early Bette Davis pictures. He left Warners during World War II to freelance, about the time the War Department issued a directive ordering studios to conserve on fabric, which meant short, tight skirts, no pleats, no cuffs, and no big collars. Government leaders realized that if the public saw pictures with movie stars wearing extravagant clothing, those fashions would become the vogue across the country. Not until after the war, when Dior introduced the New Look, did long, full skirts come back into style.

Studio executives often voiced concern over their stars' wardrobe. Jack Warner's concern for detail was such that he repeatedly complained about dresses designed for Bette Davis or boots selected for Gordon MacRae. Once Warner came out with an edict that none of his actors could wear a striped tie, arguing that a tie with any design detracted from the face on the screen. Even after the war Warner Bros. continued the no-nonsense look Orry-Kelly had established earlier, frequently economizing on wardrobe in smaller pictures. Gregory Peck remembered being fitted into one of Rod Cameron's old cowboy suits for *Only the Valiant* (1951). On most of the studio's films a new wardrobe would be made for stars, but the supporting cast and extras would have theirs assembled from existing stock. Sometimes a player was asked to create his or her own costume. Director George Stevens, for instance, gave character actress Jane Withers the freedom to select her dresses, wigs, and accessories for the role of Vashti in *Giant* (1956). "It was my idea," Withers explained, "that the older and larger she got, the redder her hair became. Stevens nearly fell over in his chair when I walked out for my final test in an emerald green dress with bright red hair and that long cigarette holder with the diamonds. He laughed until tears came. He just loved it. That was the most incredible experience of my adult years in films."

Robert Kalloch became Columbia's first contract designer, followed in the mid-1940s by Jean Louis, who created most of the fashions for Rita Hayworth. "I prefer to design for a personality," Jean Louis said. "It was frustrating and very hard to work at Columbia because they had so few stars." (Louis int., Kobal) The designer acknowledged that while he had to keep in mind the character a star was playing and adhere to essential requirements of the script, most of his creations were soft and feminine. Since Hayworth was the studio's biggest asset, he tried to make everything for her special. Harry Cohn hired a succession of girls he thought looked like Rita, training them as possible replacements, but Jean Louis rarely designed for lesser players. Character people at Columbia were dressed mainly from wardrobe; on B pictures actors might even wear their own clothes, while actresses were sometimes given money to buy something appropriate at a local department store.

Charles LeMaire was head of the wardrobe department at 20th Century-Fox, although Fox never became identified with a single designer as other studios were, nor did LeMaire create a definitive look for the studio's stars the way Adrian or Travis Banton did. Yvonne Wood and Bonnie Cashin were both in the 20th Century-Fox wardrobe department during the 1940s, along with Swiss couturier Rene Hubert. "He had the most gorgeous fabrics," contract player Vanessa Brown recalled. "I didn't ever approve or disapprove the sketches, but once in a while I had a say about a color I didn't particularly like."

Whereas head designers at most studios made annual trips to Europe, except during the war years, to survey the latest fashion trends, Darryl Zanuck was never as interested in beautiful clothes as he was in story value. Many of 20th Century-Fox's costumes even on A films, particularly during the early years, came from stock wardrobe, and actors in series like *Charlie Chan* and *Mr. Moto* wore the same attire in picture after picture. Nonetheless, actress Lynn Bari made a habit of stopping by the Fox wardrobe department to find out what pictures were coming up, sneaking a look at parts she might like to play, since scripts were sent to wardrobe before most casting decisions were made.

From 1936 until 1951 Edward Stevenson was the head designer at RKO, where recurring financial problems necessitated realistic rather than opulent wardrobes, except for the Fred Astaire–Ginger Rogers musicals and an occasional big production. The same was true of Universal, except that Deanna Durbin's clothes were lavish. Later, toward the end of the Golden Era, Universal producer Ross Hunter insisted on designer clothes for his pictures, real jewels for his stars, flowers in their dressing rooms, all the accoutrements of the big studios in their halcyon days. "Ross Hunter was doing his best to bring back glamour," Vanessa Brown said, "and he treated us elegantly."

Occasionally a stage designer would be brought in for the costumes on a Hollywood film, as Cecil Beaton was for *Gigi* (1958) and *My Fair Lady*

(1964). Samuel Goldwyn even lured Coco Chanel from France for three pictures. But for the most part the old studios relied on their own staff. "There didn't appear to be many design jobs in town," Oleg Cassini later complained. "There were only four major studios: Paramount, Twentieth Century-Fox, MGM, and Warner Bros. The others—Columbia, RKO, Roach, and so on—were too frugal to be very clothes-conscious. Costumes were an afterthought in most movies, certainly in all B pictures." (Cassini, *My Own Fashion*)

Poverty Row studios relied heavily on Western Costume Company, since 1912 the largest stock costume house in Hollywood, which initially serviced westerns but eventually moved into other areas. Republic's B pictures consistently drew from Western Costume; the studio seldom created anything original except costumes worn by Vera Hruba Ralston, Republic's pampered star. "I did nine westerns in eleven weeks," declared contract actress Peggy Stewart. "I fit wardrobe on my lunch hour." Stewart could recall only one dress Republic had specially made for her. "Everything else was rented from Western Costume," she said. "I remember I wore one of Vivien Leigh's dresses, which I loved. But Adele Palmer in wardrobe did what she could on a limited budget."

Hairstyling became another vital part of character development, and like the designers, hairdressers began their work by reading the script to determine the kind of character called for in particular scenes. The stylist then talked to the star, checking later with the director, producer, and cameraman to make sure they agreed that what was being planned was acceptable. "We used to spend days screen testing hairstyles," said Nellie Manley, who worked at Paramount for thirty-one years. "When Olivia de Havilland did *The Heiress*, we spent a week testing her hair! It had to be a very severe hairdo to help her look like an old maid. I used braids and still kept a smooth line around her face." (Manley int., Steen)

Like designers, hairstylists had to research period pictures. On other occasions the stylist was restricted by a look unique to a particular star, as in the case of Veronica Lake. Paramount hairdresser LaVaughn Speer enhanced the Lake hairdo. "Veronica had beautiful silky blonde hair," Nellie Manley remembered, "which had to be rolled just so to get that wave to hit at a certain angle over her right eye. Many times I watched LaVaughn slave over getting that wave perfect for the camera." Paramount had one big room for hairdressing, with three double-mirrored tables. On filming days, actresses reported first to their hairdressers when they arrived at the studio in the morning. "It would be nothing to have all the lady stars sitting at these mirrors at the same time," Manley recalled. "We would wash their hair, then iron it dry. In later years we used pincurls. Still later rollers came in." (Manley int., Steen) If an actress had to be on the set by eight o'clock, a hairdresser would usually start on her by six-thirty. If she was shooting outside the studio and had to leave at seven, the actress had to report to hairdressing at five-thirty.

Nellie Manley was one of thirteen women who began the hairdressers' guild, which required applicants to have at least two years' experience in a beauty shop before they were eligible for membership. Then they had to pass an examination. Gladys Witten, who worked at 20th Century-Fox, had owned a beauty shop in Kentucky with her sister before coming to Los Angeles. Witten recalled her first Hollywood interview at RKO, where she was told the field was overcrowded and to go home to Kentucky. But she stayed, taking a job selling cosmetics at Max Factor's. Eventually she was hired by 20th Century-Fox, where she worked under Beth Langston, head of Fox's hairdressing department. Witten was assigned to Linda Darnell and Vivian Blaine and later did Marilyn Monroe's hair. She curled Richard Burton's hair every morning for *My Cousin Rachel* (1952), without receiving screen credit, since that was the department head's prerogative.

Stars were expected to look their best at all times, although there were fleeting lapses into realism, sometimes covertly achieved. During preparation for *The Gunfighter* (1950), director Henry King, cameraman Arthur Miller, and actor Gregory Peck got together and pored over books of photographs of the old West. They saw that in the real West men didn't look like movie cowboys or movie gunmen and that their clothes were an odd mixture, some of which had been brought from the East. They decided Peck should wear a handlebar mustache in the picture. "Then we realized that the barbering was very primitive out West," the actor said, "and so we gave my character a bad haircut. It really did look like a cereal bowl had been put over my head. It was high up, and the sideburns instead of being long and sort of seductive and handsome-looking were very high above my ears." Since Darryl Zanuck and the company's president, Spyros Skouras, were both abroad at the time, production on the film began with no interference. Two weeks into shooting, Skouras returned, spent a day watching rushes, and immediately started bellowing over the telephone to Henry King. "Skouras hates the mustache, he hates your wardrobe, and he despises the haircut," the director told Peck. "He says that you're a sex symbol and a valuable property, and we're destroying you with this ugly haircut and mustache. He wants to know how much it will cost to reshoot the two weeks we've done." The director and actor met with the unit manager, who estimated it would cost between $175,000 and $200,000 to reshoot what they had filmed. "Could you just double that and back it up?" King asked him. Since the unit manager was an old friend of the director's, he agreed to report the inflated amount. "He phoned the figure of $400,000 over to Skouras's secretary," Peck said with amusement, "and there was more bellowing on the phone! But Skouras finally decided he couldn't afford to redo those scenes, so we went ahead with the handlebar mustache and the quaint clothes and the cheap haircut. It was authentic, and I think that was one of the things that made *The Gunfighter* successful."

At MGM, even more than at other studios, actresses looked as if they had just come from the beauty parlor at all times. "You practically had to go to the front office if you wanted something as real as having your hair mussed," Mary Astor wrote. If a player had a scene in which she'd been out in the wind or just gotten out of bed, a hairstylist might loosen a strand or two, but otherwise the actress looked perfectly groomed. In *The Clock*, for example, Judy Garland and Robert Walker spent twenty-four hours rushing about New York City during World War II, yet Garland always looked as if she had just come out of the makeup room. At Metro, Astor insisted, "all automobiles were shiny. A picture never hung crooked, a door never squeaked, stocking seams were always straight, and no actress ever had a shiny nose." (Astor, *Life on Film*) If an actor or actress was splattered with mud in a scene, somehow the mud mysteriously vanished by the next.

Actresses with a morning call at the studio customarily were awakened around four-thirty to be in the hairdressing room by six-thirty. At MGM, Elizabeth Taylor, June Allyson, Esther Williams, Jane Powell, Greer Garson, Debbie Reynolds, and Janet Leigh might all come in to have their hair done. "To go into that hairdressing room early in the morning was an experience," said MGM contract player Jean Porter. "You'd see Lana Turner come in with those beautiful blue eyes. She looked absolutely gorgeous, and she had been up half the night. She'd come in telling fantastic stories about her dates." That's where all the studio gossip was bandied about, as actresses discussed what they had done the night before, what was going on on other sets, what parts were coming up, and the scripts currently being circulated. The relationship between a player and her hairstylist generally became intimate. "When an actress comes in so early in the morning," explained Nellie Manley, "you are the first person she talks with that day. She may want to tell you her problems: She may have had a fight with her boyfriend, or her husband, or the kids might have been ill all night! They have to tell someone about it. You become a confidante." (Manley int., Steen)

Many performers learned their lines for the day under the dryer. Usually there were coffee and doughnuts for everyone, and some studios occasionally provided breakfast. Character actress Ann Doran recalled taking her coffee and doughnut into a little cubicle at Warner Bros. to have her hair done one morning and finding Barbara Stanwyck down on her hands and knees cleaning the floor. "They had messed up the floor while they were doing her hair," said Doran. "She was worried about the next person who was coming in and wanted things to be clean for them."

Actors normally arrived at the studio around seven or seven-thirty, reporting directly to the makeup room. Balding male stars, like Charles Boyer or Humphrey Bogart, had to be fitted each morning with a toupee. Supporting player Walter Abel remembered how he, Bing Crosby, and Fred Astaire all wore lace pieces with hair to cover their baldness in *Holiday Inn* (1942). "Crosby was completely bald," Abel said, "and Fred didn't have much hair, and I had a semblance, so we each had a hairdresser assigned to us."

Actresses went into the makeup department in the mornings before their hair was combed out. There, with their hair still in rollers, they chatted with players from other pictures. Big stars, at least at 20th Century-Fox, had their hair and makeup done in private dressing rooms in the stars' building, but for most performers mornings offered a chance to socialize. "It was very much a homey atmosphere," said actress Dina Merrill. "You got to know all the hairdressers and all the guys on the different shows. You'd see them every morning, day after day, and it was a very friendly, warm kind of atmosphere."

It was the makeup artist's job to see that a star's complexion appeared flawless. Arlene Dahl's beauty spot created much contention at Metro, since some producers argued it should be made her trademark, while others insisted it be airbrushed away. Eventually the beauty mark did became an asset. David O. Selznick was concerned about Rock Hudson's Adam's apple during the filming of *A Farewell to Arms* (1957), so that each morning it had to be shaded and made up. "That was the only makeup I wore in the picture," Hudson said. On the other hand, Audrey Totter in *High Wall* (1947), playing a lady psychiatrist, wore no makeup at all, since Metro felt professional women of that sort should look plain.

The head of makeup, like other department heads, usually sat behind a desk, supervising the people under him or her and performing administrative duties. Two stars might share the same makeup artist, while six character actors often stood in line waiting for one person. Actresses tended to look alike during Hollywood's Golden Era. "Our mouths were all made up the same," explained Jean Porter, "our eyebrows were very much the same, our makeup was an awful lot alike."

Female stars usually arrived in the makeup room a few minutes before seven, were handed a magazine, and were escorted to their assigned booths. The Metro makeup department, as Ann Rutherford remembered it, "smelled lovely. All of Max Factor's best scents were drifting out through the walls." (Rutherford int., Wagner) Two players might be paired in the same stall morning after morning. During the filming of an *Andy Hardy* picture, Jean Porter shared a makeup woman, Violet DeNoyer, with young Elizabeth Taylor, who was then working on *National Velvet* (1944). "I would just sit there and gaze at this gorgeous little girl," Porter said, "and listen to her talk about the horse she was riding in the film."

Makeup artists were known as a temperamental group; for example, Jack Dawn, who for years headed the makeup department at MGM, insisted that his staff have good quarters, adjustable barber's chairs, plenty of mirrors, and proper lighting. He demanded that his people be treated as artists rather than technicians, although his own method of discipline was frequently harsh. "Dawn was always firing people," William Tuttle recalled. "He'd fire you for two or three days and then bring you back." But he paid top money and attracted top people. "Jack Dawn always insisted that I never walk out on the lot with my smock on," said Tuttle, "and he always wanted me to put a coat on, wear a tie, and have my shoes shined. His view was that if you dressed like a bum, people would treat you that way. I worked for him for seventeen years, and his approach paid off. He himself was always very well dressed, almost like an executive, but he did a great deal to elevate the business in that respect."

Initially Tuttle worked for Dawn at the old Fox studio, at a time when makeup people were not highly regarded. "Many of the early makeup artists came from the ranks of actors," said Tuttle, "actors who were not too good at acting but were quite adept at changing their appearance, since they got more work that way. Eventually during the silent era, studios began hiring these people just to do makeup, and that was really the birth of the makeup profession." As an apprentice William Tuttle swept floors and scrubbed the baseboards, since Dawn was a fanatic about cleanliness. He did whatever typing there was to do, made out reports, ran errands, and answered the telephone. "I laid out all the makeup for him," Tuttle remembered, "all the colors, and that way I got familiar with what each person was wearing. There was no set form of training people for these jobs. You just sort of plunged in. There was no set period for the apprenticeship, just whenever they felt you were competent."

Dawn saw sketches Tuttle had drawn at Fox and decided Tuttle would work well with makeup, first letting him assist on tests. Then one day the makeup man assigned to *The Mask of the Vampire* (1934) failed to show up, and Tuttle was sent to the set to do what he could until someone else was hired. He ended up finishing the picture by himself. When Jack Dawn moved to Metro later in 1934, he took Tuttle along.

"Dawn was a very tough taskmaster," his protégé recalled. "He was almost Prussian in his manner and his training. You almost snapped to attention when he walked in the room. But he was a marvelous teacher. I was the first he started in the business, but he trained about twenty after me, all of whom turned out to be top people. He tried to get people with an artistic background." Sometimes politics was involved, as when an executive called and announced that he had a cousin who wanted to go into makeup.

Tuttle felt an aloofness among workers on the MGM lot that he hadn't encountered at Fox, an attitude he thought emanated from Louis B. Mayer. "Mayer managed to hire top people in each craft," Tuttle said, "and Metro took a lot of pride in their technical staff." As a beginner, Tuttle used to open the department at six o'clock in the morning and close it at seven or eight at night. He worked Saturdays and sometimes Sundays. His phone number was printed on the door of the department, so that if someone needed information at night, he could be reached, since Dawn didn't want to be bothered. "I was on duty twenty-four hours a day," Tuttle said; eventually he became department head himself.

The undisputed dynasty of Hollywood makeup had been founded during the silent era when George Westmore, a Jewish cockney, began the first makeup department in movie history at the Selig Studio in 1917. George's six sons expanded the Westmore empire, at one time or another heading the makeup departments at Paramount, Warner Bros., 20th Century-Fox, RKO, Universal, Selznick International, Eagle-Lion, First National, and a dozen other studios that flourished briefly during the Golden Era.

Monte, the oldest of the Westmore brothers, at age nineteen took a job as a busboy in the cafeteria of Famous Players-Lasky, the forerunner of Paramount, and within five years was doing the makeup for DeMille's *The King of Kings*. Perc Westmore formed the makeup department at First National, beginning with twelve character actors who happened to know basic principles of stage makeup. He remained in that position for twenty-seven years until the company was absorbed by Warner Bros. Wally Westmore headed the department at Paramount, remaining at that studio for forty-three years. Ern Westmore headed the department at RKO from 1929 to 1931, then went to 20th Century-Fox shortly after Zanuck and Fox merged companies. Bud Westmore, the handsomest of the brothers, was in charge of the makeup department at Universal for almost twenty-four years, while Frank, the youngest, started as an apprentice under his brother Wally at Paramount. In addition, the House of Westmore, the family business on Sunset Boulevard, remained for decades the most famous beauty salon in the country, specializing in wigs and cosmetics, with a lavish lobby and its own makeup plant.

"The Westmores' artistry in creating ingenious horror and aging makeups helped change the movies from a make-believe to a realistic medium," Frank Westmore said. "All the Westmores were a unique combination of sculptor, painter, researcher, anthropologist, and creative theoretician—sometimes even engineer and psychologist." (Westmore and Davidson, *Westmores of Hollywood*) Each was adept at applying liquid rubber for aging, but Wally was responsible for creating Fredric March's transformation in *Dr. Jekyll and Mr. Hyde* (1931). Later, when Wally was Paramount's department head, such opportunities became rare. "I saw Wally and Bud practically having to abandon doing creative makeup in order to administer their large departments," their brother Frank lamented. "Recruiting and training small armies of makeup artists, assigning the right artist to the right job, buying and dispensing supplies, attending daily production meetings and weekly budget meetings, reading scripts, and not only making suggestions for new makeup ideas for a film but—a Westmore specialty—dreaming up publicity gimmicks for it. Only occasionally were they able to get out in the field and use their skills on an especially important project." (Westmore and Davidson, *Westmores of Hollywood*)

Wally Westmore's department, located on the top floor of the building housing the stars' dressing rooms at Paramount, became a social center for the studio's players. Even members of the publicity office frequently dropped by, recognizing that the makeup department offered an endless flow of gossip and newsworthy tidbits.

At Universal Jane Wyatt was in for a shock in its makeup department. "I'll never forget the first time I got made up," Wyatt said. "I went in and Jack Pierce, the makeup man assigned to me, sat me down and immediately started pulling out all my eyebrows. I said, 'No, no, stop!' He said, 'Don't tell me what to do, little girl. I've made up the greatest.' I looked up on the wall and there was Boris Karloff as The Mummy, there was Bela Lugosi as Dracula!"

During his years at Warner Bros. Perc Westmore trained between sixty-five and seventy makeup people, selecting apprentices from among sketch artists, painters, and musicians, "anyone who had a touch of art in them." Eventually he did over sixty pictures with Bette Davis, winning Davis's respect and trust, in part because he explained what he was doing in detail. "Here was a gal," he said, "I don't care what the part, who would go along with the makeup I decided on. When she played Queen Elizabeth I in *Elizabeth and Essex* with Errol Flynn, I shaved her head halfway back!" (P. Westmore int., Steen) Perc had more prestige in the industry than any other Westmore, remaining at Warners until 1950.

Studio makeup artists felt that, in order to attain popularity, actors needed to be presented within the parameters of a certain image. "The mold might change from week to week," actor Robert Stack said, "but if a specific image seemed to be working at the box office, every actor was expected to fit. If his ears, nose, or other features failed to conform, the makeup boys took over." (Stack, *Straight Shooting*) As late as the 1940s a great deal of yellow makeup was still applied, much like what had been used on stages during the gaslight era. "When theaters had gas footlights, it was evidently thought that a yellow tint to the skin was needed," actress Joan Fontaine commented. "Even the Westmores hadn't got over that yet. They went into makeup before it had developed into a science."

Applying body makeup was a separate task, performed by a specialist who did nothing else, sponging players with pancake moistened by a clammy Sea Breeze lotion, covering their arms, hands, neck, and ears. Swedish actress Viveca Lindfors found the Hollywood system too specialized and individual workers too possessive of their particular jobs. "In Sweden I would have had one person helping me with my makeup, my hair, and my costume," Lindfors said, which meant fewer people on the set and more opportunity for concentration.

Once the makeup department had finished its task, actresses returned to the hairstylist, who combed out their hair or added a wig or any false hair that was needed. At Metro a row of labeled wig stands held the hairpieces to be used in each star's current production. There was usually a last-minute rush to the various sound stages; on the set a hairdresser made sure each actress's hair looked perfect, and a makeup artist added final touches. While the director lined up the first shot, a wardrobe woman might adjust a belt or the body makeup girl might dab an ear with her sponge. "It was all very plush," Rosemary DeCamp said. "Every detail was watched with jealous eyes by the department responsible," noted Mary Astor. "After every take a swarm of bees would surround me." (Astor, *Life on Film*) Hairdressers were on their feet most of the day, making certain the performer's hairdo matched in every shot. "There was hardly ever any place to sit down," said Nellie Manley. "You had to carry your equipment everywhere you went. And the hours were especially long. Your work had to please the actress, the director, the producer, and the cameraman!" (Manley int., Steen)

The big studios pampered their stars and spared few luxuries during the making of major films. But if money was seldom an object, the studios demanded professionalism in return. Craftsmen in the various fields either adjusted to the system or found themselves seeking other employment.

12

On the Set

Street sets at 20th Century-Fox around 1960. Courtesy of the
Academy of Motion Picture Arts and Sciences.

During the making of a picture most actors and actresses studied their lines the night before filming them. Sometimes a stand-in cued them on dialogue, although many seasoned performers preferred not to finalize their dialogue until they had run through it a time or two with the director. "I am one who learns on her feet," actress Lola Albright said, "so that I can coordinate the dialogue with the movement." Long monologues or difficult passages required advance preparation; in most cases actors had carefully analyzed the script for motivation and story value beforehand.

The assistant director on each production made sure that all the actors needed for the day's shooting had been notified and, in conjunction with the production manager, made certain the sets and locations were ready. The second assistant had earlier described to the casting office what types of extras the director wanted. During Gerd Oswald's tenure as a second assistant director at Paramount, he was assigned to several pictures featuring Alan Ladd. Since Ladd was short, Sue Carol Ladd, Alan's agent and wife, insisted that no extra taller than five-foot-two be hired. "When I was on a Ladd picture," Oswald said, "I had to screen the extras at seven in the morning to make sure there wasn't a tall one among them. If there was, he was out."

Most directors worked for many years with the same assistant, and many first assistants had a second assistant. It was the assistant director's job to break the script down into the number of days it would take to film it, mapping out each day's work; he also determined the number of actors required and the number of sets to be used, both interior and exterior. An assistant director served as a foreman, making certain the director had the facilities he needed and that actors were on the set when scheduled.

Often there was a third assistant, who worked primarily as a herder, supervising logistics. When Otto Lang served as third assistant to director William Wellman on *The Ox-Bow Incident* (1943), his main function was to see that the horses were properly placed and faced the right direction, making sure the number of animals and their colors matched the shots taken the day before. Each production also had a script supervisor, who checked everything the camera would photograph, to see that various takes matched previous shots. Obvious elements, such as whether an actor's tie was tied or not, were important, but so were smaller details like whether or not a button on a shirt pocket was buttoned. "Actions and gestures are most important to watch," script girl Catalina Lawrence explained. "If an actor puts his hand to his face on a certain line in a certain take, then he must always use that same gesture on the same line in all the other takes. The same goes for the use of props and any movement like the lighting and puffing of cigarettes." (Lawrence int., Steen) Since movies are customarily shot out of sequence, an exterior scene might involve an actor's walking into a house carrying a briefcase; a week later an interior sequence might be filmed that picks him up as he comes through the door. By writing notes in the script—some before filming, more as the scene was being shot—the script coordinator made sure the actor was carrying that same briefcase in the same hand, and that his clothes were the same.

Sometimes the old studios felt special advisors were necessary to insure accuracy. Throughout the making of *So Proudly We Hail* (1943), a World War II film depicting the recent fall of Bataan and Corregidor in the Philippines, a nurse and a doctor who had in actuality served there came on the set at Paramount as technical advisors. When 20th Century-Fox made *The Story of Alexander Graham Bell* (1939) with Don Ameche in the title role, Bell's daughter was present during shooting to verify historical details.

The producer and art director made countless decisions regarding sets before shooting began, consulting frequently with the director and production manager. If walls were not necessary, the producer might decide the scene would be shot in a bedroom alcove to be more cost-efficient. But bigger questions often arose that caused problems. Sometimes huge scenery had to be painted to hide modern construction or other distractions. Exteriors were frequently constructed on sound stages, which placed added demands on the cinematographer. Action sequences could be especially troublesome. Scriptwriter Charles Bennett remembered a climactic scene in C. B. DeMille's *Unconquered* (1947) in which Gary Cooper and Paulette Goddard escape from hostile Indians by going over a gigantic waterfall in a canoe, Cooper saving the day by grabbing a tree limb conveniently growing out from the bank. Days were spent deciding how to get the characters out of that situation; DeMille even had sketches drawn.

On mornings when they were shooting, directors usually arrived at the studio around eight o'clock. They'd stop at the makeup department to greet their leading lady or have coffee and a doughnut with their leading man before going down to the set. The camera crew also had been at the studio since eight, the director having told them the previous evening what his first setup of the day would be. From eight to nine the crew would prepare lighting for the scene, going through the action with stand-ins. Directors arrived to find the set surrounded by a maze of lights, reflectors, sound booms, and recording equipment, with wires and cables all over the floor. A camera crew usually included an operator, two assistant operators, perhaps twenty electricians, and six grips (who among other tasks actually moved the camera), besides the cinematographer himself.

The director would come on the set and begin discussing angles with his cameraman. Sometimes he had changed his mind on what he wanted in the initial setup. "I never let a cameraman set the camera or put on the lens," director Edward Dmytryk said. "Those were always my choices. I told them I wanted low-key lighting or I wanted high-key lighting, or I wanted it to look like Rembrandt or whatever." Other directors relied more heavily on their cameramen, making this the prime relationship in filming. "I found the one common denominator in the various studios were the crews," observed director Robert Wise. "They were almost all of equal quality."

The old studios hired excellent cinematographers, many of whom became prima donnas in their own right. Since lighting differed in each studio, definite styles existed, although the quality remained uniformly high. "The British film industry could never hold a candle to what was done here," English import Pamela Mason claimed. "And the speed! It would take sixteen weeks to do in England what it would take five weeks to do here." The great cinematographers were artists, painting with light, combining art with the science of motion pictures. Since movement was essential on the screen, the same scene would be shot from various angles and distances with different lenses, traditionally the long-shot, medium-shot, close-up technique.

Metro-Goldwyn-Mayer alone had thirteen cameramen under contract, each with his own approach, even though the studio was known for its soft, flattering close-ups and glossy look. Joseph Ruttenberg, Metro's top cameraman, won four Oscars and ten nominations, specializing in low-key lighting. Ruttenberg preferred shadows to create atmosphere, and a soft focus for women. "On still portraits," he said, "you can retouch wrinkles or blemishes on the face, but in motion pictures you don't get that opportunity. So you do it with a diffusion disc. I used to use ladies' stockings, very sheer stockings, to get the diffusion." At 20th Century-Fox, where Darryl Zanuck preferred a newsreel quality in his pictures, actors used to claim that at MGM they put a Navajo blanket over the lens.

Joseph Ruttenberg became Greer Garson's favorite cameraman because he found the secret of photographing her. "Greer had one good side, and I recognized which it was," Ruttenberg said. Metro even built sets to favor Garson's right side. Cary Grant, on the other hand, wanted his face to look dark all the time. Walter Pidgeon worried about his double chin. "Clark Gable didn't care how he looked," Ruttenberg recalled; "most of the male stars didn't care."

Veteran stars like Rosalind Russell and Lucille Ball became excellent technicians, fully aware of camera angles and the intricacies of lighting. Ball was a serious professional on a set, far from the jokester she appeared on film.. "Lucille knew a camera up and down," said Ann Doran. "She learned hers the hard way, because she was under contract at a studio where there was no training. You got your training by doing."

Like most cinematographers, Ruttenberg welcomed challenges, and MGM gave him opportunities to experiment. "Everybody from the top to the bottom was cooperative," the cameraman said. "Whatever you asked for, you'd get with no argument, as long as you convinced them that you knew what you were talking about." Ruttenberg immersed himself in the script, discussing the story at length with the director. The cinematographer had to make sure the lighting suited the action of a scene, but the director usually determined camera movements. A vital concern was to establish and maintain a mood throughout the film. "This is artistic work," Joe Ruttenberg insisted. "This isn't making an automobile or something. You have to dream up an idea every day."

The cinematographer was the boss over most of the crew. As a rule he had his own operator and regular assistants, his own gaffer (or head electrician), and his chief grip (a movie set's handyman). Major camera crews during the big studio era filmed an average of four pictures a year.

James Wong Howe, under contract to Warner Bros. in the 1940s, demonstrated a penchant for hard work and dark, moody lighting. A Chinese immigrant, Howe became the first member of a minority to achieve major success in Hollywood. Eventually he filmed 125 pictures and won Academy Awards for *The Rose Tattoo* (1955) and *Hud* (1963). Like most of the old cameramen, he preferred black and white photography. Short in stature, Howe had a quick temper and could be tyrannical with his crew.

Howe felt that any camera movement should have a purpose, dependent always on the script. "Unfortunately a great percentage of directors do not understand the use of the various lenses and have to rely heavily on the cameraman," he said. "The cameraman must photograph the actors, not how they look personally, but how they are supposed to look for that particular role." (Howe int., Steen) Sources of light should be naturalistic, Howe believed, and light in a room at a certain time of day needed to come from the proper direction. "The director, of course, thinks of his action with his actor," he declared, "and we think mostly in terms of lighting." (Howe int., McBride I)

Howe felt honesty and simplicity were the best guides to proper cinematography. When he photographed *Body and Soul* (1947), a fight picture with John Garfield, several close shots of the boxers were needed. "It was difficult to follow the action with a big camera on a dolly," Howe said, "so I put on a pair of roller skates and used a hand-held camera while the man who was my grip pushed me around the ring. The resulting effect was very exciting. It brought the audience right into the ring." (Howe int., Steen)

Leon Shamroy, a cinematographer under contract to 20th Century-Fox, was an even more bombastic personality. Shamroy was cryptic and opinionated; he used foul language and earned his title, "Grumble Guts." His passion, aside from work, was horse racing. Every spare minute "Shamy" would be on the telephone to Hialeah or one of the other tracks checking on that day's winners. One of the great Hollywood cameramen, he developed a technique of using a minimum of lights on the set. "God was the greatest photographer," Shamroy liked to say. "He'd only gotten one light." (Shamroy int., Higham)

Perhaps the most creative cinematographer in old Hollywood was Gregg Toland, who filmed *The Grapes of Wrath* (1940) and *Citizen Kane* (1941) and worked with director William Wyler on *The Little Foxes* (1941) and *The Best Years of Our Lives* (1946). A sparrow-like man physically, quick in his movements, Toland perfected the deep focus technique. "He was able to keep foreground, middleground, and background all sharp by using a wide angle lens and using more light," William Wyler explained. "This influenced my direction to some extent. Through his technique I was able to keep groups of people in the picture together and let the audience see both action and reaction from all the members in the scene at the same time. That offers much more to look at and lets the audience do their own cutting. I never had as close a relationship with any cameraman as I did with Toland."

Directors often preferred to use the same cinematographer repeatedly. For years Frank Capra worked almost exclusively with Joseph Walker, who pioneered the use of a zoom lens but was also expert in photographing close-ups of stars. A close-up required a key light, a bright spotlight of controllable intensity generally placed high and slightly to one side of the camera to bring out the contours of the actor's face. "The placement of this light can be critical and determined only by trial and error," Walker said. George Kelley, an easy-going Irishman, served as Walker's long-time assistant, whose duty it was to accurately follow focus, which constantly had to change since actors moved about. Walker spent most of his career at Columbia, where hours were long and budgets tight. An old Columbia maxim was: "By noon on the first day of a picture, you're already a day behind schedule." (Walker and Walker, *Light on Her Face*)

Stars usually had their favorite cameramen and felt more comfortable working with a familiar crew. William Daniels was Greta Garbo's cameraman for many years; Lee Garmes, known for his atmospheric backlighting, made a specialty of photographing Marlene Dietrich. "I vary my work considerably according to the story," Daniels said. "Even my lighting of Garbo varied from picture to picture." (Daniels int., Higham) Michael Gordon, the director who worked with Bill Daniels on *Woman in Hiding* (1950), felt Daniels accomplished his goals brilliantly, making a significant contribution to that melodrama. Gordon told him he wanted a lonely feeling in the film and remembered tears coming to his eyes when he saw the dailies, because Daniels had achieved it so totally.

Arthur C. Miller, under contract to 20th Century-Fox during the 1940s, took a consistently realistic approach. "The basic principle I have had in making pictures," said Miller, "was to make them look like real life, and then emphasize the visuals slightly." Never a soft-focus man, he liked crisp, sharp, and solid images. "I had the same gaffer for eighteen years," Miller remarked, "and in the end we'd just have to look at each other and we knew what we were going to do. There was no need for words; we were like a lot of dummies all through shooting." (Miller int., Higham)

Color photography posed special problems, in the early days requiring massive cameras and far more intense light, which produced scorching heat on the not yet air-conditioned sound stages. Performers sweltered in the tremendous candlepower pouring down on them, which was also nearly blinding. At each break, makeup people would rush in with ice-cold chamois to mop perspiration from the faces of performers. During the making of *Woman's World* (1954), actor Fred MacMurray brought a thermometer on the set one day when the cast was shooting a dinner scene with twelve people gathered around a table. Where they were sitting registered 115 degrees.

Dr. Herbert Kalmus, the inventor of Technicolor, had incorporated his Technicolor company in 1915 and patented the process, creating a monopoly. Initially studios had to use Technicolor equipment and Technicolor crews on the cameras. Their contract stipulated that none but Technicolor technicians were allowed in the Technicolor lab. Kalmus set up restrictive rules, insisting that a color expert, usually his wife Natalie, be present on the set whenever Technicolor was used, creating endless conflicts between the color consultant and directors, art decorators, and costumers. *Becky Sharp* (1935) was the first Hollywood film to use the three-color process, in which three strips of film were fed through the camera magazine simultaneously.

"I've always maintained that color photography should not be naturalistic," cinematographer Stanley Cortez said. "Even though a thing might be technically wrong, to me that wrong thing can be *dramatically* right. . . . You must *distort* color, play around with it, make it work for *you*, intentionally throw it off balance. You can mirror emotions in color." (Cortez int., Higham) Gradually most Hollywood directors became color-conscious, finding that in many cases color could be used to enhance the drama. In *Duel in the Sun* (1946) a grandly effective sunset was created by painting on backings. "We shot that on the stage," director King Vidor explained, "so we could control the color. We didn't take a chance by going out on a desert someplace and getting a white sky. We used color to enhance the value of the film." On musicals color was particularly effective. Gene Kelly recalled that he and director Vincente Minnelli spent hours with the Technicolor people controlling the color in *The Pirate* (1948). "If you look at a good print of *The Pirate*, frame by frame," Kelly said, "you'll see just magnificent color. It's almost like a painting."

For dramatic stories, however, most of the old directors, like the cinematographers, preferred to work in black and white. "You'll probably say I'm old-fashioned, but black and white is real photography," director John Ford insisted. "You've got to know your job and be very careful to lay your shadows properly and get the perspective right. In color, there it is." (Ford int., Bogdanovich)

Cameramen in the big studio era formed a close-knit group, organized into the American Society of Cinematographers, which had been founded in 1919. They enjoyed the camaraderie of their own fraternity and seldom mixed socially with actors. Performers, however, quickly realized that the cinematographer on a picture could be helpful, and actresses usually made a point of getting acquainted with their cameraman early on. "The first day of a film, you always go up and talk to the cameraman," declared actress Barbara Rush. "The way you are photographed is almost as important as the way you act. I've seen girls who can't act at all come off well in films, simply because they were lit and photographed so beautifully. They were so breathtaking it didn't matter if they could act." A sympathetic cinematographer could also help an actor hit his marks, so that he was always in focus, or let him know when he should take a step back if someone else in the scene was throwing a shadow on him.

On congenial sets, actors got to know crew members, and some stars—

Barbara Stanwyck, for example—made a habit of learning everybody's name and something about them. "You'd be surprised what a difference it makes if you say to a crew member, 'Did your little boy go through that tonsil operation all right?'" actress Irene Dunne said. At tiny Republic sets were uncommonly friendly. Jack Swain, a fine cameraman, won the nickname "Dreamboat" because he was such fun. "Swain would start blushing right from the bottom of his neck and just turn red all the way up his wonderful Irish face," recalled Republic actress Peggy Stewart, who spent much of her free time with the crew. "Everybody used to do everything they could to make Dreamboat blush."

Work on the set in the morning began with blocking the action—that is, determining where the actors would stand and how they would move, since it was essential they hit a series of marks chalked on the floor in succession, so that the camera would remain in focus and the lighting would be right. The first rehearsals were as much for the benefit of the camera crew as for the actors. As a rule the director would start by walking his actors through the scene with the cinematographer, then rehearse what he had decided to shoot. Stage actors soon realized they had to adjust to the camera, which required far less projection either of voice or gesture than in the theater. "You simply have to make a mental adjustment to the fact that the microphone may be two feet over your nose, and the camera may be three or four feet from you and can see every thought that crosses your mind," Gregory Peck said. "When they first train the camera and the lights on you, it's a new experience. You feel like a bug under a microscope. It's a trade that has to be learned."

Some actors never mastered the technique. "I had an awful time on my first film," admitted stage actress Kim Hunter. "The technical differences required a lot of adjustment. The silence was the worst. In films the sound department required absolutely dead silence while the camera is rolling. At first that utter silence was so 'loud' it threw me completely. It took me a long time to learn how to deal with it, to be able to draw a tight circle of concentration around myself in order to function under such unnatural conditions." Others made the transition more easily. "It wasn't all that difficult for me," actor Don Ameche said, "I think probably because of the style I had created for myself in radio. I always tried to make dialogue as completely conversational as I could, so that audiences would think this was someone talking to them. Radio was heard in the home, and I carried that approach into motion pictures."

"You have to gauge a performance by where that ear and where that eye are, where the microphone and camera are," veteran film actress Ann

Doran said. "How much you can move your head, how much you can use your hands and body depend on where the camera is. The closer the camera gets, the smaller your performance becomes, the more inside you it gets. It's no longer what you are doing with your body that tells the story, it's what you are thinking. Every major story point in any picture comes out in a close-up. And that's when you don't do a thing, you just think it. You don't act." Actress Joan Fontaine made the point in almost religious terms. "I feel that the camera is so pure and so marvelous," she said, "it's really like playing to God, because that camera is the *truth*. You don't have to justify yourself or explain anything—it sees all, knows all. So the camera has been my closest friend."

Once the director was satisfied with a rehearsal, the actors went to their portable dressing rooms on the set to check their hair and makeup, while their stand-ins took over for the final lighting. The director might run through the scene again with his principals and then shoot it. First a master shot was made, which determined location. This shot included everybody in the scene, orienting audiences to what was taking place. Next might come a two-shot, then individual shots, and finally the close-ups. In film work performers do not need to memorize an entire script before shooting starts. The first rehearsal for movie actors is a dress rehearsal, ordinarily preceding the final performance by only a matter of minutes, never more than a few hours.

Most directors during Hollywood's Golden Age worked slowly, on major productions rarely filming more than two and a half to three pages of script a day. Director Lewis Allen said he usually shot only eight setups a day, but had worked out each in advance. He knew his sets, having watched them being built and decorated, so he was confident that the set dressing wouldn't interfere with staging the scene for the camera. "I would lie in bed at night," the director explained, "and say, 'Over by the fireplace I'll have Herbert Marshall sitting on the sofa and I'll make a couple of singles. Then I'll have Joe go to the front door, and I'll dolly back with him and have him cross the room.' So when I arrived on the set, I was ready for the day's work."

Good directors felt it was important to project a point of view on the set, giving the impression of authority. The atmosphere on a set tended to reflect the personality of the director, so that nervous types had a way of making everyone else nervous. Some directors were highly disciplined, tedious in their attention to details; others might project a less meticulously polished approach. "If you show that you are in a panic, then the panic spreads all over," Vincente Minnelli observed. "So you have to sound as though you're completely self-assured." (Minnelli int., Schickel)

After the initial setup of the day was shot, actors retired to their dressing rooms while the crew prepared for the next one. The long waits between scenes were boring, especially for performers from the stage unaccustomed to working in Hollywood. If actresses knew their lines for the scene coming up, they might pass the time doing needlepoint or knitting or gossiping with their hairdresser. Others took a nap or read. "If I did a movie for ten or fifteen weeks," comedian Red Buttons said, "I would have read ten or fifteen novels." It might take an hour or more to light the next setup, time enough for actors to lose their concentration. Tony Curtis played records that he felt suited his character to keep himself in the proper mood. "It was a very boring way to make a living," commented singer Johnnie Ray. "It was also a lazy way to make a living."

Astute directors realized the importance of keeping a set alive, not letting the energy level drop. During the silent picture days studios had musicians play on the set, not just during shooting but between takes as well. Director George Cukor used humor and almost constant chatter. "It's very valuable on a set to have someone whose energy sparks everyone else," Tony Randall declared. "It becomes a contagious thing, so that actors don't go dead." Celeste Holm felt telling jokes and stories between takes was part of her responsibility. "It's silly," she said, "but it's important just to keep the bubble, particularly when you're waiting around. One thing about picture making is that you feel you're wasting your life away. So much time on the set is spent just lying about, and that's boring and destructive."

"You must pace yourself," Gene Kelly said. "You have to know when to relax. Actors have to wind themselves up, but it's an emotional wind-up." A dancer's problems are far more complex. "A dancer," Kelly said, "is like a baseball pitcher—he gets all set and boom! He rehearses an hour for one take. Then they go to another take, and he rests for an hour. Now he's cold; his muscles are cold. He has to go into another whole baseball game. In midair he has to hit marks and be at the right place for moving cameras. Also emotionally he has to be up for doing his best. Actors don't have the least idea of a dancer's problems for the cinema."

While most directors pampered their stars, others could be tyrants. John Ford, who had a kind side, nevertheless always had someone he chewed out, someone who could do nothing right. Henry Hathaway and Otto Preminger did the same, causing more than one actress to leave the set in tears. On *Roseanna McCoy* (1949), Irving Reis needled Marshall Thompson. Thompson had been engaged for six months, and his wedding had been announced for at least four months. "I was still working the day I got married," Thompson remembered, "and Reis kept me on the set until the last shot of the day. He did that deliberately. I sat around all day, just sat there, while the

crew and everybody else got mad. Finally we did my scene late in the afternoon, and everyone said, 'Oh, that was great!' to get me out of there."

Some directors believed in a lot of rehearsal time, others nearly none.
Old-time directors like William Wellman encouraged spontaneity, fearing
that more than a minimum of rehearsal would result in performances too
theatrical for the camera. If performers stumbled on words or made mistakes that were natural, veteran directors often left them as a realistic touch.
Others took more time and were precise, carefully discussing and analyzing
characters and motivation before a scene went before the cameras. Occasionally a director would rehearse an entire script as if it were a stage play,
though studio executives tended to view such preparation as a waste of time.
But if the director was powerful enough, he got his way. Edmund Goulding,
who came from the London stage, discussed and rehearsed *The Constant
Nymph* (1943) all morning around a long table, occasionally altering a line
with the writer's permission. Then he spent the afternoon shooting what the
cast had prepared that morning. Actor Robert Montgomery reversed the
process when he directed *Lady in the Lake* (1947), rehearsing every afternoon after lunch what he intended to film the following morning. "I went
to bed confident and could get a good night's sleep," recalled Jayne Meadows, who acted in the picture, "knowing what I was going to do the next
morning. On no other movie I made did that happen."

"Directing motion pictures is unlike directing plays," commented Edward
Dmytryk, who started in the business as a film cutter. Sometimes Dmytryk
had a day or two of rehearsal, but he generally found that unproductive. He
felt directing a film was "like a chess game, except instead of thinking four or
five moves ahead, a director has to think sixty-four or a hundred and fifty
moves ahead." Dmytryk believed in being flexible until the last minute. "I
wanted never, never to be fixed," he said, "always trying to improve as I went
along. I wanted dialogue to overlap, I wanted people to jump on each other's
lines the way we do in real life. If you film dialogue the way it's written, it becomes mechanical."

Many old-time action directors turned most of the rehearsal over to a dialogue coach, who got the main characters together a week or so ahead of
shooting, going over lines on the set with actors, encouraging them to listen
and react to one another. During World War II, when the government put
a restriction on the amount of film allotted, rehearsals became essential.
"Until then directors could use as much film as they wanted and selected
whatever print they wanted," Paramount drama coach Phyllis Loughton remembered. "The war eliminated that. Directors were allowed only one
print."

Every director had his own approach. Some were ushers, telling actors where to move and little else. Others were shouters, yelling at cast and crew alike; still others commanded respect by never raising their voices. Perceptive directors understood that while actors responded in different ways, all needed to be reassured. Some performers wanted instruction on every move—how to sit, how to stand, how to deliver a line. Others preferred to find characters themselves, working out details that made sense to them. Frequently directors offered suggestions sparingly, more concerned with the camera and staying on schedule. Actor Alfred Drake insisted the only direction he got from Al Green during the filming of *Tars and Spars* (1946) was, "Smile. You look better when you smile."

Great directors added creative touches that turned good scenes into remarkable ones. The dialogue between Margo Channing and Bill Sampson preceding the famous "Fasten your seatbelt" party in *All About Eve* (1950) consisted mainly of exposition, so inert that Bette Davis, who played Margo, complained to director Joe Mankiewicz. "Do you see that candy jar on the piano?" Mankiewicz asked the actress. "The madder you get the more you want a piece of candy." Davis understood and used the candy as a prop, lifting the lid of the jar and slamming it down to build an effect, finally snatching a piece of candy and tossing it angrily into her mouth. "It was a genius piece of business," Davis said. (Davis int., McBride II)

During the making of the original screen version of *The Front Page* (1931), director Lewis Milestone had a typewriter bell ring on Walter Burns's famous line "The son-of-a-bitch stole my watch," just in time to obliterate the objectionable word and allow the phrase to pass the censorship office. "We did thirty-six takes," remembered actor Pat O'Brien, "because it had to be perfectly synchronized." Studio executives considered Milestone a genius for finding a way around a difficult situation. "It was the simplest damn thing in the world," the veteran director insisted shortly before his death. "You just had to have a little imagination. The answer came to me probably a half minute before we shot the scene."

Like any artist, film directors frequently drew from their own experiences. Some scenes from *The Best Years of Our Lives* (1946) were adapted from William Wyler's own life, since he was himself a veteran recently home from the war, suffering a partial loss of hearing. "He put a great deal of himself into that picture," Wyler's widow recalled. The scene where Fredric March comes home and meets Myrna Loy, playing his wife, the two walking down a hall toward one another, Wyler took from his reunion with his own wife upon returning from Europe. "I was standing in the door of a room at the Plaza Hotel, at the end of a long hall," Mrs. Wyler explained, "and he came down the hall toward me. That's how the scene in the picture came about."

Film directors also served as an audience on the set, encouraging actors and giving them confidence. It was important that a director be discriminating yet establish that he could be trusted. Frank Capra consistently proved an appreciative audience for performers, laughing out loud between lines, knowing that his laughter could be eliminated from the sound track in the editing room. "He loved everything," singer Anna Maria Alberghetti remembered. "He would enjoy his own scenes as they were being shot." His obvious pleasure endeared him to actors, who gave their best in return. Michael Curtiz cried so audibly during Walter Huston's death scene in *Yankee Doodle Dandy* (1942) that he broke up his own take; Delmer Daves kept a box of Kleenex beside the camera when directing a sad scene. "He would cry no matter how many takes it took," character actress Rosemary DeCamp recalled, "and reach for the Kleenex, which was very inspiring for an actor. I mean we really put out."

Nothing pleased an actor more than to hear the crew applaud at the close of a scene. "That happened to me only three or four times," Catherine McLeod commented, "but it was really rewarding. Usually the crew were yawning and scratching and waiting for the coffee break, so when they did applaud, it was very flattering."

Often the director's major function was to hold an actor down, not letting him overdo for the camera, which picked up the slightest nuance. "Filming a little bit each day," William Wyler remarked, "there is a temptation for an actor to give his best each day, and his best sometimes is too much. I'd just say, 'Easy, easy. Take it easy.'" This was particularly true of stage actors used to projecting to the top row of the gallery. Wyler's biggest problem with Laurence Olivier during the making of *Wuthering Heights* (1939) was to help the actor reduce the size of his performance. "I got him to play the part not so theatrically," the director said, "but more inside. Somehow if you feel it inside, it comes out on the screen."

Some directors were better with action scenes or handling crowds, while others spent most of their time adjusting the camera. Sometimes a director's attention was usurped by meddlesome producers or last-minute rewrites. On *The Paradine Case* (1948) independent producer David Selznick rewrote the dialogue every night, giving director Alfred Hitchcock new pages before work started on the set the next morning. "We'd be handed blue pages and perhaps go off in our trailers and memorize ten or fifteen pages of dialogue before we could shoot," Gregory Peck recalled. "And the dialogue was invariably worse, not better."

If the director and his leading lady were in the midst of a love affair, friction could erupt on the set, with other actors claiming she received special attention and most of the close-ups. If a star was in love with the director and he appeared interested in someone else, the situation could become even worse. Occasionally an actor and a director didn't get along or had drastically different approaches. On the set of *The Return of Frank James* (1940), it became evident that Henry Fonda and director Fritz Lang would clash. "Fonda had a reputation of never raising his voice, never getting mad," child actor Jackie Cooper observed. "I saw Henry Fonda with veins about to burst out of the side of his head, screaming at Fritz Lang. I mean yelling and pounding and slamming and kicking, because Lang had no respect for actors at all. He wanted everybody to be a puppet. Fritz didn't give a damn about any of us. He just wanted people of a certain size and shape and wanted to move them around the way he demanded."

Professional rivalry sometimes proved disruptive, with cliques developing and actors counting how many times the director had invited others to lunch. "There is a tendency when a director is in charge of a boatload of stars," Celeste Holm said, "for each star to try to win the director's chief eye." Some directors surrounded themselves with a stock company—character actors and stunt men, as well as crew. Often these were acquaintances from the silent days, since strong loyalties existed within the old studio system.

Most film directors enjoyed molding young performers. Marshall Thompson had only one line in *They Were Expendable* (1945), but John Ford coached him for the six weeks required to make the picture. "Thompson!" the director would call, as they sat around between takes. The novice would hear his name and go rushing over. "Give me your line!" Ford would order. If they got into an elevator together, Ford would demand: "What's your line, Thompson?" They finally filmed the line on the very last day of shooting. "I didn't know what he was doing," the actor said later. "What he wanted was absolute rote, that was the whole idea, since I played a young ensign. That was Ford's method of working. He'd trick you into performing."

Since it was the quality of the work that counted in the end, successful directors were often stern taskmasters. During the making of two pictures, Gregory Peck and director Raoul Walsh became great companions. Walsh kept his set laughing, but he could also be caustic, tough, and demanding. "I learned a lot from him," said Peck. "I learned how to drink bourbon and eat steak for lunch and then go back and work for the rest of the day. If you ever let up on your energy or got sleepy or forgot your lines, he could be very sarcastic, because you were supposed to be able to have a couple of bourbons and eat a big steak and fried potatoes for lunch and then come back and work like hell until six o'clock. Walsh was that kind of a fellow. Yet underneath it all he was sensitive and loyal and kind and had an enormous appetite for literature and the arts."

"I don't think any of us thought of ourselves as artists," director Vincent Sherman said. "It was a business. Sure, there was definite artistic consideration in the preparation of the script. But we didn't go out and parade that and claim that we were artists. We were trying to make a successful picture." Nor were most filmmakers concerned during the shooting of a picture with reviews or what the critics would say. "We didn't know about four stars and *Time* magazine and whatever," director George Sidney claimed. "We weren't paying any attention to those things. We were just making movies."

In the early sound era much filming was done at night, since the studios weren't properly insulated and recording equipment picked up outside noises during the daytime. But the studios quickly improved their sound facilities, hiring the best technicians available. Douglas Shearer, brother of actress Norma Shearer, headed the sound department at Metro. Shearer was cranky but conscientious and pioneered many acoustical innovations. All the major studios, even Republic, soon had excellent sound, although there were always problems. During the making of *Moby Dick* (1930), when a rowboat got smashed by the whale, the Warners sound department couldn't figure out how to replicate the noise. Finally six peanut shells were crunched in front of a microphone; the effect was exactly right.

Actors on most sets developed an esprit, with close relationships frequently maturing into lasting friendships. Some performers were aloof, while others were companionable and sparkled with charm. At times there were rivalries, vying for close-ups, and contempt for those who appeared more photogenic than talented. But despite the friction and hard work, a lighthearted atmosphere existed on all but the strictest sets. "My darling, precious friend Gordon MacRae," sighed Doris Day, remembering her co-star from many features. "If the movie ran for three months, we laughed for three months straight. Between Gordon and director David Butler and Gene Nelson, we just cracked up constantly." When Doris and Rock Hudson made *Pillow Talk* (1959) and *Lover Come Back* (1962) together, they couldn't look at one another without laughing. "I'd look at her forehead or her nose," said Hudson. "We did terrible things to each other; with our backs to the camera we'd make faces at one another. It was perhaps acting rather juvenile, but the twinkle in our eyes, I think, is what helped make those films successful."

Bob Hope, Red Skelton, Danny Kaye, and Jerry Lewis were all notorious clowns on the set, ad-libbing during takes, although those seldom reached the screen. On her second day of shooting *A Southern Yankee* (1948) with Red Skelton, Arlene Dahl went into her dressing room to find a stink-bomb inside. All of her clothes had to be fumigated. "I don't want to talk to you about the script," Skelton said to her, "but I think you get the idea after having been in your dressing room!" Jerry Lewis constantly joked when he worked with Dean Martin, the two of them pulling all sorts of pranks, even wrecking things on the set. "They used to drive the prop man crazy," producer Hal Wallis said. "They'd shoot out the exit lights and get into fights with water guns."

Summer Stock (1950), Judy Garland's last picture for Metro, was miserable for everyone concerned, largely because of Garland's absenteeism and acute emotional problems. "When you start a picture and the bets around the lot are two-to-one that the picture will never be finished, it's not very comfortable," director Charles Walters reflected. "I would go to dailies and think, 'How dare this look like a happy picture!'" Yet Eddie Bracken, who was in the picture, remembered mainly how thrilling it was when Garland was on the set and working. Bracken was adept at picking pockets, removing watches from people without their knowing it. He announced to the cast of *Summer Stock* one day that he could take a bra off a woman and she wouldn't even be aware of it. Everybody laughed, and several voices claimed he couldn't do it. "In front of four hundred people I asked Judy Garland if she'd dance with me," Bracken recalled. "I showed her a few steps I was working on, and in front of four hundred people I took off Judy's bra, supposedly without her knowing it. Of course, Judy and I had it all planned. She put on a phony bra, and I pulled it out and held it in back of me. Judy said, 'Oh, my goodness!' and tried to cover up and ran off the set. The four hundred people thought I was a living genius. But that's the kind of person Judy Garland was. She enjoyed tricks and gags."

On the first day of filming *A Night at the Opera* (1935), the Marx Brothers sent singer Kitty Carlisle three dozen roses COD from her co-star Allan Jones, whom she'd known for years in New York. Bud Abbott and Lou Costello, on the other hand, didn't like each other and fought constantly. Between takes they spent their time gambling, playing poker for thousands of dollars. "I enjoyed watching them gamble," remembered actress Jean Porter, who made *Abbott and Costello in Hollywood* (1945) with the comedy team. "I used to watch money cross that table like I'd never seen before. I'd never experienced anything like it. They'd win and lose without even changing expression on their faces."

During filming of *That's My Man* (1947), a group of Catholic nuns from the convent where Catherine McLeod had gone to school came on the set one day to visit the actress. McLeod's co-star, Don Ameche, and director Frank Borzage found out about the visit and arranged to shoot only love scenes during the time the sisters were present. "If you didn't have some fun on a picture," actor Cornel Wilde said, "it gets so tiring, especially in those days when we had longer hours and that enormous heat!"

Making *The Wizard of Oz* (1939) was an ordeal, since Jack Haley was encased in tin, Bert Lahr wore a lion's costume weighing ninety pounds, and Ray Bolger had carefully numbered pieces of straw sticking out of his arms. "The three of us had to have fun or go crazy," said Bolger. "We did everything we could to keep ourselves from going berserk. We clowned and drove the poor director nuts. We gave him double talk and the fade off. He didn't know what we were talking about."

Big stars, particularly actresses, were accustomed to special treatment; a few eventually demanded it. Marion Davies, mistress of newspaper tycoon William Randolph Hearst, was accompanied onto the set of *Cain and Mabel* (1936) at Warner Bros. by a four-piece ensemble playing a triumphal march. Greta Garbo preferred privacy and remained in her dressing room at Metro until she was paged, with screens around the door to give her extra protection. Lana Turner arrived on the set like a queen with an entourage including her hairdresser, makeup man, body makeup girl, wardrobe lady, and someone who played victrola records for mood music. Turner disliked strangers standing around the set watching while she worked; basically she was shy. Shyness also caused Rita Hayworth to spend most of her time in her dressing room with the door shut. No one loved a practical joke more than Betty Grable, who had a raucous sense of humor, and Joan Fontaine was capable of tossing off a bawdy riddle—shocking some, delighting others. While Bette Davis could be demanding, she functioned as a member of the team and even adopted a maternal attitude toward young performers unable to protect themselves. Frank Sinatra was erratic—sometimes friendly, sometimes not. Sinatra hated to do more than a few takes and insisted on working only in the afternoons, rarely before one o'clock. Yet he arrived on the set fully prepared. "Frank would come in and he'd be wound up ready to go," recalled George Sidney, who directed the singer in *Pal Joey* (1957), "and we'd knock off two or three days at a time. We whizzed through that picture a number of days under schedule."

With all the waiting, actors spent long periods on a set fraternizing. Robert Cummings introduced Arlene Dahl to health foods during the making of *The Black Book* (1949). David Niven told funny stories during the filming of his pictures. "I could hardly wait to get to the studio just to laugh with David," said Barbara Rush. During the shooting of *Duel in the Sun* (1947),

Gregory Peck and Jennifer Jones discovered they were both from small towns and had much in common. "We had both knocked around Broadway at the same time," Peck remembered, "and we compared notes on who was the hungriest. Jennifer was very intense about her work; I was intense about mine. We were both young careerists who wanted to do good work and get to the top." Doris Day had grown up in Cincinnati watching Ginger Rogers' pictures over and over. "Ginger was my favorite movie star," recalled Day. In *Storm Warning* (1951) she and Rogers played sisters. "We had such a wonderful time," said Day. "We talked a lot and had really good conversations. I'd like to have been her sister in real life."

Greer Garson usually arranged for tea in the afternoon on her sets; Joan Blondell frequently brought homemade cookies to work. "Joan's heart was like an artichoke," actress Coleen Gray declared, "a petal for everyone." Character actor Victor McLaglen owned a farm and often brought sausages to fellow workers. On most sets there was good-natured kidding, with nicknames assigned to various people. The all-star cast of *Executive Suite* (1954) liked to tease Louis Calhern, but turned serious the minute it was time for a take. "Then at the end they'd start ribbing again," director Robert Wise recalled. On *The Searchers* (1956) actor Ward Bond took most of the teasing, since he tended to be a braggart. In the middle of one scene the camera suddenly stopped. "Ward had arrived on the set, unplugged the camera to plug in his electric razor, and started shaving," explained Harry Carey, Jr.

Gradually actors who stayed in the business adjusted to the fact that movies were shot out of sequence, and they learned to make story points stand out for the audience as well as for other characters. Performers developed a sense of where to stand, how to watch the key light to make sure someone else wasn't casting a shadow over their face, and ways to handle colleagues determined to upstage them.

Rosalind Russell became a quick study, often learning lines in her dressing room. "I never saw Rosalind study," claimed her husband, producer Frederick Brisson. "She could take a script, read it three or four times, and learn everybody's part. On the set when she found they had done rewrites, she would go into her trailer, look at that piece of paper, maybe two or three pages, and play the scene as if she had learned it a week ago." Some performers—Jack Carson, for instance—had a great facility for remembering details, so that they matched in various takes. Most others were oblivious to such matters, unless the director called specific points to their attention.

Seasoned actors in most instances were helpful to newcomers, initiating them into the secrets of picturemaking. Gene Kelly felt lucky to have worked with Judy Garland on his first musical, *For Me and My Gal* (1942), and later

Jerry Lewis taught Dina Merrill how to make physical comedy work during *Don't Give Up the Ship* (1959). Character actress Agnes Moorehead coached young Laraine Day in *Keep Your Powder Dry* (1945), and western star Dale Evans claimed she learned more about movie acting from George "Gabby" Hayes than from anybody else. "It's all in the eyes," Hayes used to tell her. "If you don't believe it in the eyes, you don't get the point across."

When Marshall Thompson admitted to Mary Astor on his first picture, *Blonde Fever* (1944), that he was nervous, the actress whispered, "Give me your hand and we'll do the scene together." Astor put her hand in his, and her palm was wet with sweat. "We all get nervous," Thompson recognized later, "it never dies. But that encouraged me; from then on I understood a lot more." Jane Wyman, a recent Oscar winner, was patient with young Rock Hudson throughout the shooting of *Magnificent Obsession* (1954), even though her inexperienced co-star sometimes required thirty or forty takes. When Hudson thanked her for her kindness at the end of the picture, Wyman said simply, "It was handed to me by somebody, and I handed it to you. Now it's your turn to hand it to somebody else." (Hudson int.)

Not all performers were so accommodating. Some were temperamental, others lacked skill, a few were downright lazy. Joan Crawford insisted her sets be kept at a certain temperature, colder than most of her colleagues liked. John Barrymore late in his career had trouble retaining lines, while other stars suffered from severe emotional problems, sometimes manifested in alcoholism or drug addiction, creating hardships for everyone. The cast of *Sabrina* (1954) divided into two camps, the Humphrey Bogart faction mixing as little as possible with the William Holden group. On *Oklahoma!* (1955), Gloria Grahame proved miscast and troublesome. "I'd never worked on a picture before where one person alienated everybody in the entire crew and cast," Grahame's co-star Gene Nelson declared.

Occasionally actors recognized from the beginning that they were working with an impossible script. Such experiences could be torture. If a director hadn't wanted an actor but had had him assigned to his picture anyway, the atmosphere on a set could grow tense. "One of the challenges of film-making," said actor Lloyd Nolan, "was just to keep your sense of humor, because you could really bump into some foul balls. Making a film is a long, tough, tedious business." Sometimes pictures were successful despite internal bickering, but such experiences might be miserable for those involved. "You got to the point where you could open a stage door and know whether it was a happy company or a company in trouble," said Metro publicist Esmé Chandlee. "The vibes hit you right away."

Many actors liked to get a shot within the first three or four takes, convinced they became mechanical or stale after that. Some even thought they were best on the first take and preferred to keep rehearsals to a minimum. Judy Garland, an accomplished performer and a quick study, did the "On the Atchison, Topeka, and the Santa Fe" number for *The Harvey Girls* (1946) in one shot. The number had been planned weeks in advance; the performers walked through it once after lunch, with Garland seeming not to listen. But director George Sidney got what he wanted on the very first take. "Judy was amazing," said Sidney. "She never forgot anything." On the other hand, the waltz scene in *The Merry Widow* (1952), choreographed by Jack Cole, involved 112 pairs and took almost a full week to shoot. "We rehearsed that number forever," Fernando Lamas remembered. "Lana Turner was a very good dancer, and that waltz became spectacular. But it was complicated. Studios could afford to do that in those days."

Movie work was not only exhausting, it could be dangerous, especially before the formation of the Screen Actors Guild. Real bullets were sometimes used in combat sequences, and a punch in the face was a hazard of the trade. Lucille Ball claimed she was "bitten in the can" by a crocodile during the making of *Roman Scandals* (1933) for Sam Goldwyn. "Luckily," she added, "the crocodile was about a hundred years old and had no teeth." Character actor Lloyd Nolan almost lost an eye working on *Lady in the Lake*. In that picture he was to be shot in a window well by a man coming down a fire escape. "When he fired his gun," Nolan recalled, "something had to break the glass. So they had a sharpshooter, out of sight of the camera, with a pellet gun. He would shoot where the guy aimed his pistol to break the glass. The damn glass came right out at a ninety degree angle, straight at me."

Other situations were more frightening than threatening. Actress Virginia Mayo stood in awe of James Cagney when they were paired for *White Heat* (1949). "Jimmy was such a great talent that everybody deferred to him as the star," Mayo said. In one of their scenes together, where Cagney had to treat her roughly, the actress grew truly frightened. "I swear I didn't have to do much except remember my lines and be scared," she recalled. "He really did frighten me in that scene." During filming of *The Mummy's Hand* (1940), Peggy Moran, dressed in a satin nightgown, was carried around in a dark cave by the monster, an actor she never met without makeup. The man playing the mummy had to report to Universal Studios at four in the morning so he could be wrapped, and by the time Moran arrived at six, he couldn't talk. "He looked awful!" the actress declared. "And I had to be carried around in the dark by this guy whose face I'd never seen, nor had I ever heard his voice. I was literally scared to death of him. So I wasn't acting in those scenes. It was real."

Coleen Gray remembered watching actor Richard Widmark throw Mildred Dunnock down the stairs in *Kiss of Death* (1947), but rather than being frightened, she was impressed. "He was a New York actor," said Gray. "Victor Mature and I were from the west coast, so I felt inferior to Widmark, because I hadn't had any stage experience." Theater actors frequently complained that movie work lacked honesty, since everything could be redone, and sound, including dialogue, could be altered in the editing room. "I found there was a lack of spontaneity," dancer Ray Bolger said, "because you were restricted. You had to work within a certain area. I was never happy with my motion picture career." Gregory Peck, on the other hand, adapted easily. "I liked screen work from the beginning," said Peck. "Gradually I stopped returning to the stage, and I think it was just because I loved making films. It seemed to be what I was born to do. I found the whole intricacy of picture-making intriguing, along with the teamwork required—the number of contributions that all have to be first-rate in order to come out with a good finished film that's complete and hits its target. The human element is very intricate. The photographer, the sound men, the prop man, the costume designer, the screenwriters, the director, and the producer all have egos, and they all have to blend somehow and make a whole."

Occasionally an actor might be working on more than one picture at the same time, or filming one and rushing off for costume fittings or to make tests for another between scenes. Sometimes special skills had to be acquired. Bonita Granville took ice skating lessons for *The People vs. Dr. Kildare* (1941). Robert Alda, Cornel Wilde, and Catherine McLeod had to learn to play the piano convincingly for *Rhapsody in Blue, A Song to Remember* (both 1945), and *I've Always Loved You* (1946) respectively. If Alda, playing George Gershwin, hit an F instead of a G in close-ups, an observer from the Warner Bros. music department pointed out the mistake and shooting stopped, even though he was playing a dead keyboard to a prerecorded sound track. McLeod was coached by pianist Arthur Rubenstein, who provided the music for her picture while she played a piano with no strings. "I learned sixty-four pages of the Rachmaninoff concerto," McLeod said. "I'd go home every night and practice. I had taken piano for eight years, so I knew what I was doing—sort of." In *A Song to Remember*, Cornel Wilde played Chopin; the final segment found the composer on a concert tour to raise money for his downtrodden countrymen. The sequence took eight days to film and focused mainly on Wilde at the piano, requiring eight to ten hours a day. "By the fourth day," the actor said, "there was blood on the keys, because my fingers were bleeding."

Betty Hutton spent six months on a sound stage at Paramount learning to fly like a trapeze artist for *The Greatest Show on Earth* (1952), starting on an eight-foot ladder and working her way up. "Betty became a flier," associate producer Henry Wilcoxon said. "We didn't need any doubles; she did it herself." Douglas Fairbanks, Jr., rehearsed for weeks on his stunts for *Sinbad the Sailor* (1947), coming to gymnastics less readily than his athletic father had. "It was sort of like an amusing ballet," the actor explained, "and I didn't want to have doubles. My coach, who was about my age, knew what I could do and what I couldn't. We'd work it out, so that I could perform the stunts with a semblance of naturalness and yet be relatively graceful." Westerns required riding and shooting and all kinds of physical skills. "I learned to run at a horse and vault over his rump into the saddle," Gregory Peck recalled with amusement. "I learned to step off a horse that was running and how to take a fall on a horse without getting caught underneath him. I practiced roping buckets and trash cans for hours and hours and hours and finally learned to rope steers on horseback. It was all kid stuff really."

Special treatment was given child actors, partly because the law demanded it. Unwelcome on a set were stage mothers, particularly meddlesome types who sat behind the camera and attempted to interfere with a director's instructions. Margaret O'Brien's mother was usually with her daughter during the 1940s, as was Elizabeth Taylor's, and Taylor frequently brought along her pet chipmunk, Nibbles, as well. "Elizabeth always had a pet," publicist Emily Torchia recalled, "but Nibbles went with her everywhere."

By the time child star Jane Withers made *The Farmer Takes a Wife* (1935), she knew her way around a movie set, even though she was only nine years old at the time. That picture was Henry Fonda's first film, and he was terrified. "I felt the vibrations," Withers remembered later, "so I took him aside from everybody else." The child explained how everyone there wanted to help him and assured him he had nothing to worry about. Finally she took his hand and said, "Lord, this is a new young man to the film industry. Please guide him and let him know how special it is. He's going to make a lot of folks happy. Please, God, take real good care of him." Fonda smiled and gave the girl a bear hug. "That was the beginning of a very special friendship," Withers said.

During the Great Depression extras on pictures earned three dollars a day and were given a box lunch, more money than they could make waiting tables or working in a store. People would stand outside the studio gates near the casting offices, hoping an assistant director would choose them for a day's work. On giant spectacles there were times when five hundred to a thousand extras might be needed for a single scene, all herded onto the set like cattle. Assistant directors were constantly aware of time, conscious that if crowds were involved, overtime became all the more costly. Some extras were shrewd enough to realize that if they could maneuver themselves into getting their faces in front of the camera, they might be brought back for another day's pay. For most extras, though, it was exciting just to get inside the studio and see the stars. As a young man, character actor Gale Gordon worked as a dress extra on an early Greta Garbo picture. Years later he remembered the excitement of being on a Garbo set. "She was the most beautiful thing I ever saw in my life," said Gordon. "She wore these beautiful gowns and was absolutely ethereal. Just to watch this magnificent, beautiful creature floating around was a great thrill for me. We'd all look at the chairs with the names on them and think what it'd be like to sit in one with a name on it. You wouldn't dare go near one or touch it; you'd be thrown out. There was a great class distinction in those days."

Birthdays of stars usually called for a celebration on the set, with an enormous cake and champagne or ice cream brought in around four o'clock, while the publicity office took charge. During Christmas season a tree might appear on the set and gifts would be exchanged. Most directors permitted a cocktail break if filming was to continue past five in the afternoon, although studios frowned upon their workers drinking before then. A wrap party was customarily given on the last day of a major production for the cast and crew and all the department heads associated with it. Normally the commissary catered such parties, while the producer or director might arrange for a combo or entertainment. Occasionally a star would take fellow workers to dinner or present crew members with an appropriate gift as a memento.

At the end of each day's shooting an assistant director made out a production report, detailing the number of scenes completed, the number of camera setups used, the film footage exposed, and other items for a log of that day's work. Every morning the producer knew exactly how much the picture had cost to that point, since a cost sheet had been delivered to him. The assistant director's job was to make sure that no one went on overtime unnecessarily and to account for every hour the picture fell behind schedule. Once the project was completed, final reports had to be turned in to the production office before the film was readied for post-production.

With the last shot the intimacy that had developed during the making of a picture vanished, as the people involved went their separate ways. "This business seems to be one of forming and reforming new families as you go from one project to another," said screenwriter Hal Kanter. The teamwork and support system that had existed over the course of filming, the bonds that had grown as a result of the pressures, intensity, and mutual objectives, dissipated almost overnight. "A movie set is a very closed little world that's created every time a picture starts," Margaret Tallichet Wyler observed, "and then at the end of the movie the bubble bursts and it's gone."

"Each time a picture is finished," Coleen Gray said, "it's like an amputation. You have instant friendships and you become very close with people in this two-month period or whatever it is. Then the day it's finished, it's over. You have to go on to something else." Gray considered *Riding High* (1950) her most satisfying professional experience. "I felt worse when that picture was over than any other," she said. "It was so hard that I left town. I went down to La Jolla, where I used to work as a waitress, and I drove around in the rain, just kind of sorting out my thoughts."

Gregory Peck felt actors tended to lose sight of reality during the making of a film. "The make believe becomes reality," Peck explained, "and the real world recedes into the distance. You stop reading the news for the most part, and you get totally wrapped up in these little charades that we put on film. You almost have to get the kind of belief into it that makes them seem real."

Although the work was hard, most picture people loved their jobs. "We couldn't wait to wake up in the morning at five-thirty and go to work," declared actress Ruth Hussey. "I lived to go to the studio and make believe," Lizabeth Scott said. "My work was my hobby, because it was playful, it was joyous, it was fulfilling, and it was gratifying. It was a difficult life, but there was something happy about it. Making pictures you're a child, perhaps that's it."

It seemed to be the relationships that mattered most to Hollywood workers. Despite the friction, tensions, disappointments, and egos, most studio workers were craftsmen dedicated to doing a job to the best of their abilities. What counted in the end was the satisfaction of having completed a job as well as circumstances permitted, and the knowledge that some of the finest talent in the business had contributed to a team effort.

13

On Location

Old dining hall at Goulding's Lodge in Monument Valley, Utah, used as a set in John Ford's She Wore a Yellow Ribbon. *Courtesy of the Academy of Motion Picture Arts and Sciences.*

In the early days Hollywood filmmakers were not particularly concerned about filming on location. "A rock is a rock, and a tree is a tree," producer Abe Stern declared. "Shoot it in Griffith Park!" (Vidor, *Tree Is a Tree*) Many of the first silent movies were photographed in the streets around the studio, and even *The Great Train Robbery* (1903), Edwin S. Porter's classic western, was filmed in New Jersey rather than the West.

Even after the formation of the big studios, foreign locations were generally simulated on the backlot or one of the larger sound stages. Metro-Goldwyn-Mayer's lot contained sections of English, French, and Asian villages, as well as New York, New England, and New Orleans streets. At 20th Century-Fox, where the cutting department later stood had been a section of Hyde Park, while part of a steamship sat where the new administration building was eventually built. "You'd turn a corner," studio film editor Marjorie Fowler said, "and you were just in another world. They had a New York street, a Chicago street, and a Normandy village." The Fox backlot was full of permanent sets, an adventure to walk through. "You'd go into a French village that had been part of a war set and was bombed out," recalled Marjorie's husband, film editor Gene Fowler, Jr. "It was like you were moving from time to time."

When 20th Century-Fox made *The Robe* (1953), the studio reproduced Jerusalem on its backlot. Paramount re-created Bataan and Corregidor on its lot for *So Proudly We Hail* (1943), and Metro built both the bridge and the London railroad depot used in *Waterloo Bridge* (1940) on a sound stage. *Red Dust* (1932), although set in Indochina, was filmed on Stage 6 at MGM. "Stage 6 became a jungle with a hut in it," said actor Gene Raymond, "and it stank to high heaven. We had to have rain, and all of a sudden the rain created mud. Then they put the hot lights on it, and the moisture steamed up. So it was not a pleasant picture for anybody, the technical staff included."

Although the scenes in Bastogne were shot on Metro's backlot, most of *Battleground* (1949) was filmed on Stage 30, where weather conditions could be controlled. A forest of pine trees was brought in, along with fog machines and several kinds of snow devices, which could produce falling snow, dirty snow, and clean snow. The scenes looked realistic because they were covered with fog most of the time, and audiences didn't detect that the cyclorama in back was a fake. Director William Wellman wanted to see the actors' breath to indicate the extreme cold and ordered a huge air-conditioning system to bring the temperature on the sound stage down to the thirties. "When you came into that stage," screenwriter Robert Pirosh declared, "you put on a field jacket."

An advantage to working on a sound stage was that weather, lighting, and seasons could be controlled. *Desire Under the Elms* (1958) was shot entirely at Paramount, where all four seasons could be shown on the same set. "In later years we would have gone on location to Vermont or New Hampshire or wherever," director Delbert Mann said, "and found a way to overcome the change of season problem, achieving a greater sense of realism." Working on a sound stage or backlot also was cheaper, required less advanced technology, and could be coordinated by the studio manager. Sets often looked unconvincing, however, unless skillfully camouflaged, and second unit footage shot on location frequently didn't match what had been done in the studio.

To keep audiences' attention off the backgrounds of Metro's *The Three Musketeers* (1948), director George Sidney cast Gene Kelly in the role of D'Artagnan, allowing him to choreograph the duels almost as ballet. "I wanted those duels to have a kind of lyrical quality," said Kelly. "I knew we could get something that the action fellows couldn't do, that only a dancer could do. That was the challenge—to broaden it out and put the camera on it like a dance number. I wanted audiences to see the performer kick and fight and laugh without the quick cuts, a phony punch in the jaw, or a quick cut to a foot kicking or a false mannequin." The *Three Musketeers* dueling sequences were filmed in Busch Gardens in Pasadena and along the beach at Malibu, rather than in back of Luxembourg Palace and along the French coast, but the action was good enough that few viewers noticed. Sidney even used the golf course in back of the Culver City studio. "One day I had Frank Morgan as the King of France and all the horses going off," the director remembered, "and here came a fellow along in his undershirt playing golf. So we had to sit there and wait till this guy got through." But the picture cost less than a million dollars and looked expensive. "There were holes in those sets we used," said Sidney, "but no one saw them. We had colors projected here and there and a little fog machine going. That was exciting, to create as you went, sometimes covering up something. We worked these things out; that's what made it fun."

Sometimes settings outside the studio, but nearby, were used to simulate foreign locations. The White Cliffs of Dover scene in *Sylvia Scarlett* (1936) was photographed north of Malibu Beach. For the silent version of *Ben-Hur*

(1926) the Circus Maximus was re-created on a tract of land about four miles from the MGM lot. During the shooting of the chariot races the studio hired ten thousand extras for three dollars a day plus a box lunch. "I was the forty-fifth assistant on that," laughed Joseph Newman. "I handled the mobs and gave out the box lunches." Paramount's *Untamed* (1940) required a great blizzard, which was actually shot in an ice house in downtown Los Angeles during the middle of summer. "We used to have to go into a decompression room before we went out into the heat, because it was below zero in there," actress Patricia Morison recalled. "They had ice machines with propellers grinding up ice and pelting it at us." The plane crash at the beginning of *Lost Horizon* (1937) was also filmed in an ice house.

Occasionally a camera unit would be sent on location to take shots that would be brought back to the studio and projected onto a screen; the actors would then perform in front of the screen, coordinating their movements with the background action supplied by the second unit. In *The Snows of Kilimanjaro* (1952) Gregory Peck had to shoot a rhinoceros; someone had gone to Africa and filmed the killing of a rhino for 20th Century-Fox. "I pointed my rifle at the charging rhino on the screen," said Peck, "and when the red light blinked off camera range, I pulled the trigger and down went the rhino." Another time *Kilimanjaro* cast members were in dugout canoes, floating in a tank amid artificial pussy willows and reeds, representing an African riverbank. The actors were shooting at crocodiles projected on the screen, while a prop man was rocking the boat and a wave-maker was stirring the water. "It was all very artificial," Peck concluded, "but that's the way it used to be done in those days. At that time the studios still hadn't come round to the idea of sending film units wherever the action took place."

On program pictures, stock footage would be intercut with what had been filmed in the studio. Most of the studios owned ranches in the San Fernando Valley, where the Columbia backlot also was located, several miles from the main studio. Broadway musical comedy star Vivienne Segal remembered her four pictures for Warner Bros. during the early 1930s with dismay; she was bitten by red ants making *Golden Dawn* on the Warner ranch, and the studio had to call a doctor. Lucille Ball, usually a trouper, could recall complaining only once during her early days as a Hollywood show girl—on RKO's *The Three Musketeers* (1935), when the studio kept sending her out to its ranch during the summer dressed in layers of petticoats, velvet gowns, and plumed headdresses. When she fainted from the heat, Ball made the location manager send her home in the middle of the day.

Much outdoor filming on routine westerns took place either at Iverson's

Ranch on the outskirts of the San Fernando Valley or at Corrigan's Ranch near Chatsworth. Actress Peggy Stewart recalled that the first two days of Republic's B westerns were normally shot at Iverson's, where the stunt work was filmed. "We didn't have a whole lot of dialogue on location," Stewart said, "just yelling at the stagecoach or something like that." There wasn't much direction on inexpensive westerns, since the emphasis was mainly on action. Actors were expected to come to work knowing their lines; the crew was invariably fighting against time, and sunshine was too valuable to waste. "We tried to make most of our pictures in the summertime," cowboy star Gene Autry remembered, "because of the long days."

Sometimes crews would go on location to nearby Victorville or Barstow for desert shots or to Lone Pine for mountains. Vasquez Rocks, near Palmdale in the high desert, proved another prime location, and Kernville offered accessible hills, boulders, and scrub oaks.

If the weather cooperated, nearby location work could be pleasant, with jokes and fun; romances often developed on location. Gene Autry felt he came to know his crew better working outdoors than he did in the studio. After the unions were formed, film crews were no longer required to work on Saturday or Sunday, but weekends on location were just like any other workday, which shortened schedules but made a long week. Outdoor shooting usually necessitated reflectors for proper lighting, and if much action was involved, duplicate costumes were made in case something happened to them while workers were away from the studio. Sometimes aspiring directors were given a chance on location to prove what they could do behind the camera. Freddie Brannon, for example, who started out as a prop man at Republic, eventually moved up to being a B director. "They'd break a director in just the same as the horses from Hudkins Stable would break the drugstore cowboys in," Peggy Stewart said.

Republic rented horses from Hudkins, near Warner Bros., although other studios used different stables. Wranglers hung out in the vicinity, and the nearby Hitching Post became their favorite bar, where cowboy extras gathered between assignments. "I enjoyed the wranglers," said actress Catherine McLeod, who made several Republic westerns. "They were decent, honest, funny men. They taught me all kinds of rope tricks and things like that. They were also quite protective and watched out for me: 'Honey, we'll be right behind you.' I was athletic, so I enjoyed the outdoors."

In some instances actors and actresses had to do their own riding and even take risks. Actress Marie Windsor remembered taking riding lessons at the stables for at least two weeks prior to a picture. "The cowhands and Bill Elliott taught me how to sling a gun and twirl it and put it in my holster,"

Windsor said. During the making of *My Pal Trigger* (1946), western star Dale Evans decided to do her own riding in a scene without consulting producer Armand Schaeffer, despite the fact that the scene called for an English saddle and Evans had never ridden with one before. "They wanted to use my double, but I decided I was going to do it myself," the actress recalled, "which was foolish. I got up on that horse, and I took a good deep seat and a handful of mane, and I rode that thoroughbred wide open behind a camera car with rocks and gravel flying. I was scared but exhilarated." When the producer heard what Evans had done, he stormed out and forbade her to do that again on one of his pictures, since schedules and budgets could have been ruined. From then on Evans let her double take over the more dangerous riding scenes.

Stuntmen were involved in most action sequences filmed away from the studio, particularly on westerns but also on slapstick comedies. Stuntmen and doubles had been practically unknown during the early years, although the antics of the Keystone Kops and others sometimes involved considerable danger. Insurance companies were reluctant to cover silent picture actors, since jumping from rooftop to rooftop and falling down stairs were all part of a day's work. Later on, when stuntmen routinely doubled for actors, a second unit with a separate director and crew often set up at a ranch, while the main company worked inside the studio. During the 1930s a top stuntman earned twenty-five or thirty dollars a day for fourteen hours of work. Studios hired them on a picture by picture basis, although the best ones worked steadily. There weren't many recognized stuntmen (no more than thirty during the Depression years), but there were quite a few cowboys who were good horsemen. They did their own equestrian stunts, were willing to work as extras, and hoped for an occasional line or two of dialogue, which earned them more money.

Most stuntmen specialized in fights or horse stunts, although working in cars grew increasingly important. Preparation was essential, and stuntmen worked closely with the prop department and special effects people. In fights they wore arm pads and, when necessary, knee pads and shin pads. If they were performing a hazardous feat, they wore a lace-up vest of three-quarter-inch rubber, extending from the shoulders down to the tailbone. In cars most stuntmen made sure the seats were bolted down and that they had roll bars, so that when they rolled over, they rolled on the bar. "If I was ever doing a stunt that was a turnover or anything like that," veteran stuntman Gil Perkins said, "where the gas tank was liable to ignite, I did it with half a gallon of gas in the tank. I had the rest drained out."

Perkins eventually worked in over two thousand pictures, doing high wire stunts, falls off buildings, high dives, water stunts, everything except airplane and motorcycle stunts. He became an active member of the Screen Actors Guild and served as a stunt coordinator, for which he received screen credit. The coordinator read through the script and noted all the stunt sequences, figuring out how to accomplish each feat and arranging for the necessary equipment.

Yakima Canutt, perhaps the greatest of all the Hollywood stuntmen, became a legend, famed for his stunts in *Stagecoach* (1939). In the John Ford classic Canutt leaped to the back two horses pulling the coach, then to the next two horses, and then the next two, finally dropping underneath the galloping teams. Other noted stuntmen included Dave Sharp, Duke Green, and later Dean Smith, as well as such stuntwomen as Helen Thurston and Alice Van, the championship trick rider who doubled for Dale Evans. Stuntmen customarily taught each other how to rig a car to turn over, how to do a saddle fall or a rearing horse fall. But despite precautions, accidents did happen. Bob Morgan (actress Yvonne DeCarlo's stuntman husband) lost a leg filming the railroad sequence in *How the West Was Won* (1962), and earlier a stuntwoman in an Eddie Cobb western had drowned in Colorado when she was caught by quicksand in a stream. Eventually the Stuntmen's Association of Motion Pictures had between 130 and 150 members and issued rules and regulations for their protection.

Even famous movie horses—Trigger, Champion, Topper—were doubled for dangerous stunts. Until 1940 it was not unusual for horses to be killed in movie falls because of a technique known as the "Running W" that was used to pull the horses' feet out from under them. "A length of airplane cable was tied to a stationary object, then attached to the four legs of the horse by means of small rings on hobbles," explained stuntman Chuck Roberson. "The rider then spurred the horse into a canter, and as they reached the mark, the horse's feet were snapped out from under him. The result was often a dead horse." (Roberson, *Fall Guy*) In three takes of one scene for *Northwest Mounted Police* (1940) nine horses were killed. At that point the Humane Society stepped in and a law was passed making the "Running W" illegal. Horses were then trained to fall by having the riders turn their heads when they wanted them to go down.

At best, location work could be hazardous. Rock Hudson remembered doing his own stunts on his first western at Universal, *Winchester '73* (1950), in which he played an Indian. "I didn't even know there were stuntmen," the actor claimed. "I was doing horse falls at a full gallop with just a loincloth on, in gravel and rocks. There I was with no padding. Like an ass I did it twice. But I didn't know. I was eager, and the director took advantage of my stupidity."

The more important the picture, the more likely the company was to go on location, although before 1950 location work was restricted mainly to Hollywood westerns. The country around Tucson, Flagstaff, and Yuma in Arizona became popular for filming, as were Durango, Colorado; Moab, Utah; and Gallup, New Mexico. George Stevens's *Shane* (1953) was shot in the Grand Tetons, near Jackson Hole, Wyoming; Otto Preminger's *River of No Return* (1954) was made in the Canadian Rockies. Some locations were miserable, with heat in the summer, sandstorms in the desert, rain that made work impossible.

Actress Julie Adams remembered making *The Man from the Alamo* (1953) in the summertime near Agoura, about twenty miles from Hollywood. "It was one of those action films where you sit around for hours and hours waiting for them to line up the covered wagons and horses," Adams recalled, "and it was hot! One day when they finally got to my close-up, it was so hot my nose began to bleed. They just closed up the show; that was it for the day." *Hombre* (1967) was a difficult picture for Barbara Rush and everyone else associated with it. For six months they were on location in the San Maria Mountains near Tucson; every morning, Rush said, they got into landrovers and drove up to the mountains through dirt and dust. But earlier conditions in filmmaking had been worse. Director William Wyler made his first all-talking picture, *Hell's Heroes* (1930), in Death Valley during July and August. "At that time the camera was in a big box," Wyler recalled. "The camera had to keep moving because the nature of this story was such that the men who were being pursued had to keep moving. When they stopped and talked, we had to bury a microphone in a cactus. And we had to push that box without making any noise. It was tough, with a guy on top of the box and a microphone hanging over and guys pushing the box in the sand in 110 or 120 degree heat. One time we opened the box and the cameraman had passed out, because inside it was 150 degrees."

During the filming of *The Outriders* (1950) in Kanab, Utah, actress Arlene Dahl's naturally red hair bleached almost blonde from the sun. The studio had to rinse it with a red color that looked like a vegetable dye, and eventually the star was sent back to Los Angeles one weekend to have her hair rinsed so she would again be a redhead. Joel McCrea headed the picture's cast, and the Kanab location wasn't easy for him or anyone else, but it was particularly hard on the fair-skinned Dahl. "I had a terrible time because I burned underneath my makeup," the actress remembered. "The sun was that hot. So I had to have an umbrella, carried by a prop man or somebody, wherever I went. If I wasn't in the shade, I had to be under an umbrella."

Makeup call on location was generally at seven A.M., with actors and actresses sometimes sitting on a rock while it was applied. Keeping the makeup on in summer heat was a problem, as was the boredom of waiting around for the next scene. "Nothing could be further from the truth than any notion of glamour and excitement in regards to moviemaking," wrote Native American character actor Iron Eyes Cody. (Cody, *Iron Eyes*) As time went on, people's nerves became jangled. Fights often broke out, especially if somebody brought out a bottle of liquor.

To ease the tension, director John Ford had Danny Borzage play folk tunes on his accordion between shots, especially "Red River Valley." After the day's work was completed, film talk was banned, and Ford encouraged games, usually dominoes or pitch. Endless practical jokes were played, usually on actor Ward Bond, the most reactionary and opinionated of Ford's stock company but a close friend of the director and most of the company. Although Ford liked people who drank, he insisted on abstinence during the making of a picture, and he dealt with flagrant violations severely.

Stuntman Chuck Roberson worked for Ford on *Rio Grande* (1950), the last of the director's famous cavalry trilogy. "The first day on location in Moab, Utah, with Old Man Ford was something akin to my first day in the army," Roberson wrote. "Moab was just a little, one-horse Mormon town in the middle of nowhere. There were no motels, so we bunked in army tents with wooden-planked floors and side flaps that could be raised each day to let the sand blow through." (Roberson, *Fall Guy*) More often the company would be spread out at various hotels around a town. On *She Wore a Yellow Ribbon* (1949), another segment of Ford's cavalry trilogy, the director's daughter, Barbara, shared a small house at Goulding's Lodge with actresses Joanne Dru and Mildred Natwick. "I remember the Indians there," said Natwick. "They had a camp. At night I could hear them singing and dancing, just for their own amusement. It was quite spooky, but interesting."

John Ford was the first to use the landscape to lift the western movie into legend. With *Stagecoach* the director first used Monument Valley, which straddles Arizona and Utah in the Navajo Reservation, a spectacular area with gigantic monoliths rising from the desert floor. Ford used Monument Valley as a metaphor for a moral universe, a primeval garden of natural dignity and innocence invaded by civilization. The landscape, grand and savage, became an integral part of the epic struggle between pioneers and the wilderness, as its cathedral-like buttes and mesas tower over the stagecoach winding its way across a vast panorama. With civilization's progress and order also came destruction and moral decay, as the invaders enter a majestic land already filled with chaos, violent outlaws, and marauding Indians.

In *Stagecoach*, the valley "is not simply a valley, but a valley melodramatized," Tag Gallagher wrote, "and the coach is not simply a coach, but the historic mythos of 'the West.'" (Gallagher, *John Ford*)

No one had ever used Monument Valley before in films, since it was located in one of the most inaccessible parts of the country, a hundred miles or more northeast of Flagstaff, with a single dirt road, no bridges, and no telephones. Ford began shooting *Stagecoach* in late October, 1938, and headquartered his company at Harry Goulding's trading post, the only lodging available, where big western-style dinners of barbecued ribs or steak were served each evening. After dinner there was usually a game of pitch in Ford's quarters, played with silver dollars, since the director liked to hear the sound of the heavy coins as they were tossed into the pot. The isolation of Monument Valley freed Ford from rigid studio schedules, allowing him to take advantage of a desert squall, dramatic natural backlighting, or an impressive cloud formation.

Ford returned to Monument Valley six more times (for *Fort Apache*, *She Wore a Yellow Ribbon*, *The Searchers*, and *Cheyenne Autumn*, among others), making it the Valhalla of western films. He befriended the Navajos, paying them union wages at a time when Indians rarely earned more than fifty cents a day. He studied their language, participated in their sports, and eventually was adopted into their tribe and given the name Natani Nez, meaning Tall Soldier. "I tried to copy the Remington style," Ford told Peter Bogdanovich, referring to *She Wore a Yellow Ribbon*. "You can't copy him one hundred percent, but at least I tried to get in his color and movement, and I think I succeeded partly." (Ford int., Bogdanovich)

Other filmmakers became enamored with the West during location assignments and thrived on being outdoors with the elements. Douglas Fairbanks, Jr., was barely sixteen when he made *Wild Horse Mesa* (1925), but the experience was one he never forgot. "I would ride with the Indians out on the Arizona desert," said Fairbanks, "and help them while they rounded up hundreds of head of wild horses. I used to be romantically fascinated with seeing them race across the desert in the evening with their long hair let down. They'd strip down to their breechcloth and ride bareback on their horses, streaking across the desert. It was something that would have been painted by Remington or Russell, very romantic."

Since the cast and crew of most westerns consisted mostly of men, the few women in the company often found location work lonely. "I got sick and tired of being the only girl with over a hundred guys," script supervisor Catalina Lawrence complained, "and still never having a pass made! At night the guys would always play cards around the hotel. I would stack the cards, stack the chips, and go out and get the drinks for the fellows. They treated me just like I was a kid sister running around for them." (Lawrence int., Steen) During the weeks that *Blood on the Moon* (1948) was filmed near Sedona, Arizona, stars Robert Mitchum and Robert Preston spent much of their time teasing leading ladies Barbara Bel Geddes and Phyllis Thaxter. (Wise int.)

Studio rivalries were active even hundreds of miles away from Hollywood. Errol Flynn and Miriam Hopkins detested each other, yet Warner Bros. cast them as the leads in *Virginia City* (1940). During location shooting in Flagstaff, the two fought continually. When the script called for them to play a love scene, off the set they weren't speaking to one another. The night before the scene was to be shot, each used a typewriter in their respective rooms to rewrite the lines to their own advantage. "It took us three days to unscramble that," scriptwriter Robert Buckner said.

Working great distances from the studio created problems, not the least of which was feeding as many as 250 people every day on the set, particularly once guild regulations required hot lunches. Director Henry Hathaway ran into trouble filming *True Grit* (1969) in Texas when the script called for corn dodgers and no one knew how to make them. Hathaway, who insisted on authenticity, had a vague notion of what he wanted and described his idea of corn dodgers over the telephone to Paramount commissary head Pauline Kessinger. Commissary workers made them out of corn meal and cooked them in deep fat. Some had red coloring added, Kessinger recalled, because they were supposed to have blood on them. About six hundred corn dodgers were made in the Paramount kitchen and flown to Hathaway in Texas.

Technically, location work proved more difficult for the crew than working in the studio, and logistical problems at times were nearly insurmountable. Paramount's *California* (1947) was shot in northern Arizona near Cameron. Production coordinator Gerd Oswald was stationed in Flagstaff; his job was to find extras, get them costumed, and arrange transportation for the hour and a half ride out to the location. Making *Red River* (1948), director Howard Hawks's company had to transport around Arizona fifteen hundred head of cattle, as well as the cast and crew, adding thousands of dollars to the cost. The studio had not figured that expense in the budget, and Hawks went about $800,000 over making the picture.

Usually it was lucky for a company if skies were clear or if clouds were photogenic—fluffy cumulus or dramatic thunderhead clouds, for example. But sometimes clement weather presented difficulties, especially for Technicolor. Director King Vidor recognized during the filming of *Northwest*

Passage (1940) that he didn't want beautiful blue skies, which were incompatible with the hardships the frontiersmen in his picture were experiencing. Vidor realized that the colors had to be controlled. "That was quite an experience," the director said. "The blue skies were just too colorful; we didn't want the film to be too gaudy. That's okay for musicals, but not for drama."

Vidor's company, which filmed *Northwest Passage* mostly in Idaho, required two trains loaded with equipment. Thirty miles from Boise they had ninety boats, thirty or forty tents, Indian teepees, and a full crew of carpenters. The Metro company took over an unused resort for living accommodations, but the working environment remained difficult. "Some of us had tick fever," Vidor said. "Everyone was afraid that they would be bitten by ticks. I remember we used to come in at night and put a white sheet on the floor, undress on the white sheet, drop all the clothes on the floor, and see if there were any ticks that showed up on the sheet." (Vidor int., Dowd and Shepard)

Under such adverse circumstances keeping even tempers became difficult. During the shooting of 20th Century-Fox's *Two Flags West* (1950), a film about a Union captain and a former Confederate colonel on patrol out West, the company worked near the San Ildefonso Indian reservation and lived in a camp originally intended for construction workers. For ten days they were plagued by sandstorms, but the actors maintained their sense of humor, referring to the picture as "Two Fags West." Usually there were compensations—travel, establishing close friendships, evening camaraderie with the company—and after years of location experience studio production departments had perfected efficient techniques, despite the fact that constant adjustments were necessary.

Westerns were not the only movies that required location shooting, although in the early days studio heads were reluctant to send their crews far from home. Metro's *Treasure Island* (1934) was filmed on Catalina Island, where the company spent six weeks battling the weather and trying to keep yachts, speedboats, water skiers, and airplanes out of the picture. A year later MGM's *Mutiny on the Bounty*, also filmed around Catalina, faced even greater problems, including fog every morning until eleven or twelve o'clock. Production head Irving Thalberg flew over one weekend to meet with director Frank Lloyd, who complained that he was getting only a couple of hours of work done each day. Thalberg told him to film what he needed if it took six months. The *Bounty* company remained on Catalina eight weeks, then came into the studio, and finally returned to the island for another two weeks to finish the picture; background shots with doubles had been made in Tahiti six months earlier.

Warner Bros.' *Dive Bomber* (1941) was filmed at the Naval base in San Diego and on an aircraft carrier at sea, with director Michael Curtiz seasick much of the time. When he was well, Curtiz was constantly telling the captain to turn the other way because the smoke from the ship was getting in the way of his camera. The trench warfare in Lewis Milestone's classic *All Quiet on the Western Front* (1930) was filmed on a hill near Newport Beach. "There's no film to this day where movie battle scenes are more realistic than *All Quiet*," declared actor Lew Ayres. "With its black and white photography, it's almost like a documentary. I have seen many picture books on World War I, and strangely enough, occasionally I will pick out shots that were taken from *All Quiet on the Western Front*, right from our set. They simply put them in because the realism of that film was so tremendous."

Most of the outdoor sequences for *The Adventures of Robin Hood* (1938), with Errol Flynn at his swashbuckling best, were photographed in Bridwell Park, near Chico, California, and Paramount's *Frenchman's Creek* (1944) was shot on the Russian River. *Tarzan and the Huntress* (1947), featuring Johnny Weissmuller in the title role, was filmed in Santa Anita, in a park that later became an arboretum. The park had a river running through it, complete with lily pads and hanging vines, but the set decorator added tropical plants and peacocks to make Tarzan's jungle more cinematically complete. Even Metro's *The Good Earth* (1937) was shot in California; filming Pearl Buck's novel in China was never considered. Later, *The Keys of the Kingdom* (1945) and *Anna and the King of Siam* (1946) were both photographed entirely on the 20th Century-Fox backlot.

The sand dunes near Indio, California, became known as Hollywood's Sahara, and many of the "tits and sand" potboilers were filmed there. Since Indio was a morning's bus ride from Los Angeles, the studio didn't have the expense of hotels, only meals; the crew raced back to the city the minute the sun went down. Warner Bros.' *The Desert Song* (1943), however, was shot near Gallup, New Mexico, where director Robert Florey arranged some excellent desert scenes for his star, Dennis Morgan.

Parts of Metro's *The Yearling* (1946) were filmed on location in Florida, and Gregory Peck remembered spending two and a half days there working on two pages of dialogue, trying to keep the deer in camera range amid blistering heat. The cast did the scene seventy-two times, exhausting everyone. MGM shot the rest of the picture indoors on Stage 15. The studio grew corn on the backlot from seeds so the set decorators could bring progressively more mature corn to the sound stage each day to show how the field in the movie had grown. "You really couldn't believe those scenes hadn't been taken on a farm," publicist Ann Straus commented.

Sometimes there were obvious locations—Sun Valley, for instance, where the ski scenes for dozens of pictures were filmed. In *Thin Ice* (1937), Darryl Zanuck's ski instructor, Otto Lang, doubled for skater Sonja Henie in some of her trickier moments on the slopes, wearing the Norwegian star's costume. "Sonja Henie was about five-one or five-two," said Lang, "and it was really an ordeal for me to get into those clothes. The only way it could be done was to split the seams open in back and then hold them together with safety pins. Wearing a wig, I could only be photographed from the front and from far away."

Viva Villa! (1934), Metro's film with Wallace Beery as Pancho Villa, was shot in the interior of Mexico, where the work proved extremely difficult. "The climate was hot and dusty," recalled cameraman James Wong Howe. "We had to haul in our water. Many people got sick. For living accommodations they had to bring out a train with a kitchen coach attached. The cook was Mexican, so we ate mostly Mexican dishes, which in this case were terrible. We were told not to go out nights because there were bandits! We had to sit in the train and play cards or chew the rag." (Howe int., Steen) By the time 20th Century-Fox photographed *Captain from Castile* (1947) in Mexico, conditions had improved. The Fox company stayed four months, during which Indians came down from the mountains to work as extras in the Samuel Shellabarger epic about the Spanish conquest. Since most of the Indians spoke no English, interpreters had to tell them what to do.

Elia Kazan's *Viva Zapata!* (1952) was filmed just across the border, along the Rio Grande. Cecil B. DeMille took a second unit deep into the south of Mexico to find jungle footage for *The Story of Dr. Wassell* (1944). Rather than travel to Africa, director Zoltan Korda re-created the big game hunt for the Hemingway story that became *The Macomber Affair* (1947) in Mexico. The company drove to the location every day in taxicabs, over sixty miles of dirt roads.

But until the 1950s Hollywood filmmakers usually stayed as near their studios as possible. For Metro's remake of *Show Boat*, even in 1951, director George Sidney went to Natchez, Mississippi for only five or six days to shoot atmosphere, while the showboat itself was built and floated on MGM's Lot 3. Sidney's company spent twenty-eight days in the studio filming the musical, rearranging schedules when the boat caught fire. Veteran director Henry Hathaway, on the other hand, though he admitted he could easily have made *Brigham Young—Frontiersman* (1940) on the 20th Century-Fox backlot, insisted that the studio send him to Lone Pine, Big Bear, and the Utah desert for six weeks, realizing that his film's reputation would be substantially enhanced. "The locations made it a big picture," Hathaway later said.

Hollywood filmmakers effectively used urban locations as early as 1928, when King Vidor shot the sidewalks of New York City for his silent masterpiece, *The Crowd*. As Vidor described the experiment, "We designed a pushcart perambulator carrying what appeared to be inoffensive packing boxes. Inside the hollowed-out boxes there was room for one small-sized cameraman and one silent camera. We pushed this contraption from the Bowery to Times Square and no one ever detected our subterfuge." (Vidor, *Tree Is a Tree*) The director's aim was to capture the movement and grit of city life in almost a documentary fashion; he also filmed long shots of Coney Island and took his crew up to Niagara Falls.

In the mid-1940s several semidocumentary movies used actual urban locations. *The House on 92nd Street* (1945) was filmed on the streets of New York, inside real houses and offices. "I never shot one thing in the studio," director Henry Hathaway said. "It was a true story." The film's producer, Louis de Rochemont, worked closely with the FBI, while even the actors didn't know the picture was about the atomic bomb. *Boomerang* (1947) was also photographed entirely on location, in Greenwich, attesting to de Rochemont's commitment to realism. Louis de Rochemont was a major figure in the documentary approach, pioneering techniques in film production that revolutionized Hollywood.

Kiss of Death (1947), another true story, brought Henry Hathaway back to New York; he found working on the city's streets exciting, though some of his actors posed problems. Victor Mature, the picture's star, spent most of his nights carousing and much of the day on the set sleeping. "I bought him a couple of new suits," claimed Hathaway, "and I finally found him in the men's toilet, lying on the floor asleep—in the new suit I'd just bought for him!" The director had a reputation for being tough with actresses, a tendency that surfaced while filming a scene with newcomer Coleen Gray. When Hathaway grew angry with her, he made a biting remark that caused Gray to cry. She ran upstairs in the house where they were working, and Hathaway soon followed to comfort her. "He couldn't have been sweeter," Gray said, "so I dried my eyes and went down and did the scene the way he wanted it. He was an absolute lamb all the rest of the picture."

Later car chases were filmed around New York, usually at night, across the Manhattan Bridge, in Queens, and through the streets of Brooklyn. Whereas Metro's *On the Town* (1949) introduced New York locations into screen musicals, *West Side Story* (1961) expanded the concept; both transliterated Broadway shows into cinematic productions. Making *West Side Story* the company realized quickly that it wasn't realistic to have a gang of teenagers suddenly dancing on film. The prologue was elongated for the movie, allowing the teenage boys to break into dance slowly, almost doubling the introductory section.

By the time *West Side Story* was made, location work had become commonplace, almost mandatory. Metro's *Seven Brides for Seven Brothers* (1954) included some beautiful footage filmed in Oregon, although doubles were used for the dancers. Much of 20th Century-Fox's *Carousel* (1956), including musical numbers, was shot with its stars in New England, and some of the circus scenes in DeMille's *The Greatest Show on Earth* (1952) were performed before real audiences in Philadelphia and Baltimore. Portions of Stanley Kramer's *My Six Convicts* (1952) were photographed at San Quentin, as were key sequences in Robert Wise's *I Want to Live!* (1958). While most of the outdoor scenes for Warner's *Giant* were filmed in Texas, the ranch house in the picture, never more than a shell, was built in Burbank and shipped on a flatcar to Marfa, where it was held in place by four telephone poles and cables. "That was kind of a desolate place," actor Rock Hudson remembered, "but the warm nights were wonderful, with really clean air. There wasn't sky enough for all the stars. But we used to have to go to Alpine to go swimming. There was a pool at the country club there, a forty-mile drive from Marfa."

During the shooting of a scene for *A Place in the Sun* (1951) at Lake Tahoe, several extras were waiting to be filmed sunbathing on the shore, some under umbrellas, while director George Stevens was on a camera boat photographing a raft on which Elizabeth Taylor and Montgomery Clift were seated. When he was ready to start the action, Stevens yelled to his assistant director, Gerd Oswald, and Oswald notified the extras on the shoreline to begin whatever they had been told to do. The director started the scene, only to have it interrupted when an airplane flew over. Stevens bellowed, "Cut!" and upbraided Oswald for letting the plane go by. "Mr. Stevens," the assistant director pleaded, "you must be kidding." But Stevens was furious: "If you were an adequate assistant director, you would know the exact flight schedules and would not let me start this scene knowing that it was going to be interrupted by a plane." Oswald smiled imperceptibly. "Mr. Stevens," he announced, "I do have the flight schedule. The next flight listed here was not due for another two minutes." Stevens apologized and Oswald went back to consulting his flight schedule.

Around 1950 Hollywood discovered Europe. With the postwar surge in international travel and the advent of television, movie audiences would no longer be satisfied by studio replicas of foreign locales. Movie companies at this time had a great deal of money frozen in European countries; the only way to use it was to film pictures there. American citizens living and working abroad for a two-year period enjoyed a tax advantage, an arrangement that appealed to many stars. Labor costs were cheaper in Europe, so that filmmakers began working around the globe, adding authenticity to their product by shooting in Paris, London, Madrid, Rome, or Tokyo. Eventually it became easier to sell a script to be filmed in Europe than one that could be made in Hollywood.

Metro's *Quo Vadis* (1951) cost seven million dollars, an exorbitant sum for that time. To keep the budget as low as possible, the studio decided to film the epic in Italy, where labor was cheap. Director Mervyn LeRoy devoted a year to preproduction planning and for three weeks on location had thirty thousand people and twenty assistants working for him. "We had trouble with the lions," LeRoy said, referring to the sequences in the Colosseum. "We got lions from every circus in Europe—young ones, old ones—and all the lion tamers we dressed up as the Christians. But that was a tough picture."

Ivanhoe (1952) with Robert Taylor was filmed at MGM's studio in England. Producer Pandro Berman estimated that he saved $650,000 by making the picture at the British studio, and he had been spared the usual interference by the Culver City hierarchy. Berman found better casting among English character actors, and he could take advantage of the British countryside as background. Once *Ivanhoe* proved successful at the box office, Metro decided on *Knights of the Round Table* (1954) as a follow-up. Since there was trouble with the English unions, the big battle scenes for the King Arthur story were shot in Ireland, using the Irish army as extras.

Warner Bros.' *Captain Horatio Hornblower* (1951) was filmed in England, although the Hollywood actors performed their British roles with an American accent. "The idea of an American movie hero's playing with an English accent just wasn't considered in those days," said Gregory Peck, who undertook the Hornblower role. Shooting schedules were more leisurely in Great Britain; there was usually time for the cast to do some sightseeing. When actress Mildred Natwick worked on John Ford's *The Quiet Man* (1952), she was before the cameras only about twelve days out of the six weeks she spent in Ireland, giving her time to vacation and see the countryside.

Billy Wilder's *A Foreign Affair* (1948) was filmed in Berlin when animosities from World War II were still strong, especially since some of the local population thought the American crew had come to make an anti-German propaganda picture. But when Henry Hathaway directed *The Desert Fox* (1951), he was allowed to photograph the scenes in Germany he needed for his story of Field Marshal Rommel, and he actually used Rommel's house in the movie. Twentieth Century-Fox's *Night People* (1954) was shot during festival time in Munich, where the company stayed at the finest hotels and enjoyed the finest food.

William Wyler's *Roman Holiday* (1953) was among the first American pictures to be made in Rome, where the company filmed in the streets for six months; Jean Negulesco's *Three Coins in the Fountain* was also photographed in Rome. King Vidor shot *War and Peace* (1956) mostly in Italy and *Solomon and Sheba* (1958) in Spain.

Toward the end of his career Cecil B. DeMille spent nearly two months in Egypt working on *The Ten Commandments* (1956), directing scenes in which as many as eight thousand people took part. "We engaged twenty assistant directors who spoke both Arabic and English," DeMille wrote, "briefed them each day on the next day's work, and sent them, in costume, into the midst of the Exodus, each one responsible for directing a certain segment of the great moving mass of people, animals, and wagons." (Hayne, *Autobiography of DeMille*)

Metro's *Valley of the Kings* (1954) was also filmed in Egypt, and Alfred Hitchcock's *The Man Who Knew Too Much* (1956) was shot in Marrakesh. For *The Swiss Family Robinson* (1960) the Disney Studio sent its cast and crew to Tobago, and MGM's *Something of Value* (1957) was filmed in Kenya. One Sunday morning during the making of *Something of Value*, the cameraman, the director, and stars Rock Hudson and Sidney Poitier were taken out into a field by a white hunter to meet the Mau Mau. "We were told to stand quietly and not move, or we would be shot," Hudson recalled. "When they found that they were safe, the Mau Mau came out of the trees and the bushes and surrounded us. It was terrifying! I never saw such hatred as in those black, piercing eyes."

Negotiations with foreign governments frequently proved troublesome on location work, particularly when large numbers of extras were involved; the reception of movie companies could fluctuate with shifts in diplomatic relations. The length of time involved on a location assignment sometimes was exasperating. William Wyler returned to Rome in 1958 for *Ben-Hur*, spending eight months there working on the epic six days a week, with practically no social life. "I used to walk around alone about five or six o'clock on dark, cold Roman evenings," Mrs. Wyler remembered, "thinking, 'How long?'" But her husband's experience was mild compared with the disasters involved making 20th Century-Fox's *Cleopatra* (1963). "We rehearsed four months on that procession that brought Cleopatra into Rome," said choreographer Hermes Pan. "I had seventy-five black dancers and seventy-five white dancers, as well as animals. It was just like a nightmare."

Seasonal changes occasionally posed problems with schedules on location. On *Lust for Life* (1956), Vincente Minnelli's crew had to race to capture on film the ripening wheat fields in France so they would match Van Gogh's color palette. The company traveled to all the places Van Gogh had painted—to Belgium where he worked in the coal mines, to Paris, to the fields outside Arles—capturing on film the light and atmosphere he had created in his paintings.

Gradually Hollywood cameras became eyes to the world, as more and more productions moved out of Los Angeles. After the big studios began their decline, actors and directors frequently selected the projects that offered them opportunities for travel, since everything was paid for and accommodations were first-class. Metro unit publicist Emily Torchia remembered fondly the European locations to which she was assigned: "I'm grateful to MGM, because I could never have traveled the way I did on my own. In those days it was always deluxe. It was a great thrill for me to be in Europe for five or six months at a time." Eventually major studios included someone in their foreign production units who had a grasp of the European market, as the revenue from outside the United States in some cases exceeded what film companies could expect from domestic sales. "I love location shooting," said director Delbert Mann, who launched his film career just as the old studios were beginning to retrench. "I enjoy the problems, the coping with weather and the physical problems, trying to retain the focus on people and their relationships in front of different settings which have freshness and newness to them." Mann, who came to Hollywood from television, represented a younger breed of filmmaker, typical of the generation that rose to prominence after the big studios had discovered the world.

14

Film Editing

Editing equipment at Metro during World War II arranged to look like a cannon. Courtesy of the Academy of Motion Picture Arts and Sciences.

Film editing is the only professional skill indigenous to the motion picture industry. The film editor has the power to shape, improve, even rewrite the basic story. "Editing is to cutting as architecture is to bricklaying," former film editor Edward Dmytryk observed; "one is an art, the other a craft." (Dmytryk, *On Film Editing*) While many Hollywood pictures have been saved in the editing room, others have been ruined by unskilled cutters. Whether a film editor is an artist or a craftsman depends on the editor, but in any case editing remains at the heart of the motion picture process.

Not long after Edwin S. Porter began to experiment with the intercutting of simultaneous and related action around 1903, film editing lifted motion pictures from a recording medium into a dramatic art form. D. W. Griffith soon discovered that a filmmaker could expand or compress time and space to meet his needs. Then Griffith began shooting close-ups to enhance the impact of a player's reaction; he created a rhythm that made his pictures flow smoothly, building climaxes to stirring crescendos. Perhaps reflective of the national temperament, the pace of American movies from the outset tended to be faster than that of European films. Director William Wyler once remarked that a European film might open with clouds that dissolve into a shot of cloud patterns moving slowly along the ground, which in turn dissolves into another shot of the clouds. An American film more likely would open on the clouds, out of which flies an airplane that shortly explodes.

Vital though film editing became to the industry, in the early years of the big studio system anyone who called himself or herself a film editor would have been considered a snob in Hollywood. Film editors were known as cutters, a term a later generation grew to despise. Throughout the 1920s and early 1930s self-effacement seemed part of an editor's job, and many chose the profession largely because it permitted them to work imaginatively without risking public scrutiny of their personal worth. "To wield the power of creation without bearing ultimate authority or responsibility was a perfect formula for a cautious personality," wrote Ralph Rosenblum and Robert Karen. (Rosenblum and Karen, *When the Shooting Stops*) Some editors during the major studios' early decades were given a great deal of responsibility, others very little, but all were expected to play a subordinate role and were viewed as technicians. "In the Hollywood I grew up in," Edward Dmytryk claimed, "'Art' was a dirty word. 'Empiricism' was the watchword of the day, though probably not one film worker in twenty had heard the word and not one in a hundred could define it." (Dmytryk, *It's a Hell of a Life*) Billy Hamilton, who headed the editing department at RKO for a time, routinely allowed his assistants to do the editing, while he sat with them and patiently corrected their mistakes. Young Robert Parrish, after showing Hamilton a sequence one day, asked his boss if he wasn't afraid somebody on his staff would learn too much, become an editor, and take his job. "Editors, my ass," Hamilton fumed. "You don't edit film, you cut it, the same as you cut cloth or sheet metal or cheese." (Parrish, *Growing Up*)

In 1938 the Wagner Labor Relations Act was passed, and the unionization of film workers progressed quickly. In an attempt to raise the status of the editing profession, the term officially became "film editor," although less knowledgeable studio executives still considered editors near the bottom of the pecking order within the technical fields. In 1950 the American Cinema Editors was organized, the professional organization to which most of Hollywood's working editors belong. With Elmo Williams's acclaim two years later for reworking *High Noon*, the Gary Cooper western that Williams turned into a classic, members of the guild became more aggressive in trumpeting the role of editors in press interviews, so that the craft enjoyed increased publicity and slowly won prestige and respect.

Elmo Williams and his department head, Harry Gerstad, won Academy Awards for the skillful editing of *High Noon*. Based on a story by John W. Cunningham called "The Tin Star," *High Noon* was written by Carl Foreman, directed by Fred Zinnemann, and produced by Stanley Kramer. "I loved the story from the start," Elmo Williams declared. "I thought it was one of those rare stories that just had the right cinematic ingredients to make a successful film." In the picture Gary Cooper plays a middle-aged sheriff, recently wed to a Quaker wife. He is deserted in his hour of need by an entire town when an outlaw gang the departing sheriff had sent to prison returns seeking vengeance. Director Fred Zinnemann supervised the initial cut, which producer Stanley Kramer saw and disliked so thoroughly he threatened to take his name off the project. After the initial screening Williams asked if he could have a few days to work on the film. "I think I can make something out of this," the editor told Kramer.

Carl Foreman had written peripheral stories, all involving characters coming to Cooper's aid, into the screenplay . "The first thing I did," Williams said, "was to eliminate those stories. Every time I left the town and Cooper the suspense of the show fell apart." Next Williams trimmed Grace Kelly's role as Cooper's bride. "I told Fred Zinnemann at the time he was making his initial cut that he was overbalancing the film in favor of Kelly, that he was enlarging her part too much. She was the moral force behind the film, but that was all. Fred was enamored with Grace and didn't really agree."

"To me the picture was the kind of story that a grandfather might tell his grandson," Williams explained. "So I approached it as a folk tale and edited the show along those lines. I could see the grandfather talking to his grandson and maybe making the story a little more dramatic than it was, building it up, to hold his grandson's interest." Williams cut the picture from around two hours to eighty-six minutes. Stanley Kramer liked the revised version, but insisted it was too short. In those days to obtain first feature billing, a movie had to run at least ninety-one minutes. "So I went back and padded the film out," Williams recalled, "mainly Katy Jurado's role, her relationship to Cooper and with Lloyd Bridges. It worked fine." Williams took the rhythm of the picture from Gary Cooper's way of walking and his slow drawl. "What I wanted to do with the end was to play lonely, simple sounds against the sudden, loud, staccato gunfire. So I have a fly buzzing around in the office, I have the scratch of Cooper's pen as he's writing his last will and testament, I have the ticking of the clock—single, isolated sounds, always with Cooper working and listening for the train."

Eventually director Fred Zinnemann and screenwriter Carl Foreman resented the unprecedented publicity Williams received for turning *High Noon* into a western classic. Producer Stanley Kramer later denied the contention that Williams's editing had made the picture a success.

Perhaps the most celebrated piece of editing in an American film is the breakfast table sequence in *Citizen Kane*, in which the disintegration of Kane's first marriage is vividly depicted. Robert Wise, a young film editor (later director of *The Sound of Music*, then only a year older than *Kane*'s boy genius, Orson Welles) had replaced an older cutter who hadn't suited Welles, who was both the director and star of the picture. Wise found the experience "gratifying, maddening, and exhilarating. Working with Orson, things were never on an even keel for very long. He could be outrageous in his behavior and do or say things that would make you so mad you wanted to tell him to take the picture and shove it. But before you could do that, he would come up with some brilliant idea that would have your mouth gaping open." Wise worked on the breakfast montage for weeks, interworking the images and dialogue. "The scene was in the script," Wise said, "and Orson shot it. But the actual timing of it, what made it work, was all done in the cutting room. Mark Robson, my assistant, and I spent long hours working with the speed of those quips that went from one person to the other, experimenting with where the in-coming voice should start, starting the in-coming voice earlier to see how it would work, getting the whole rhythm of the sequence. That rhythm was built right in the cutting room, just trying and experimenting and working to improve it." Welles, as director, gave Wise instructions but left his editor alone to do his work. "We knew we were getting something special and extraordinary in the film that was coming in," Wise observed later, "but we wouldn't have projected it as the tremendous classic the years have shown it to be."

The backgrounds of the Golden Age editors demonstrate the swift, haphazard way in which the American film industry grew. Edward Dmytryk, later a successful director, started as a teenager at Paramount as a projectionist, transferring to the cutting department not long after the introduction of sound. Dmytryk's break came with *The Royal Family of Broadway* (1930) when his technical knowledge proved useful to that picture's co-director, George Cukor, recently arrived from the stage. Elmo Williams had never considered entering the picture business when he enrolled in the University of California at Los Angeles; he wanted to become an artist. He was working as a carhop at the corner of Wilshire and Westwood Boulevards, where Merrill White, an editor at Paramount who had cut several of the Ernst Lubitsch pictures, was a frequent customer. Williams had waited on him three times when White asked if he'd like to go to England as his assistant. The boy said yes, and White paid his passage and bought him appropriate clothes. White was scheduled to edit a series of pictures in London, directed by Herbert Wilcox and starring Anna Neagle. "I used to just hang around the cutting room watching Merrill," Williams recalled years later. "I had nothing else to do. I was fascinated by the hot splicer. When nobody was around, I used to go and play with the splicing machine and taught myself to splice, taught myself to rewind, and then taught myself how to assemble film." Merrill White's teaching method was to let his protégé get into trouble, then help him work his way out. After he'd been in London for nine months Williams cut his first film. Soon he took over the editing of the Herbert Wilcox pictures, with his mentor acting as supervisor.

Daniel Mandell, who became Samuel Goldwyn's primary editor, started his career as an acrobat in vaudeville and later performed with the Ringling Brothers Circus. Like Elmo Williams, Mandell had never thought of working in motion pictures. He was simply searching for something to do when Metro offered him a job. "I grabbed at it like a drowning man catching a straw," he said later. Mandell didn't have a high school education, but learned his trade from his friend Arthur Ripley, the cutter who got him started in the business. Mandell worked for a time as an editor at Universal, where he went from picture to picture, mastering the craft. Then Sam Goldwyn, for whom he edited *Wuthering Heights* (1939), *The Pride of the Yankees* (1942), and *The Best Years of Our Lives* (1946), offered him a job earning more money. "I learned a lot from my experience in vaudeville," Mandell said. "I guess you develop a sense of how audiences will react to certain things. That you can't teach anyone; that comes from experience." (Mandell int., AFI)

William Reynolds, who won an Oscar for editing *The Sound of Music* (1965), grew up in Elmira, New York. A devoted movie fan, Reynolds was an English literature major at Princeton. After college he got a job working in the prop department at Fox Studios shortly before the merger with 20th Century. He soon realized he was interested in editing, so he began making contacts and managed to work his way into the editorial department as an apprentice. Reynolds said it was easier then because there were no unions.

During the 1920s several women graduated into film editing from jobs as negative cutters, script girls, or secretaries. Barred from more prestigious crafts within the old studio system, women could demonstrate their technical skills and assert their creativity in the editing room. "Trained from childhood to think of themselves as assistants rather than originators," Rosenblum and Karen declared, "they found in editing a safe outlet for their genius— and directors found in them the ideal combination of aptitude and submission." (Rosenblum and Karen, *When the Shooting Stops*) Elmo Williams felt women made excellent editors. "They are sensitive and as a rule have more patience than men do," he claimed. "Women don't mind all the fiddly little details that you have to deal with; they're very thorough."

Anne Bauchens, who edited all of Cecil B. DeMille's pictures from his remake of *The Squaw Man* (1918) through the sound version of *The Ten Commandments* (1956), as a young girl had been William DeMille's secretary. Born in St. Louis, Bauchens had aspired to become an actress and worked as a telephone operator to pay for lessons in drama, dance, and gymnastics. Arriving in Hollywood from New York with her boss Bill DeMille, she soon became his brother's trusted film editor, winning an Academy Award in 1940 for *Northwest Mounted Police*, the first woman to win an Oscar in that field. Bauchens worked with DeMille on his scripts before any footage was shot, suggesting transitions; she later devoted sixteen to eighteen hours a day to a project once it had reached the editing stage. On *The Ten Commandments*, DeMille's last picture, the seventy-five-year-old Bauchens labored for seven months. "She is still the best film editor I know," the director insisted. (Hayne, *Autobiography of DeMille*)

Margaret Booth, who served as editorial supervisor at Metro-Goldwyn-

Mayer, screened and advised on the structure of every MGM movie from 1939 through 1968. Booth began her career at Metro during the silent era and learned her craft by watching director John Stahl. Basically Stahl cut his own pictures in the early days, but he allowed Booth in the projection room, where he showed her different techniques of cutting. He pointed out why he went to a close-up: "Always play it in the long shot unless you want to punctuate something," Stahl told her. At night Booth would stay at the studio until two or three in the morning, practicing with the outtakes from a picture and learning to edit on her own. Working on silent pictures Booth became aware of rhythm. "You learned to cut film like poetry," she said. (Booth int., *Focus on Film*)

With the arrival of sound, Booth was frequently sent on the set to assist directors who had recently come from the New York stage and hadn't yet learned to make camera setups. She edited several of the Clarence Brown pictures and appreciated the director's lack of interference; she rarely saw him in the cutting room. Booth would edit portions of the film, then screen them with Brown, making modifications in response to his critique. Irving Thalberg came to rely heavily on Booth, giving her control of MGM's total output, a position she held for thirty years.

Adrienne Fazan, who eventually worked under Margaret Booth at Metro, started in the picture business in the late 1920s at First National. She became an assistant editor for Alexander Hall in the Colleen Moore unit at First National. She began working at MGM in 1930, cutting films in the shorts department. "It was wonderful training," Fazan later reflected. (Knox, *Magic Factory*) Outspoken as well as creative, Fazan graduated to feature pictures and in 1951 was nominated for an Academy Award for *An American in Paris*; in 1958 she won the Oscar for *Gigi*.

Dorothy Arzner, who entered the cutting room during the 1920s after typing scripts, eventually became one of Hollywood's few women directors during the big studio era. Arzner used to hold film up to the light, pull it through her fingers, and cut negative in her hands. "I was a very fast cutter," she said. "I cut something like thirty-two pictures in one year at Realart, a subsidiary of Paramount." (Rosenblum and Karen, *When the Shooting Stops*) Publicist Gail Gifford recalled how her friend Viola Lawrence at Columbia followed her husband into film editing, before it became necessary to go through the unions. He taught his wife the craft and taught her well, as evidenced by the number of successful pictures she edited.

Barbara McLean, whom Darryl Zanuck eventually made head of 20th Century-Fox's editing department and who edited all of Zanuck's own productions, launched her career in her father's film laboratory in Palisades

Park, New Jersey. She came to Hollywood in 1924, and Sol Wurtzel soon gave her a job at the old Fox studio on Western Avenue. McLean worked as Allen McNeil's assistant on *The Mighty Barnum* (1934) and with McNeil co-edited *The House of Rothschild* (also 1934). In 1935 she won an Oscar nomination for *Les Miserables* and subsequently edited or supervised hundreds of pictures. Attractive and dark-haired, known in the trade as Bobbie, McLean was nominated seven times for Academy Awards and won an Oscar for *Wilson* (1944). Creative, imaginative, and expert in her art, McLean was also quiet, efficient, and cooperative. She worked extensively with director Henry King, repeatedly demonstrating a solid dramatic grasp, a knowledge of what could be done with film, and a keen awareness of story values. Often she viewed a movie over a hundred times before the final cutting. McLean once said women were better cutters than men "because every woman is at heart a mother. A woman uses the scissors on a film like a mother would, with affection and understanding and tolerance." (McLean int., AFI)

Film editors agree that their craft requires a sense of story, structure, and pacing. "Rhythm counts so much, the pauses count so much. Everything must be rhythmic," insisted Margaret Booth. "It's the same as when people speak or dance. You can tell right away when it's wrong." ("Film Editors Forum") Daniel Mandell acquired his instinct for timing through acrobatics, while several other editors benefited from a musical background. Barbara McLean studied music as a girl and was convinced those skills gave her an awareness of tempo that helped her in editing. She could cut a musical, for instance, and have it always on the beat. William Reynolds concurred that timing is vital in editing and that it varies with action pictures and comedies (which move quickly) and dramatic scenes (which may need to be paced more slowly).

Most editors indicate that the director establishes the rhythm of a film, but Elmo Williams noted that the pace for him was set by the leading actor or actress. "They have a certain rhythm in the way they speak, the way they walk, the way they move," Williams said. "Since they're on the screen most of the time, that rhythm kind of establishes a rhythm for the picture."

Sometimes the editor wants an off-rhythm cut to create a feeling of disturbance. During his years at RKO, Elmo Williams edited *They Won't Believe Me* (1947), starring Robert Young. In the film Young's character is cheating on his wife and buys his girlfriend a diamond bracelet. The lovers meet in a restaurant, but instead of Young's character giving her the bracelet, they quarrel. Young goes home with the bracelet still in his pocket, to discover that it's his twentieth anniversary. His wife's family is there, dressed to go to the theater and celebrate. Convinced Young's character has forgotten his anniversary, the family begin to needle him. Suddenly Young remembers the bracelet and hands it to his wife, arguing that he had been detained at the office.

"The sequence didn't quite work," Williams recalled later. "I don't know whether the actors didn't quite know how to handle it or what, but I cut the sequence way out of rhythm on purpose. I would nip the end of somebody's dialogue, and then I'd leave a long pause at the end of another sentence. It helped a lot, because it irritated the audience. They didn't know what was wrong, but it supplied the irritant that was missing in the playing of the scene." Unfortunately Williams's department head was a purist who objected to the timing's being off. "We got in a big argument about it," Williams said. "He had the right to fire me, so I had to do it his way. I smoothed it all out, and the sequence was dull in the end."

Successful screen editors seek hidden meaning in a scene. "If you've got a choice between a dramatic point or mechanical perfection," veteran editor Gene Fowler, Jr., advised, "always choose the dramatic point and to hell with the mechanics." Knowledge of the flexibility of film is also important. "You can make film do anything you want it to do," Robert Parrish remembered Billy Hamilton's telling him. "It will lie for you or tell you the truth. It can make you cry or laugh or inspire you or confuse you. You just have to learn how to work with it so that it says the things you want it to say." (Parrish, *Growing Up*) Knowing what to leave out becomes essential. "Don't fall in love with the film," Gene Fowler cautioned, "but make it do what you want it to do." Billy Hamilton said, "The cutting room is the court of last appeal. It's the last chance to put out the best possible movie." (Parrish, *Growing Up*)

The film footage that reaches the editor's hands represents a combination of many talents. Elmo Williams argued that what the editor does with that film is closely akin to writing. "Story lines and characterizations that were sound on paper may sag on celluloid," Williams said. "Or the reverse may be true. Minor bits can become so vital to the story as it unfolds on film that they must be featured." The editor can change performances, shade characterizations, even alter the balance so that the story is significantly changed. "Editing is not as flexible as writing," said Williams, "because you're stuck with whatever is on the film, but you can still rewrite stories by changing the emphasis."

While technical proficiency is mandatory, Williams maintained that a good film editor also must have sensitivity, since he or she is constantly playing with audiences' feelings. "I always listen to my own emotions," he said, "and shape a film so that I react to it emotionally." A smooth, slick picture is simply not enough. "An editor has to feel what emotion is there," Williams said, "and tailor the film to get as close to that emotion as possible."

As with most of the other motion picture crafts, aspiring film cutters by the 1930s found breaking into the trade difficult. The studios employed their own people and were not open to outsiders. All of the department heads had their own teams, and it was often difficult for newcomers to feel accepted. "In my early days at RKO I was made to feel like an outsider and received no cooperation," Elmo Williams remembered. "The other editors used to talk to me, but it took a long time before we became friends."

The apprenticeship system for editors was long and difficult, but for those determined to learn the craft the payoff was excellent training. Ambitious apprentices usually became assistants and, if they showed talent, they were eventually promoted to film editors—an arduous route. "I have known several promising young men who have abandoned the cutting rooms because they were unwilling to spend seven or eight years at menial labor before getting permission to put scissors to film," Edward Dmytryk noted. (Dmytryk, *On Film Editing*)

Although trade classifications later became more specialized—sound effects editor, music editor, animation editor, trailer editor, feature film editor—in the early years editors did everything—cut negative, cut their own music and sound effects, lined up optical work, and supervised whatever miniatures were involved. Hours were long, the pay substantially below that of directors, writers, and cameramen. Gene Fowler, Jr., earned $80 a week in 1935 as an assistant editor at Fox, although the situation improved with the passage of the Wagner Act. By the late 1940s Elmo Williams made $350 a week at RKO, when $305 was considered the top rate. Editors were not under contract to the studios they worked for, but moving from one to another without the sanction of executives could result in blackballing.

Cutters had nothing to say about which films they worked on during the big studio era; they took whatever project was assigned. Sometimes a producer or star might request a particular film editor, especially once a cutter's reputation was established. Elmo Williams became known as a film doctor while he was still at RKO, and he was often asked to salvage sick pictures. "I got a little upset about that," he said. "It's pretty rare that the film that has to be doctored becomes a big hit." While it was gratifying to make a success out of a failure, editors only got credit for the really successful pictures. Within the major studios there was a division between the editors who worked on B films and those who worked on A productions, although it was possible to move up if a shortage developed in the A category. Assistant editors were usually assigned B pictures first, before moving on to A assignments.

Final cuts varied markedly from studio to studio, and on a purely technical level each editor developed his or her own approach. "It is probably safe to say that no two cutters will cut a film or even a moderately lengthy sequence in exactly the same way," Edward Dmytryk observed. (Dmytryk, *On Film Editing*) The finer the technique the less noticeable the editor's contribution, since the viewer should perceive only a natural, seamless flow of movement. Most editors work closely with the film's director, beginning any new assignment by reading the script, breaking it down and noting questions or problems. If they felt there were holes in the script, in the story or in the character, they could file a report with their department head, since the department head attended meetings with the producer and writers.

Studio directors often wanted their editors on the set. When they were, editors had more input into the creative process—suggesting setups, offering advice that might save time or money, making sure everything essential was covered on film, reminding the director about tempo. Barbara McLean used to spend three or four hours on the set with every director she worked with, noting their approach, particularly during the first weeks of production. At one point Paramount had a policy that an editor should be on the set at all times, even though some directors considered the cutter's presence a threat to their authority and arguments resulted. Most editors, however, contended that a director should have his way, and those who worked with the same ones repeatedly could almost second-guess the director and interpret the story the way he wanted.

The editing process usually begins with the cutter's compressing the raw footage into a rough cut, which will then gradually be refined. The editor's creativity comes in the selection, pacing, and arrangement of material. Excessive dialogue is eliminated, action is tightened, performances are heightened or toned down, flaws are covered, perhaps even allowing a scene to unfold around a bad actor's back. "The editor's first cut is a pretty pure version of the film," said Marjorie Fowler. "She or he starts cutting a film with the values that are there, what one presumes the director has tried to say through his film." An editor's skills are partly innate, partly acquired, as is the technical facility for handling film and the associated equipment.

One of a film editor's constant concerns is to make certain, after seeing each day's rushes, that all the shots needed for a sequence have been furnished. Cutters are realistic, more aware of the film that comes from the lab than the pages of script given to the director. Each day's rushes are spliced together on reels, then coded. "This took first priority," Marjorie Fowler remembered from her years as an assistant cutter at 20th Century-Fox. "The one set routine was getting the dailies ready for running." Since the director is still busy shooting the picture, the editor makes the rough cut of a film largely on his or her own. "You're not trying to make your version of the film," emphasized William Reynolds. "You should be trying to make the director's version and what you understand that he wants." But with extensive coverage the editor has a considerable amount of leeway; there are an infinite number of decisions on when to use close-ups, long-shots, or reactions. If the coverage is incomplete or the film is bad, the editor may devote much of his or her attention to covering up errors. Elmo Williams still looks back on *Nocturne* (1946) at RKO with horror. "I had to do an awful lot of trickery to make it work at all," he said.

In the early silent era editors simply determined breaks by jerking film through their fingers in front of an overhead light. The Moviola, an editing machine invented in 1919, came into wide use around 1925; it consists of a peep-show viewer that provides a better sense of what the edited film will look like on the screen. Picture and sound are run through the Moviola on two separate pieces of film. The editor can run the two separately or in sync and make separate splicing marks on each. Editors cutting a film spend hour after hour before their Moviola, watching the action and listening to the dialogue, always keeping the big screen in mind and periodically checking their cut in the projection room. The Moviola allows the editor to start and stop quickly, move forward or backward, and find the exact frame for a cut.

With the advent of sound the requirements of synchronization initially put a cramp in fluid editorial style. For a year or more, until technical obstacles could be overcome, stilted, static pictures were commonplace. The problems with keeping the picture and sound track in sync, and of organizing "trims" (the remains of various shots) for possible use later, resulted in the creation of the position of assistant cutter, which quickly grew in importance. "I remember the first time I ever came in the cutting room," Gene Fowler, Jr., said. "I saw this morass of film, bins full of film. I didn't know what it was or how anybody could possibly keep track of anything." Trims have to be filed by the assistant so that editors can readily locate whatever footage they might need, and an elaborate coding system was devised at the old studios, with logged code books so that everything became easily accessible.

The clapper snapping shut at the beginning of each take indicates the sync. The assistant editor runs both picture and sound track through a synchronizer, keeping them exactly even. Then the assistant takes both into the coding room and puts the same number on the beginning of the picture and the beginning of the sound track. The coding machine prints a consecutive number on every foot on both picture and track so that the editor can synchronize the two easily.

When Gene Fowler started working in the editing department at 20th Century-Fox he was first taught how to splice, a job no one else wanted. In the big studio era pieces of film were spliced with an acetone-base cement. "We used a heated Bell and Howell splicing machine that we worked with our hands and feet," recalled Robert Parrish. "A piece of positive film would be put in place, emulsion side up. We'd scrape the emulsion off, down to the celluloid base, apply some cement with a little brush, and then clamp the next piece onto it, celluloid to celluloid. These two pieces were held firmly between two hot metal plates for a few seconds and then the spliced-together film was rolled onto a metal reel. Each splice took about fifteen seconds to make." (Parrish, *Growing Up*)

Much of the film editor's work was performed in isolation, in bleak, windowless rooms. Elmo Williams preferred editing at night, since there were fewer distractions than during the day. "I liked to just lose myself in the film," he said. "So I did all of my editing after dinner, sometimes working until two in the morning."

Most cutters were conscientious; the good ones were meticulous craftsmen who exercised taste in choosing the best of several takes, constantly weighing and resolving problems. "At the beginning of a scene," explained Bill Reynolds, "you very often start with a master shot to show the audience where you are and who's there. But at a given point when something particularly important happens or a line of dialogue comes up, you may start to use close-ups. You use them for significant reactions of people to what's being said or what's happening. There's no rule about when to use them; it's just your own gut feeling about it."

"I wish the word 'editing' had never been invented," Verna Fields lamented. "People feel there is some kind of friction between the director and the editor, because the word 'editing' implies correcting, and it's not. In French the word is *monteur*, which is what it is: you're mounting the film." (Fields int., McBride I) Although the rough cut of a picture is largely put together by the editor and his or her assistants, once filming is completed, the major editing generally takes place in close association with the director and sometimes the producer. "Once the filming is over," Bill Reynolds said, "you start going through the film with a fine tooth comb with the director. It becomes a give and take."

Elmo Williams's procedure was to make a loose first cut, then run it for himself in the screening room. Conscious of how important music could be to the dramatic impact of a film, Williams haunted local record shops looking for music appropriate for certain scenes. "I never screen a film for a producer or a director until I put a temporary dub on it," he said.

Most directors wanted to view dailies with their editor to select the takes to be considered. Director Henry King usually asked his editor to join him for lunch. If the picture was being made on location, the editor would consolidate portions of the film already shot, then take it wherever the movie was being filmed, so the director could see how his work was coming along and if anything was missing. When Henry King was photographing *Captain from Castile* (1947) in Mexico, Barbara McLean flew down several times for editing conferences. During the shooting of *The Westerner* (1940) in Tucson for Sam Goldwyn, Daniel Mandell was on the set most of the time. If the editor was busy with another picture, phone conversations with the director on location might suffice. While Elia Kazan was shooting *Viva Zapata!* (1952) in northern Mexico, Barbara McLean was finishing up another assignment, so she talked to the director every day after she'd seen the rushes.

Occasionally the editor was asked to solve a problem the director couldn't. During the making of *The Pride of the Yankees* (1942), for example, director Sam Wood discovered that Gary Cooper, who played Lou Gehrig in the picture, was right-handed whereas Gehrig had been left-handed. Daniel Mandell came up with an editor's solution. "It's easy," he said. "All you have to do is put the letters on his shirt backward, have him hit right-handed, run to third base instead of first, do everything in reverse, and we'll flop the film over." (Mandell int., AFI) Other solutions were less obvious. When Lewis Allen was directing Gail Russell in *The Uninvited* (1944), he realized Russell was gorgeous, but no actress. Allen consulted Doane Harrison, who was cutting the picture. "The only thing you can do with this girl," Harrison told the director, "is to get it in bits and pieces. Just do it three or four lines at a time, and I'll splice the pieces and make it work." (Allen int.)

Some directors were notorious for shooting very little film, giving the editor only footage they liked. Adrienne Fazan found Vincente Minnelli to be stingy when she cut *An American in Paris* for him. "He never gave me enough close-ups," Fazan complained. (Knox, *Magic Factory*) Other directors mistrusted editors, purposely giving them only what they wanted in the final cut. "I got where I would only shoot what I needed and what I wanted," director Charles Walters said. "I liked my own flow and sense of rhythm and choreography. I didn't trust cutters." Joseph Mankiewicz preferred to cut as he shot. "I'm not giving that cutter one frame that I don't want on the screen," Mankiewicz often said. (Holm int.) John Ford, on the other hand, seldom concerned himself with editing, convinced that what he wanted to convey was unmistakably on film.

After discussions of the editing, even the most powerful directors were unwelcome in the cutting room while the actual work was being done, and producers usually left editors alone unless they had specific objections. A producer might question the length of a scene that was strictly atmospheric, or a joint decision could be reached to add a sepia tint to a scene. "In a black and white show you can add sepia for warmth," Elmo Williams pointed out. "Sometimes they used to add blue tint for moonlight. That can help the mood of a scene just as a piece of music does." If the director and producer disagreed on the interpretation of a script, the editor could become embroiled in discord. "A gift for tact and diplomacy, the ability to be the recipient of confidences from both sides without betraying either," film editor William Murphy wrote, "sometimes seem as necessary as skill in editing itself." (Murphy, "Film Editing") Daniel Mandell recalled that Samuel Goldwyn and director William Wyler weren't always in accord. "I was in the middle," said Mandell. "I had to please both of them. I used to cut it the way I thought it should be and then have them look at it, and if they disagreed about anything, I'd just sit there and let them fight—and keep my mouth shut unless they asked me."

Occasionally a star would come to the cutting room, maybe with a box of chocolates, in an attempt to convince the editor to put in a couple of extra close-ups. "Some actors count the number of close-ups you make of them," said Elmo Williams, "and if you make one of this star, you've got to make one of that one, whether you need it or not." Actors and actresses often were fearful about what could happen to their performance in the cutting room, and they were crushed if their part was lessened. Actor John Saxon had hoped his role in John Huston's *The Unforgiven* (1960) would lead to an Academy Award nomination. When the film was released, more than half of his performance was cut. Actors from the stage found it particularly disturbing to have so little control over their final performance. "Your performance goes into the can," commented actress Julia Meade, "and it's up to the director and the editor to put it together." Sometimes a part would be discarded in the cutting room simply because it overshadowed a player the studio was trying to build into a star. "You're all on the cutting room floor," director Woody Van Dyke announced to actress Signe Hasso, after making her part bigger in *Journey for Margaret* (1942). "You were too good," the director told Hasso. "Your part took too much interest away from Margaret O'Brien." (Hasso int.)

Since the stars of the Golden Era were established personalities, film editors didn't have to spend a great deal of time building characterizations. Gary Cooper projected a likable quality and won the sympathy of audiences almost immediately. Wallace Beery could chew tobacco and be a dirty old man and even play a villain, but he had a heart of gold and audiences knew it as soon as he walked on the screen. "In editing you had to keep that characterization consistent," Elmo Williams said, "but you didn't have to build it."

Most of the big studio editors found musicals less difficult to cut than dramatic pictures. "With musical numbers, you go by the music," said Adrienne Fazan; "it's very easy. The dancers and the director take great pains to match action." (Knox, *Magic Factory*) Editors simply had to know when to cut on the downbeats. Sometimes technical problems arose prompting the music editor to ask for a scene to be lengthened or shortened.

After a movie has been filmed there is always a certain amount of "looping"—dialogue to be replaced for various reasons. By the time *Giant* (1956) was finished, actor James Dean was dead. Parts of his drunk scene as Jett Rink had to be looped, since Dean had played it too drunk and sections of the dialogue couldn't be understood. Warner Bros. brought in actor Nick Adams to re-record portions of that speech. Frequently editors ordered inserts—brief footage like a lantern being tossed into a haystack or shots of a prop that will clarify or add mood to a scene; sometimes editors directed those inserts themselves.

Of the big studio heads Darryl Zanuck was the most intimately involved with the editing process. Zanuck respected editors, was appreciative of what they could do, and was eager to see the first cut of every major 20th Century-Fox film. If a retake was necessary, Zanuck needed to know as soon as possible, to keep actors available and to make sure the set wasn't torn down. Zanuck fancied himself an expert film editor and demanded final say. "He wasn't a good editor," Gene Fowler, Jr., insisted. "That's a myth. Zanuck was a fine constructionist. We always claimed he cut with an ax." Zanuck often eliminated motivation, emphasizing action, but he could pinpoint defects and knew precisely what was necessary to salvage a weak script. "He could take a picture that was a bomb," assistant director Gerd Oswald claimed, "and he would take the middle and put it at the end, the end and put it in front, mix the whole thing up, and all of a sudden you had a decent picture."

Zanuck possessed great powers of concentration. He had the editor sit next to him in the cutting room, and when a problem arose, he'd touch the editor on the arm, indicating a change needed to be made. Sometimes the atmosphere grew tense. Zanuck always sat in an enormous, leather wingback chair. "All you could see were puffs of smoke from his tremendous cigar coming up like Indian smoke signals," producer Herbert Bayard Swope, Jr., recalled. "If Zanuck coughed, you died."

Usually the Fox production chief brought along a half dozen of his stooges, all of whom had designated seats. Since there was no intercom, Zanuck's chair had a button on the side allowing him to signal the projection booth when to start, stop, or focus the film. On the other side of his chair was a series of buttons that sent an electric shock to the seats of his various henchmen. Whenever Zanuck was feeling playful or thought things were getting dull, he'd push one of those buttons. "One of the guys would raise to his feet yelling and screaming," Gene Fowler recollected, "which they all thought was funny." When the picture ended, Zanuck usually called for soft drinks. "Everybody would uncap their bottles and throw the caps on the floor," Gene Fowler said. Zanuck would begin by asking everyone's reaction, as he paced up and down with a broken polo mallet. "He would line the caps up," recalled Fowler, "and as he'd walk by he'd whack them into the screen and dent it."

But most of the evening's task was taken seriously. Zanuck's capacity for remembering the exact spot in a movie he felt needed to be expanded, eliminated, or changed was legendary, and he was rarely, if ever, reluctant to make decisions. If the picture was in trouble, he would order it run reel by reel. "We'd know we were in for a good night's session then," said Fowler, "maybe not out until three or four in the morning." How much Zanuck relied on his stooges remains a point of controversy. Some observers felt he regarded their opinions highly, as representative of the average moviegoer; others discounted their influence. "I don't think he ever made changes in a picture based on anything they said," William Reynolds said. "I never observed that he talked to those people much in the projection room after a screening. Maybe his contact with them came afterward."

Early in the 1940s the head of the editing department at 20th Century-Fox was Hector Dods, an Australian who had taught Darryl Zanuck polo. "He never learned anything about editing and never wanted to know anything," Gene Fowler claimed. "The only thing he did was make assignments." Fowler himself learned a great deal when he came into the department from watching Allen McNeil, an old cutter from silent film days who had worked years before with Mack Sennett but had never quite learned how to marry sound to pictures. "He would cut the picture silent," Fowler explained, "just going by the lip movement, having seen the dailies with sound on the screen. But Mac had the magnificent faculty of looking at anything on the screen and telling you how it could be better, what the weak spot was, what was necessary or what was unnecessary." McNeil had learned his craft through experience, and knew how to survive in the old studio system, always carrying an empty film can under his arm wherever he went. "He didn't want anybody to think he was loafing," Fowler said.

On the eve of World War II there were roughly twenty cutting rooms in the long editing building at 20th Century-Fox. The building had been designed specifically for that purpose; along one side of the hall were the A editors, who worked with Zanuck, and along the other side were the B editors, who worked under Sol Wurtzel. The B editors had earlier been housed at the Fox studio on Western Avenue. Around 1940 editors on the A side received $125 a week, while editors on the B side got $110. Payday was every Wednesday.

Marjorie Fowler began in the studio's editing department as an apprentice in the coding room, finding only two other women there at the time. "One had lovely, long red fingernails and high heels," Marjorie Fowler remembered, "and was rumored to be the girlfriend of Hector Dods." Since film editors used hot splicers then and acetone to clean the blades, a woman's long fingernails and perfect nail polish indicated she wasn't doing much work. "About the second day I was there," Marjorie Fowler continued, "this woman came up to me and said, 'Now I don't want to see you carrying any film cans around here and making trouble for us.' I was determined I wasn't going to be like that." The other woman in the department was Barbara McLean, who eventually became Zanuck's top editor.

Once the men grew accustomed to her, Marjorie Fowler found everyone in the department pleasant, but she soon discovered that the life of a film editor was like being inside a cocoon. "It was just like a little world all to itself," she said. "I thought it was marvelous, as long as I didn't have to go on the set." Dailies were run every afternoon for the producer and usually again at night for the director. Editors enjoyed both steady employment and job security, as long as their performance was judged satisfactory. Once the mechanics of their trade were mastered, much of the work was routine.

In his heyday at Metro-Goldwyn-Mayer, Irving Thalberg demonstrated a keen eye for film construction, often finding values no one else did. "He won me my first Academy Award," actress Helen Hayes declared. Thalberg was at Bad Nauheim, a spa in Germany, when *The Sin of Madelon Claudet* (1931) finished shooting. Its script by Hayes's husband, Charles MacArthur, had been taken from *Lullaby*, an old play, but when Metro previewed the picture under that title, it was such a bomb the studio shelved it. "Irving came back and wanted to know what had been going on during the months he had been away," Helen Hayes remembered. He specifically asked about the picture MacArthur had written for her, only to be told it had failed completely. Thalberg insisted on taking a look. "It only needs a couple of retakes," Thalberg told MacArthur after screening the film, "a scene here and a scene there. You rewrite those and we'll retake them, and that's it—you've got a picture." Neither MacArthur nor any of the lesser studio executives saw the point, but the retakes were shot—on Sundays, since Hayes by then was making *Arrowsmith* for Samuel Goldwyn. "Sam Goldwyn was furious and nearly fired me when he found out," she said. "Irving renamed the picture *The Sin of Madelon Claudet*, which he thought would draw people at the box office, and it was a smashing hit. I won an Oscar for it."

When Thalberg died in 1936, Margaret Booth was appointed supervising editor at MGM, a position she held for three decades. She viewed the dailies of every picture made at the studio, looked at all the first cuts, and gave the editors her reaction, offering suggestions. "I think you have to go by the style of the director," Booth commented. She found the atmosphere at Metro to be cooperative, with several editors working together on a project to meet a rush deadline, each cutting two or three reels of the same picture. "It was really a great family studio," Booth said. (Booth int., *Focus on Film*)

Most workers stayed at MGM for decades. Fred Y. Smith remained in the editing department there for eighteen years, beginning with *Little Nellie Kelly* (1940), his first assignment for the company. Adrienne Fazan was at Metro even longer. She found producers and directors at the studio agreeable, especially producer Arthur Freed. "It was easy working on the Freed pictures," Fazan said. "When he hired a director on a picture, he gave him the responsibility and trusted him." (Knox, *Magic Factory*)

RKO enjoyed little of MGM's stability, although the constant turnovers in management didn't affect the RKO employees. During the late 1930s Jimmy Wilkinson headed the editing department there, assisted by Carl Hunt and two secretaries. "Jimmy Wilkinson didn't really care two hoots what kind of cutting was done as long as he protected himself," Elmo Williams declared. "He was a typical department head that thought of Jimmy number one. I never heard of him telling anybody how to cut a picture." Wilkinson was head of the department when Robert Wise, later a successful director, entered the RKO cutting rooms as an apprentice. "It just so happened that he needed a young, eager kid with a strong back who could carry prints up to the projection booth for the studio executives to run," Wise remembered years later, "someone to check prints and that kind of thing. That's how I started. I became an apprentice, a sound effects editor, a music editor, an assistant film editor, and finally an editor." As the editor of *Citizen Kane* (1941), Wise won international recognition. Robert Wise and Mark Robson, another future director, taught Robert Parrish, a new boy in the RKO department, how to splice, clean reels, file trims, and keep his mouth shut in the projection room. "Then they taught me to . . . do synchronization," Parrish wrote, "to lie to producers, and, finally, to begin to edit sequences." (Parrish, *Growing Up*)

Like the other major studios, RKO was compartmentalized, and editors labored under a strict set of rules. "Things were very tightly controlled in those days," said Elmo Williams. "We had to send in a report every night to the department head on what progress we'd made, what was the footage, how much we'd cut that day, and any problems we'd encountered. We caught hell if we didn't get those reports in." RKO employed around twelve editors and generally as many assistants. There was a trailer department, consisting of two people, and about eight people in sound editing. Each department had its own budget, and the department heads worried about going over and being reprimanded by the production boss. "All of them used to pad their budget by anywhere from ten to twenty-five percent," Elmo Williams claimed, "just to keep from being put on the hot seat."

Later the head of the editing department at RKO was Billy Hamilton, among the best cutters in Hollywood. Hamilton had been a boxer and a stunt man and had survived four or five studio administrations at RKO. "His cutting room was decorated with Petty-Girl cutouts from *Esquire* magazine and a sign that read, 'In this room ART is spelled with an F,'" Robert Parrish remembered. (Parrish, *Growing Up*) Hamilton drank a great deal and sometimes came into the cutting room in the morning in a shaky condition, only to assign work to his staff and hastily exit out the back gate to the Melrose Grotto, where he drank until noon and then returned to the editing room ready for an afternoon of serious work.

Until the advent of sound, Paramount hired mostly women cutters, but that situation changed quickly. "Feminism wasn't in vogue yet," commented Edward Dmytryk, "and the idea was that women weren't up to the technical aspects of cutting sound." Paramount acquired a fine group of cutters when First National sold out to Warner Bros. in 1929. Warners brought in their own people and eased the First National editors out; therefore first-rate editors were looking for jobs. George Arthur, who headed the editing department at Paramount, acquired Doane Harrison, Stuart Heisler, Hugh Bennett, and Billie Shea at that time.

Young Fred Y. Smith was one of the few First National editors that Warner Bros. kept on. Originally hired by First National as a projectionist, Smith spent as much time as possible in the cutting room learning from editor Hugh Bennett. He spliced film, made up cue sheets, wound film, and finally convinced Bennett he was capable of editing film. The first feature Smith cut was *Sweet Mama* (1930) with Alice White. Later he worked at Warners on several of the Busby Berkeley musicals and observed how the director "would only photograph the angles which he wanted to see on the screen. He didn't cover a scene from any other angle." (Smith int., Steen) Berkeley's technique, called "camera cutting," proved fast and inexpensive but sometimes created havoc in the editing room, since the cutter had no alternative material to cover mistakes. By the mid-1930s Warner Bros. maintained a staff of fifteen to twenty editors. Jack Warner never involved himself with the details of editing the way Darryl Zanuck or Irving Thalberg did, relying instead on his production heads.

Editing a picture normally takes twice as long as shooting one, sometimes longer. Marjorie Fowler felt five months was the ideal amount of time for a cutter to work on a film. "I've been on one as long as nine months," she said, "but I think that is bad. Your motivation deteriorates, and you lose objectivity." All of the old studios maintained a sound library, containing sounds of every description on an optical track. Since those negatives were repeatedly used, the quality became scratchy and noisy. Editors therefore recorded special sound effects for the better pictures. Eventually the composer would be brought in for discussions with the editor about the score, how much music would be needed, and its texture for various scenes. While the finished cut of a movie remained a team effort, the editor and the director during the big studio era generally exerted the greatest influence in shaping a final release, unless the producer was unusually powerful. "It's quite amazing that any film succeeds," Elmo Williams reflected, "because any one of the key people along the line can push the trolley off the track."

During Hollywood's Golden Age once the sound effects and music had been added to the finished cut, a film was usually taken out for a "sneak preview," its first test before a random audience. Sometimes the sound and picture strips hadn't been put together yet. Several theaters in the Los Angeles area were equipped with double-head projection, which allowed the separate strips to be shown in synchronization. "When a preview location was chosen," wrote Robert Parrish, "we would pile into big black studio limousines and drive to the theater about an hour before our film was scheduled to start." (Parrish, *Growing Up*) Occasionally cutters had to race to be ready. Margaret Booth recalled taking the first two reels of a film with her in one car to the screening, while the rest was being completed back at MGM. "The theatre would start the picture while the cutters who were working on the last two reels were still in Culver City," Booth said. "They would have to go practically on two wheels to get to the preview." (Booth int., *Focus on Film*)

While the picture was being shown, the editors and studio executives would watch and listen to the audience. Daniel Mandell related how he'd note the place in the film where people started coughing or where they seemed to get restless and what they laughed at. Usually there would be a sidewalk discussion with executives after the preview, followed by a meeting at the studio the next morning to decide what needed to be done. Darryl Zanuck preferred that everybody involved gather back at the Fox studio that same evening, while the film was fresh in his mind. If a picture was in trouble or simply not very good, it might be previewed as many as five times.

After a preview, stamped cards were handed to the audience as they left the theater. Each person was asked to rate the picture and make comments. Pencils were provided in the lobby, and participants could either fill the cards out there or mail them in later.

Writers and directors occasionally felt a picture had been ruined by placing too much emphasis on the preview audience's response. Paramount's *Chicago Deadline* (1949), based on Tiffany Thayer's book *One Woman* about an investigative reporter who puts together the pieces of a dead girl's life, was one such instance. Alan Ladd played the reporter, Donna Reed the girl. The problem was that Reed's character was dead throughout the entire movie. Writer Warren Duff finally came up with the idea of having her appear in a series of flashbacks in Ladd's mind, essentially dream sequences. Since big studio producers considered fantasy to be certain death at the box office, Paramount's executives decided to cut the dream scenes after they had been shot and play *Chicago Deadline* as a straight melodrama. Over director Lewis Allen's protests almost a half hour was deleted from the picture, and it was previewed in Pasadena with the fantasy sequences omitted. The audience seemed to love the picture, turning in rave comments. "Warren Duff and producer Robert Fellows and Alan Ladd and myself were fit to be tied," Lewis Allen later said. "But what could we do? We were just employees. That was one of the troubles with pictures in the old days; lots of them were fiddled around with by the studio."

Film editors with ambition often became involved in the inevitable studio politics; maintaining friendly relations with producers, directors, and executives could further one's career. But others preferred the seclusion the editing room provided. With the major studios turning out fifty or so films a year, it was easy for cutters to be engulfed by the assembly-line process, since pressure to deliver product was constant. Most editors did their work conscientiously, socialized occasionally, particularly after the American Cinema Editors was formed, but had little contact with anyone outside their own department except those involved with a specific assignment. "For film editors the major studios in those days were islands unto themselves," said Gene Fowler, Jr. "The MGM people didn't know the Fox people, the Fox people didn't know the Columbia people, the Columbia people didn't know the Warner Bros. people."

Some with unusual talent or drive found the relative obscurity of the editor's cubicle unsatisfying. Edward Dmytryk, Robert Wise, Mark Robson, Dorothy Arzner, and Stuart Heisler, among others, eventually became directors, earning more money and status. Elmo Williams won acclaim within the industry as a second unit director and later produced a few pictures. But in each case, they continued to work closely with their film editors and understood the technical aspects of filmmaking as few others did. While studio directors occasionally complained about editors who were unimaginative mechanics, other editors became recognized as virtuosos who injected film episodes with an emotional impact beyond that supplied by the actors and director.

15

Around the Lot

Columbia Studios at Sunset Boulevard and Gower Street during the 1940s. Courtesy of the Academy of Motion Picture Arts and Sciences.

From the outside, a Hollywood studio appeared drab, a motley assortment of buildings that looked like airplane hangars, warehouses, and government office buildings, surrounded by massive walls of concrete. With guards and rigid security systems, the studios resembled prisons or fortresses. Inside, during the big studios' heyday, a rush of activity continued throughout the day, every sound stage busy, people scurrying around, filled with excitement even when the work was tedious. Each studio had its own main thoroughfares and side streets, and between two thousand and six thousand employees. "It was a factory," comedienne Nancy Walker said of Metro-Goldwyn-Mayer in the 1940s, "but a very glamorous one." Throughout World War II every sound stage in Hollywood was in demand, with production units awaiting their turn. "If you were behind schedule," actress Jean Porter commented, "you were holding up the next company. That's how busy it was."

Old Hollywood had class distinctions and even a caste system, but social lines were crossed in the working environment, and a rapport among studio workers of all ranks developed over a period of time. The dressing room building for the stars at Warner Bros. was a secluded two-story structure, facing a tennis court and surrounded by giant shade trees. Each star was assigned a commodious apartment containing a living room, bedroom, kitchen, and bath, all decorated and furnished to taste. If a character actor, such as Claude Rains, attained sufficient success, he was moved from the building where featured players dressed to a star's suite, in keeping with his elevation in salary and status. The first floor of Paramount's three-story building called Dressing Room Row contained the dressing rooms of the studio's biggest stars — Bing Crosby, Mae West, Carole Lombard, and W. C. Fields, among others. The second floor housed the lesser stars, while the third provided space for supporting players. In the mid-1940s, Tyrone Power and Betty Grable occupied the largest dressing rooms on the first floor of the stars' building at 20th Century-Fox; Gene Tierney and Linda Darnell were across the hall from one another on the floor above. Power's quarters had oak paneling and leather chairs. Liquor was not permitted in the dressing rooms at any of the studios, although infractions took place. Harry Cohn posted an older woman in the dressing room area at Columbia to report what was going on there. "She was friendly and nice," actress Jeff Donnell remembered, "but she knew everything that happened, and her line was direct to Harry Cohn if anything untoward occurred."

As in most jobs, griping was almost universal. Complaining and trying to outwit the front office bound studio workers together. Directors insisted they didn't have enough time to make their pictures properly or that the studio wouldn't spend enough money. Actors referred to Warner Bros. as "San Quentin" and to 20th Century-Fox as "Penitentiary Fox." Actress Lynn Bari, under contract to Fox, claimed that Darryl Zanuck had his own private grapevine: "I always felt like Big Brother was watching. I really think they had a system there of keeping an eye on you so you didn't say the wrong thing."

Maria Montez, under contract to Universal, was perennially angry over money or roles or the studio executives' lack of consideration. Something of a primitive in her personal habits, Montez refused to walk the distance to the bathroom, using the wastebasket in her dressing room instead, a practice that was chattered about by studio gossips. (DeCamp int.)

Other performers resented their studio's "potato clause," which stipulated that their weight must be kept at a certain level. Columbia's contract with western star Charles Starrett stated that he must stay between 190 and 200 pounds. "They didn't want a fat cowboy," said Starrett. Tyrone Power, on the other hand, was slender and was expected to work out in the gym at 20th Century-Fox regularly to build himself up.

Stars mingled with studio workers and made newcomers feel welcome. Arlene Dahl recalled meeting Gary Cooper on her first day at Warner Bros.; he invited her to lunch. Young Metro player Laraine Day found Spencer Tracy remote when they were cast in the same picture, but she remembered Walter Pidgeon as charming, allowing her to use his dressing room on the set when he wasn't working. Marlene Dietrich and singer Rosemary Clooney had made a record together before Clooney signed her contract with Paramount. Dietrich was a veteran on the lot and knew everybody there. "Marlene sent notes to the various departments with me," Clooney recalled affectionately, "so I was treated very specially, because Dietrich made a point to see that I was." Dietrich coached the young singer for her role in the musical *Red Garters* (1954), since Clooney's part satirized Dietrich's character in *Destry Rides Again* (1939). "I had to learn to roll a cigarette with one hand," Clooney remembered, "and I couldn't get the tobacco to stay in. I had Bull Durham all over the Paramount lot just practicing. Finally we cheated; we put another cigarette inside the paper."

Over a period of years, lasting relationships developed. Actor Marshall

Thompson and young Marsha Hunt played siblings in Metro's *The Valley of Decision* (1945); years later they still referred to themselves as brother and sister whenever they ran into one another. Thompson dated both Elizabeth Taylor and Jane Powell while the girls were attending the Metro school. "I was out of school by that time," the actor said, but the studio was "like a home away from home." Even on days when he wasn't working, he would often go over to the studio to visit friends. Young players starting out sometimes became roommates, forming friendships in the process. If performers working on the same picture lived close to one another, they occasionally carpooled to work. Often New York actors didn't own a car, so someone who lived nearby picked them up in the morning and drove them home at night. Actress Ella Raines recalled dropping by the Beverly Hills Hotel for director George S. Kaufman when they were making *The Senator Was Indiscreet* (1947) together. "George didn't drive," the actress explained, "and this was his first film, which he approached exactly as he would a play." Raines enjoyed her association with the veteran stage director, but not all such experiences were so pleasant. "I got to drive Denise Darcel to work," actress Vanessa Brown remembered of making *Tarzan and the Slave Girl* (1950). "What a pill! At five-thirty in the morning I'd pick her up and she'd be eating an orange. She'd tell me about her diet and then read me her press clippings."

As in most families, some members were well loved, others tolerated, still others admired but not particularly liked. Most beginning players were eager to grow professionally—taking lessons, watching other people work, making tests, visiting set after set in the hope of talking to directors and producers, or at least important crew members. "I enjoyed making tests," character actress Ann Doran said, "and I learned even more from doing them than from making pictures. You'd do exactly the same scene with four different people, and it really became four different scenes."

If Cecil B. DeMille was shooting a spectacular sequence or Esther Williams was about to start one of her water numbers, the word would circulate around their respective studios, and people from all over the lot would come and watch. On days when "cattle calls" went out for chorus girls on musical numbers, male workers found an excuse to be in that vicinity. "The chorus girls obviously had to dress in tights or shorts to show their figure," Edward Dmytryk recalled from his days at Universal, "and men from all over the studio would be standing there leering."

"You would walk down the main street at Metro," publicist Esmé Chandlee said, "and you would hear Mario Lanza singing on Stage 15 or Kathryn Grayson's voice coming from up above where she was practicing with the Maestro. I remember I was walking down the street once with some visitors, and there were some other people behind us. This woman said, 'I've been on the lot for an hour and I haven't seen one star.' Right ahead of me was Judy Garland with Roddy McDowall. Judy turned around and winked at me, and the woman was looking right at her." But even professionals made mistakes when it came to identifying talent. Actress Catherine McLeod recalled running into a "little blonde with dirty tennis shoes" and being so unimpressed with her that she advised the aspiring player to get out of show business altogether. The blonde later became Marilyn Monroe.

Director Irving Rapper described Warner Bros. as "a family feud in the Democratic Party. We were always feuding, always getting suspensions, but I must say, it was a family." If cliques and pettiness existed among veteran studio workers, newcomers sometimes felt shut out. "A strange face was murder," makeup man William Tuttle said of his time as a messenger at Metro. "I would go down to the stock room to pick up some office supplies or Kleenex, and I could stand there forever. The store clerk would wait on everybody before he'd wait on me. I was an outsider. After I'd been there awhile, they finally accepted me."

Among the in-group, gossip was a constant source of amusement, for each studio functioned much like a small town. Everybody knew who was sleeping with whom and when marriages and romances had run their course. One famous Metro producer rented a house behind the studio for his frequent dalliances. "All of the secretaries could look out the administration building," said actor Marshall Thompson. "They knew it was his house, and they had binoculars to see which starlet walked in next." When the same producer fell in love with a lovely New York dancer, he ordered a mammoth dressing room decorated for her, with the walls padded in pink satin. "It was like something from *Marie Antoinette*," Jayne Meadows recalled. "I remember going into her dressing room and just staring." On the other hand, Metro contract player Richard Ney felt his marriage to the studio's grande dame, Greer Garson, almost ruined his acting career. Convinced that MGM executives were giving him parts only because he and Garson were married, Ney eventually asked for a release from his contract.

During the latter half of the 1930s, motion picture studios, like other American industries, began experiencing unionization. Much of the rapport within the various crafts henceforth came through guild activities, which sometimes entailed bargaining with studio representatives. After 1925, with the formation of Central Casting, even screen extras were organized, hired by telephone. Central Casting enabled registered players to call in for work any time. Studio casting offices notified the central bureau daily of the types of extras required and the kind of wardrobe expected for the following day's shooting. Over several months extras working regularly at the various studios got to know one another.

Producers and directors not only tended to remain loyal to actors and actresses who had worked for them over the years, but went out of their way to hire the same bit players. An esprit also developed among producers and directors. Producer Henry Blanke and director Michael Curtiz, who often worked together at Warner Bros., used to yell at each other even when discussing where they would have dinner that evening. "It was just the way they were," said Ann Doran. "They were both volatile men."

But rivalry and thoughtlessness also existed. At Paramount during the late 1940s independent producers Hal Wallis and Cecil B. DeMille jockeyed with each other for screening room 5 to view their dailies. "If they were both in production, the question was who had priority on that room," music executive Bill Stinson recollected. "Everybody thought that DeMille would, because of his longevity. But Wallis always got it; he was powerful." During the time Richard Powers was head of Metro's music department, someone came into the publicity office one Monday morning and asked, "Did you see what happened in the music department?" Since Esmé Chandlee happened to be covering that department, she went down and found Powers's piano out on the street. "When he came in at ten o'clock that morning," she said, "that's the way he learned he was through."

Some vital operations of a studio were isolated from the mainstream. Special effects became such an area, supervised by mild-mannered mavericks. While most movies utilize special effects in some form—to create rain or snow, or to make sure spears and arrows land where they are supposed to—others require major effects to simulate a plane crash or a train wreck, perhaps even the burning of Rome. Usually major effects are created in miniature and intercut on the screen with footage of actors performing on full-size sets. Process shots, involving stationary sets in the foreground with scenery and action projected onto a screen in the background, grew in importance after the early 1930s. Generally a studio's special effects department worked closely with the art department, since matte paintings and other artwork were often needed to create or enlarge an illusion, but their operation was specialized, not part of the day-to-day routine of most workers.

Until the end of the silent era special effects were mostly described as trick photography, a term old-timers continued to use. Around 1910, Norman Dawn became one of the first special effects consultants in Hollywood, devising a glass shot process that made possible shooting action with performers either in front of or behind backgrounds painted on glass. Montage work, sequences showing the evolution of time or a transition from one subject to another, gradually became more sophisticated and subtle after the late 1920s, especially in the hands of such montage geniuses as Slavko Vorkapich, who arrived in Hollywood in 1928 after studying painting in Paris. Even as a student Vorkapich visualized compositions that moved. He saw D. W. Griffith's *Intolerance* during World War I and decided motion pictures were the art for him. Believing in the choreography of natural movements, Vorkapich developed his craft, conscious of rhythm and imagery in expressing a theme cinematically. The goal of montage, he said, is to "generate general themes, not a story. It's either a time lapse or a camera mood, but it must have visual expression and must have movement."

Willis O'Brien became the pioneer of stop-motion animation, establishing his reputation with *The Lost World* (1925), in which he contrived clashes between man and prehistoric monsters. O'Brien achieved his effects by using models, placing them in the necessary positions, and adjusting them slightly after each frame of film was exposed, so that the models appear to be moving on the screen. A man of great skill and patience, he lifted model animation to an art form, achieving his greatest success with *King Kong* (1933) and winning an Academy Award for his effects in *Mighty Joe Young* (1949), both for RKO. O'Brien's model of King Kong was only eighteen inches tall. Using stop-motion animation to cause an elbow or knee to bend a fraction of an inch at a time, a fifteen-second shot might require days of work; the effects for *King Kong* took over a year to complete.

O'Brien's principal disciple, Ray Harryhausen, had seen *King Kong* as a teenager. "*King Kong* influenced my career from the beginning," Harryhausen said later. "*Kong* is reflected in all of my work—every picture I make." Harryhausen worked with Willis O'Brien on *Mighty Joe Young*, then went on his own with such monster films as *The Beast from 20,000 Fathoms* (1953) and *It Came from Beneath the Sea* (1955). Of O'Brien, Harryhausen said, "He was my hero for many years." But Harryhausen developed his own philosophy and perfected his own approach. "It's important that you believe in the fantasies you are creating on the screen," he said. "Otherwise you end up with a tongue-in-cheek attitude, which for me is death to a fantasy picture." (Harryhausen int., *Film Comment*)

Linwood Dunn, who became head of the photographic effects department at RKO, arrived in Hollywood in 1926 as a cameraman with a serial unit. Two years later the out-of-work Dunn was playing in a dance band when he received a call from RKO offering him a couple of days' work in special effects, a job that led to permanent employment. "I really lucked out," Dunn said. "RKO was the smallest of the majors, and because it was smaller we did all the various types of effects in one department. The big studios like Fox and Metro had a separate department for everything—one for matte paintings, one for optical printing, one specially for miniatures, another for background projection—but at RKO everything came through the one department." (Finch, *Special Effects*) Dunn worked on such diverse classics as *Cimarron* (1931), *Flying Down to Rio* (1933) and other Astaire-Rogers musicals, and *Citizen Kane* (1941).

Most actors and actresses worked very little with studio special effects people. Cinema enthusiasts, such as jazz singer Mel Tormé, who were curious about all aspects of filmmaking, occasionally found their way into the special effects quarters to observe, but they were the exception. "What really interested me were miniatures," Tormé said of his Metro years. "The technical end of movies still absolutely fascinates me." For most studio employees, however, the special effects department remained a mystery.

Filmmakers readily acknowledged the vital role special effects played in achieving the realism of motion pictures, and studio executives usually appropriated ample funds for effects crews to create their magic. In 1932, for example, Carl Laemmle, Jr., approved the construction of a $50,000 sound stage at Universal, to be used exclusively by the studio's special effects department. John P. Fulton was responsible for the effects in Universal's *Frankenstein* (1931) and *The Invisible Man* (1933). Later Fulton replaced Gordon Jennings as head of the special effects department at Paramount.

Jennings had worked on many of the Cecil B. DeMille productions at Paramount, including the train wreck in *The Greatest Show on Earth* (1952) and the destruction of the Temple of Dagon in *Samson and Delilah* (1949), but he died during preproduction on the remake of *The Ten Commandments* (1956). Jennings had taken over the Paramount department in the mid-1930s, supervising a number of semi-autonomous subdivisions: rear-projection, optical printing, miniatures, and matte photography. Farciot Eduoart, who had his own rear-projection unit at the studio, worked like a military tactician. Eduoart would show up with much pomp to shoot a New York street out the back window of a taxi, not a novel or demanding task, and one he had done for forty years. He, like others in the old Hollywood system, had systematically built his own empire within a unique studio subdivision.

The head of special effects at Metro was Arnold "Buddy" Gillespie, who supervised the earthquake in *San Francisco* (1936), the tornado in *The Wizard of Oz* (1939), the oil well fire in *Boom Town* (1940), and the chariot race in both versions of *Ben-Hur* (1926, 1959). Unfailingly inventive, Gillespie created the swarm of locusts in *The Good Earth* (1937) by filming coffee grounds settling in water, then reversing the film and combining that with footage of actors. The San Francisco earthquake, the evacuation of Dunkirk by Britain's "mosquito fleet" in *Mrs. Miniver* (1942), the snow avalanche in *Seven Brides for Seven Brothers* (1954), and the sea battle in the second *Ben-Hur* were all achieved with miniatures. "The trick department had to know just how many frames to the foot we were going to film to make things look real," said MGM cameraman Joseph Ruttenberg. "Miniatures are very tricky to do right. In miniature photography you have to grind your camera about 60 percent faster. You have to know mathematics to figure exactly how to photograph miniatures to look real." For the Wicked Witch's army of flying monkeys in *The Wizard of Oz*, 2,200 piano wires on an overhead trolley supported miniature monkeys and moved their wings up and down. "It was an awful job to hide the wires," said Arnold Gillespie. "They had to be painted and lighted properly so that they blended into whatever the background might be." The miniature replica of Rome set afire in *Quo Vadis* (1951) was built so that it wouldn't burn. "It covered an area nearly nine hundred square feet," Gillespie said. "Copper tubes with nozzles were distributed into every building so that fuel could be fed and controlled. We used pilot lights to start the fire." (Gillespie int., Steen) Other fires, such as the one in *Boom Town* or the remake of *Cimarron* (1960), were created with process shots, in which film of actual fires was projected onto a screen and actors performed in front of it as the whole sequence was rephotographed.

Around 1935 special effects departments had become firmly established in all the major studios, although their organization varied from company to company. Routine effects, like fire in a fireplace, would generally be handled by the property department, while on complex assignments special effects crews worked more closely with stuntmen than actors. Byron Haskin, who headed the special effects department at Warner Bros. for nine years beginning in the mid-1930s, described the effects department there as "a studio within a studio." Haskin had his own sound stage (Stage 5), his own cameramen, his own special effects directors, and between 125 and 130 people working under him. "I could have made features in there that they would have known nothing about," Haskin claimed. "I even had my own laboratory and sound department." Whether a production called for special effects or not, Haskin charged all units a minimum of $4,000, insisting that every film made on the lot had to share in his department's upkeep. Any time a production company encountered technical problems, filmmakers assumed Stage 5 could come up with a solution. "The Warners special effects department was one of the wonders of the picture business during its time," Byron Haskin noted. "We were sort of an emergency outfit—anything that had difficulty to it was ours." (Haskin int., Adamson)

Other studio employees were less vital. The bigger studios had dozens of unnecessary people on their payroll, including what Metro film editor Adrienne Fazan called "running around boys and girls," assistants to assistants. (Knox, *Magic Factory*) On the third floor of the Thalberg Building at MGM, at the end of a corridor, were a number of small offices housing a peculiar fraternity known as the Iceberg Watchers. Writer Herman Mankiewicz claimed this ancient tribe had been hired by Louis B. Mayer to keep a watch on Washington Boulevard, just below their office windows. Should they sight icebergs moving toward the studio, they were to report to him immediately. "A more prosaic view," producer John Houseman wrote, "was that these men were pensioners from the old, ruthless days of the business who knew where the bodies were buried and who were kept on the payroll to insure their silence." (Houseman, *Front and Center*)

Studio messengers carried film, interoffice memos, and scripts all over the lot. Outside mail poured into the central mailroom and was distributed to various departments twice a day by youngsters usually working their way through school or eager to break into the business. Pay was low, but studio executives often were considerate and encouraging. "It was fun walking down a street and getting a smile or nod from such familiar faces as Fredric March, Joan Blondell, and Jimmy Cagney," said Stuart Jerome of his days as a messenger at Warner Bros. When Jerome started at Warners in 1938, twenty-four messenger boys were employed in a department swarming with activity from eight in the morning until seven at night, headed by John Frederick Pappmeier. "By mid-afternoon of the second day," Jerome said, "my feet were so painfully swollen that I could barely shuffle along, forcing me to discover various hideouts that even our veterans weren't aware of."

Stuart Jerome estimated that 75 percent of the mail delivered by the Burbank post office to Warners was addressed to actors and actresses, an average of fifteen hundred to two thousand a day. The lot was divided into six routes, which took delivery boys to sound stages, offices, and dressing rooms and meant they learned most of the secrets. "We knew who suffered from headaches, heartburn, and hemorrhoids," Jerome wrote, "who wore dentures and hairpieces; we knew who were the boozers, drug addicts, and degenerates." (Jerome, *Crazy Wonderful Years*)

Metro printed a studio newspaper, edited by Bill Wright in the publicity department, that dealt mostly with human interest stories. "I wrote a fashion column on the secretaries," publicist Emily Torchia recalled. "Secretaries to producers made $65 a week in those days, and they dressed nicely—gabardine suits mostly. We didn't wear slacks; if we had, we'd have been sent home." For important announcements, usually about salaries, all of the studio's employees would be summoned to an assembly. During one period when MGM was suffering financial reverses, a platform was built in the middle of the company's main street, where everyone was asked to gather at noon to hear a speech by Nicholas Schenck, president of Loew's Inc., Metro's parent company. Schenck gave the assembled workers a pep talk, telling them how sacrifices were necessary to put Metro back where it needed to be. Another time the New York office sent out a group of efficiency experts, who insisted every worker on the lot write out a detailed description of his or her job. "Then they came and interviewed you," publicist Ann Straus said. "When it was all over, they fired a number of people and put a stapler on everybody's desk. That was to make us more efficient!"

When the Fox Film Corporation was taken over by Chase National Bank, shortly after the Wall Street crash, many people were laid off. "Some brilliant man came along," makeup artist William Tuttle recalled, "and decided we should have a charitable plan where every week the people who were working would drop money in the kitty in each department for those who had been laid off. It was called the Five Friends Plan, but we called it the Five Finger Plan. That didn't last long." Eventually everyone on the Fox lot, including the stars, took a pay cut.

Warner Bros. also had periodic financial troubles, during which times everyone would be called together on a sound stage and Jack Warner would speak. Actor Ralph Bellamy recounted one such session during the early 1930s. "I was working on the backlot and got there late," said Bellamy. "But I remember tears dropping from Jack Warner's eyes as he spoke. He said, 'You've got to take a 50 percent cut in salary or the business is dead.' Everybody loved to hear Jack make a speech because he got off the subject and ranted and raved and stamped his feet and made little or no sense, except in this case. He left it that if we didn't take a 50 percent pay cut, the business was down the drain." Bellamy claimed Warner Bros. realized millions from the salary reductions.

At Metro, at noon sharp, all the sound stage doors would open, and workers would swarm out into the streets on their way to the commissary for lunch. "The main street of MGM was a wide street," said drama coach Lillian Burns, "and in those days when those stages broke for lunch, it was like 42nd Street in New York." L. B. Mayer believed that a family who ate together stayed together and insisted the MGM commissary be commodious and serve good food, modestly priced. He also didn't want his workers leaving the lot at noon and coming back late, having enjoyed too many drinks during lunch. To simulate a family atmosphere, Mayer required the commissary to serve his mother's chicken soup with matzo balls as a specialty at thirty-five cents a bowl. Another specialty was Metro's apple pie; workers claimed it was the best they ever tasted.

There was a seating at noon and another at one o'clock, since the entire Metro family couldn't be fed at the same time. For those two hours the commissary was noisy with the clatter of dishes, the scraping of chairs, greetings from table to table. For newcomers the parade of stars could be awesome: Clark Gable, Robert Taylor, Greer Garson, Lana Turner, Spencer Tracy, Esther Williams, and the rest—sometimes in costume, other times not. Strict protocol existed on seating; most of the stars sat against the back wall. "I remember Ava Gardner breaking all the rules and going in the cafeteria with the truck drivers and grips," said dancer Leslie Caron. "That was considered just abominable." (Knox, *Magic Factory*) Extras in costume would stare enviously at their favorite stars while waitresses who had worked there for years brought their orders. "I was in the commissary the day Clark Gable came back from the war," said publicist Esmé Chandlee. "He walked in, and everybody got up and applauded." Male stars generally sat at the same table, with a dice cage on one end. "Each star would come in and turn it over," recalled talent scout Al Trescony, "and mark down the number of dice and wait until everyone else had done the same. The loser would pay for all the lunches." If a star had an interview, he or she sat apart with the journalist.

Tables were reserved for the various departments. The MGM commissary had a producers' table, a directors' table, a publicity table, a special effects table, a makeup table, writers' table, and so forth. If writers had a problem with a story, they could talk about it during lunch. "You'd go back to your office," William Ludwig recalled, "and maybe an hour later another writer would come in and say, 'I was thinking about what you said, and I've got an idea for you.' Everybody helped everybody. It was very collegial." Some imposing people were seated around the writers' table—at various times Dorothy Parker, Herman Mankiewicz, Sid Perelman, Robert Benchley, Donald Ogden Stewart, Eddie Chodorov, and Scott Fitzgerald. A perennial wit was Harry Kurnitz. One day Joe Mankiewicz, Herman's younger brother, ordered bouillabaisse for lunch. As it was set down before him, Kurnitz looked at it and quipped, "Joe, you know, if you stepped in that on your way to work, you'd go back and change your shoes!" Another time Mankiewicz ordered rare roast beef. When the meal arrived, Kurnitz looked at the plate and snorted, "You know, I've seen cows hurt worse than that get better." (Ludwig int.)

Each dining group developed its own rapport. Elizabeth Taylor, Janet Leigh, and Arlene Dahl, then youngsters just starting out, knew each other from classes and frequently sat together. "The feeling of competition was not really very strong," Dahl said, "because we had a big Daddy looking after us and developing things for us, or so we thought. Still, it was a good idea to have producers and directors see you at lunchtime." Actresses were careful of their appearance, never knowing when they might be considered for an important role. "I remember Lana Turner and June Allyson looking as if they'd just stepped out of the shower," observed publicist Emily Torchia. "Lana would come into the commissary with her hair brushed back, that thick, thick hair. It was sort of a mahagony when she first came, then we lightened it. She had very milky skin, and she'd wear white starched blouses, or pink, or pale blue. The boys always wore nice shirts or turtleneck sweaters. They all wanted to look like Gable."

Sometimes couples who were dating paired off. Singer Mel Tormé had met Ava Gardner during a recording session with bandleader Artie Shaw, whom Gardner later divorced. Tormé and Gardner started dating when he was making *Good News* (1947) at Metro and she was making *The Hucksters* (also 1947). "It was great fun to sit around in the commissary with Ava and be envied by all the guys," Tormé said later. "We went steady almost all the time I was making *Good News*."

Other times an entire company would sit together while they were shooting a film—the Freed unit at one table, the Pasternak unit at another, although there was always table hopping. Actress Jean Porter said she usually sat at the music table because she enjoyed the company of John Green and Bronislaw Kaper. "I guess it was the fun and laughter I remember most," said Porter.

For new arrivals the experience could be terrifying. "I didn't want to walk in there," Janet Leigh admitted. "The commissary was so big, and I felt the whole place was looking at me. It took me a long time to get over that, if ever." Actress Betty Garrett, a recent sensation on Broadway, claimed she'd walk in, take a look around, see all the stars sitting with their cluster of friends, and go back out and eat in the coffee shop. "It was a lonely life that first year," Garrett said. "I just used to wander around the lot."

Louis B. Mayer ate in Metro's private dining room just off the main commissary, reserved for studio executives and their guests. Big stars occasionally ate in their dressing rooms, and character actor Lloyd Nolan recalled how impressed he was that Van Johnson, at the height of his celebrity, brought his lunch to work in a pail. If a company was filming out on the backlot, as director George Sidney did on the Wild West Show sequences for *Annie Get Your Gun* (1950), lunch might be served there. For most studio employees the lunch hour was a time they could socialize, exchange ideas, and cement a team spirit. "We could all band together," said Metro writer Robert Pirosh, "the way you might in the army. We'd gripe about how stupid the front office was and not feel so alone. If you're out battling as a freelance artist, which I was for much of my life after Metro, it's different."

At Paramount, commissary head Pauline Kessinger, beloved by workers all over the lot, became the studio's den mother, on hand to greet guests and regulars as they walked through the dining hall door. "Pauline was great at handling all of those crazies," film executive Howard W. Koch declared, "making them feel they were the most important of all, which is difficult." The Paramount commissary developed a reputation for serving excellent food and maintaining an inviting atmosphere. Kessinger had worked for Paramount since the silent days, taking a job as a waitress, mainly to see the stars, at the studio where her husband was a grip. "My husband, Coley, didn't want me to work," she said. "But he went off on a location assignment, and while he was gone, I found a job. I told the manager I could only stay two weeks." Then the cashier quit, and Kessinger stayed on. Eventually she became assistant manager, and for almost thirty-five years was commissary head. She bought the food directly from the wholesale houses, always operating at a deficit, which the studio expected. "Our food cost ran high," Kessinger said, "because we offered a large menu. In most restaurants it's normal to run around 30 percent for wages. But we were running about 64 percent, which was union scale. That's why we lost money."

Since the commissary served the Paramount stars breakfast in their dressing rooms, Kessinger arrived at work by seven-thirty, sometimes not leaving the lot until after midnight. Twice a day the commissary staff provided coffee on each set—five or ten gallons, depending on the size of the company. "I would go to a production meeting every day at one-thirty," Kessinger recalled, "where the business manager would tell me how much coffee and how many doughnuts he wanted on the sets the next morning. Coffee was our lifeline. We made money on that, because the company paid for it." Kessinger supervised a staff of around eighty—cooks, waitresses, dishwashers, and busboys. Waitresses changed the color of their uniforms about every six months. "We had a captive audience," Pauline Kessinger explained, "and people got tired of seeing the same old dresses. So about every six months we'd get new uniforms for the girls." At least once a month she would call a staff meeting, in which service would be the primary topic of discussion—"just to keep everyone on their toes," Kessinger said.

The studio's stars knew they had arrived when a dish was named after them in the main dining room. Visitors could order a Lamour Salad, Turkey and Eggs à la Crosby, Tomato Bendix, Caulfield Potatoes, Fitzgerald Mushroom Soup, or Pie à la Lake, all in honor of current stars. "Clara Bow loved chow mein," Kessinger recalled. "We didn't make chow mein in the commissary, so I sent out to get it for her. One day the chef told me, 'Let me have a hand at making this chow mein.' He made it so well we carried it on the menu every Wednesday after that. Lucille Ball used to come over from RKO especially to eat our chow mein." No liquor was served in the commissary. The dining room itself was elegant, with chandeliers left behind from the making of some movie. "The set dresser just forgot to take them down," Kessinger said.

Young stars sat at the Golden Circle, a large round table in the middle of the room, covered with a gold cloth, where the newcomers could be seen by producers and directors. Maybelle Dillon was their waitress. Patricia Morison remembered sitting there and watching Cecil B. DeMille and his entourage make their grand entrance, with DeMille in puttees and jodphurs, carrying a riding crop. "Here comes Jesus Christ and the Disciples," someone would joke. Carolyn Jones, who later played Morticia on "The Addams Family," laughed when she remembered the madcap times when she and Peter Baldwin were sitting at the Golden Circle. "We created such havoc for poor Pauline," Jones confessed years later. "We used to make a joke of throwing imaginary hand grenades at one another across this giant table." One day Baldwin tossed a prop grenade toward Carolyn Jones. "I wasn't paying much attention to him," she said, "and it terrified me for an instant. I batted it away, right at Cecil B. DeMille. It hit his table and knocked over the salt. I was fit to be tied." But DeMille sent somebody over to ask if the young lady who had thrown the hand grenade would like to be introduced to the manager of the Brooklyn Dodgers.

For most studio employees the lunch break was a time of relaxation. "Bing Crosby always came in with a song on his lips," Pauline Kessinger recalled, "singing to the waitresses or to me or somebody." When Dean Martin and Jerry Lewis were a team, they were full of pranks, particularly on Saturdays. "They would grab trays and carry them through the dining room and drop them," said Kessinger. "I saw Jerry one time go over to a table set up for eight people and jerk the tablecloth out from under all the food, throwing everything on the floor. They always paid for the damage, but they were determined to have their fun." Since the writers' table at Paramount, as elsewhere, was among the more gregarious and amusing places, costume designer Edith Head usually sat there, playing "The Word Game" with Billy Wilder and Charles Brackett. Head and Wilder were the recognized champions, although even the losers enjoyed a lively time. "You really looked forward to lunch," screenwriter Hal Kanter said.

Paramount's star Betty Hutton made a practice of eating in her dressing room, since her energetic approach to performing generally left her exhausted. Bob Hope frequently did, too, when he was busy with meetings. "I usually had lunch in my dressing room," Lizabeth Scott said, "because then I could catnap for ten or fifteen minutes, or make telephone calls and catch up on my outer life. I'd go to the commissary on my days off, when I had lots of time and could sit there and observe everybody." Others brought their lunch, except when they had an interview or a special appointment.

Paramount, like most other studios, had a private dining room for studio executives, special parties, and interviews. "We entertained all kinds of heads of state with elaborate luncheons," Pauline Kessinger remembered, and there was generally a kick-off luncheon for pictures about to go into production. For *Son of Paleface* (1952), starring Bob Hope, Jane Russell, and Roy Rogers, a luncheon was held in the executive quarters. "Trigger, Rogers's horse, came and kicked a hole in the wall," Kessinger said.

Pauline Kessinger also supervised a backlot cafe and a coffee shop, which stayed open all day. "Gary Cooper was sitting at the end of the counter having a cup of coffee with me," she recalled, "when the 1933 earthquake hit. He grabbed me by the hand, and we went through the kitchen, with him pulling me behind those long legs of his. We went out under the water tank."

Before the unions required hot food at noon for workers, the commissary put up box lunches for production units shooting on nearby locations. "We would sometimes have as many as a thousand or fifteen hundred lunches to prepare," said Kessinger, "and we'd have a crew come in at four in the morning to put up the box lunches. Sometimes there'd be four or five locations filming at once, and we had to keep all of those lunches separated for the drivers to pick them up."

Occasionally food was required for filming. When a scene called for caviar, the commissary generally substituted blackberry jam. "One star wouldn't eat the blackberry jam," Pauline Kessinger recalled, "so we had to buy fresh caviar for her for several days of shooting, at $55 a pound." For the Golden Calf scene in *The Ten Commandments* the commissary prepared quarters of lamb, hundreds of pounds of ribs, and loaves of what studio workers called "Bible Bread." "We must have had a hundred baskets of fruit," Kessinger added, "because that scene went on for several days, and they didn't dare use the food after it sat under the lights."

The commissary at 20th Century-Fox, called the Cafe de Paris, was run by Nick Janios, a trusted friend of Darryl Zanuck. It was decorated with an art deco mural of the world's major cities and featured pictures of the studio's stars. There was also an executive dining room and a sun room where the writers ate. "They had a wonderful black man, Emmett, who walked around with a great rolling cart with a wonderful silver dome and the most marvelous entrees," remembered Marjorie Fowler. Since the studio was adamant that workers should not leave the lot at noon, administrators were willing to lose money on their commissary in order to offer enticing dishes at reasonable prices. Twentieth Century-Fox prided itself in serving better food than the more regal Metro did, but MGM arranger Connie Salinger was unimpressed: "That's fine," Salinger commented. "They should photograph their food and eat their film!" (Blane int.)

Zanuck entered the executive dining room at Fox every afternoon promptly at one-fifteen and almost invariably ordered a corned beef sandwich on dark rye bread and iced coffee with whipped cream. "He was always wiping off his little mustache," Herbert Bayard Swope, Jr., recalled. Producers were entitled to use the executive quarters, but Philip Dunne refused to enter even after he had achieved the requisite status. "I didn't like the monologue at lunch," said Dunne, "which was what you got from Darryl." Since crews only got a half hour for lunch, many of them ate at the hot dog stand on the lot, which sold sandwiches, candy, and soft drinks.

The head of Warner Bros.' commissary was Katherine Higgins, a thin, severe-looking woman who retained much of her Irish brogue. Producers, stars, and directors lunched in the Green Room under large, glamorous portraits of the studio's current stars; there was an adjacent cafeteria for grips, electricians, technicians, and extras. Jack Warner enjoyed gourmet meals in a separate paneled room that boasted a private cook and a butler with white gloves. Studio delinquents with sufficient prestige, such as Humphrey Bogart, Errol Flynn, and Dennis Morgan, took their meals across the street at the Lakewood Country Club, sometimes returning to work in a haze after several cocktails.

Although the commissary at Warners witnessed jokes and laughter, the lunch hour also had a serious side. "Many a deal was made in that dining room," studio publicist Bill Hendricks said, "and there were interviews going on all the time."

Harry Warner decided at one point that he would show the industry how to run a studio commissary efficiently and doubled the prices. All the writers at Warners looked at the revised menu and walked out. Soon after, they received a note from Jack Warner complaining that they had been five minutes late returning to work. "I had a stop watch," said scriptwriter Melville Shavelson, "and I would gather all the writers at the gate where the cop was clocking us. I'd give them the cue, and we'd go outside, run across the street, and eat as fast as we could. We'd come back and stand in front of the cop for maybe twenty or twenty-five minutes until it was exactly sixty minutes. Then we would come back in." Two weeks later they received a note from the story department that read: "Dear writers, come back—all is forgiven." They walked into the Warners commissary to discover that the waitresses had put up big signs: "Welcome home, writers." The malcontents were given a standing ovation and sat down to find the prices back to normal. "We came in the next day and picked up the menu, and every price had been raised a nickel," Shavelson recalled. "Two days later everything was up another nickel. In a month prices were right back where they had been."

At RKO the commissary had two prices for food—one for producers, stars, and directors, another for studio workers, including film editors. Many RKO employees preferred to go to nearby Lucey's or Oblath's, both popular restaurants with the picture crowd. Smaller Columbia didn't have a commissary for actors and crews, although there was a hot dog stand on the premises, run by a one-legged man named Henry. Columbia had an executive dining room with seating for about ten, where Harry Cohn ate. The dining room was an important addition to the mogul's power structure, since he conducted business there and collected bets after weekend football games and horse races. "Cohn invited me to lunch one day," said actor Alfred Drake, "and he had all the big shots there. He insulted almost everyone—needlessly, I thought. These were talented people, but that was Harry's way of maintaining his championship."

Universal offered one dining room for everyone, with two adjoining rooms. "It was all very warm and casual," publicist Gail Gifford said. The Sun Room, where the executives and contract players congregated, was slightly more expensive than the outer area, but the atmosphere there was generally friendly. Everyone seemed to know what everybody else was working on, and status largely depended on the success of one's latest project. "If you walked into the Sun Room and everyone said, 'Sit here,' you knew you'd done something that everybody had heard about and approved of," said drama coach Estelle Harman. "But if the mildest little thing happened, maybe a screen test that didn't go well, you'd walk into the commissary and suddenly everybody would be looking the other way."

At the end of each picture a party was customarily given at the larger studios, with food and ice sculptures and plenty of liquor. For *Sunset Boulevard* (1950) the Paramount commissary was converted into a nightclub, complete with dance floor. "We even sprayed the place with perfume," Pauline Kessinger said. To celebrate the completion of a Mitch Leisen picture, a clambake was held on Paramount's Stage 18. "It was a *real* clambake," Kessinger said. "We brought in sand and had the oysters and the corn and the lobster, all wrapped in gunnysacks, and baked them in this hot sand over bar-beque coals. Mitch Leisen was a big party-giver. He drove onto the set in a spring wagon, with the wagon loaded down with gifts for everybody, even for the waitresses and me."

On the day before Christmas all work stopped in the studios by noon, when each department gave its Christmas party. Feuds were laid aside as people made their rounds, exchanging gifts, extending season's greetings, and toasting everybody from the night watchman to the president of the company. "Mainly it was an excuse to get drunk and chase the secretaries," film editor Elmo Williams said. "I used to have a couple of drinks and go home." Others stayed as long as the liquor flowed. Universal put on a Christmas show for employees and their children, featuring the studio's young contract players. "That was great fun," Rock Hudson said. "The carpenters and electricians would build the scaffolding, and a whole sound stage would be filled with kids. Even employees from other studios could bring their kids if they wanted."

When the day's work was done, Hollywood studios resembled empty factories, much of them facade. Actress Carolyn Jones recalled how studio head Y. Frank Freeman kept four large boxer dogs in a compound at Paramount, just inside the DeMille gate. "At night around seven o'clock, when practically everyone was gone, they'd close the gates and let those dogs loose," said Jones. "They played with film cans like they were Frisbees. Guys would sail the cans in the air, and those dogs would run and catch them. I remember seeing them chomp on them, and the film cans would just crumple in their mouths. They were patrol dogs." What a few hours before had been a glamorous, exciting hive of activity suddenly became a shell, haunted by the memory of the voices and bustle, a ghost town much like the one Hollywood itself would soon become.

16

Social Life

Arthur Farnsworth, Mary Astor, Bette Davis, and Manuel Del Campo (left to right) socializing at the Ambassador Hotel's Cocoanut Grove. Courtesy of the Mary McCord/Edyth Renshaw Collection on the Performing Arts, Hamon Arts Library, Southern Methodist University.

"The idols of the screen are an international royalty," Leo Rosten wrote in 1941, "whose dress and diet and diversions are known to hundreds of millions of subjects." (Rosten, *Hollywood*) Around the globe Hollywood became Tinseltown, a land of dreams and luxury. For the American public, raised on an ethic that emphasized success, material wealth, and social mobility, Hollywood embodied a national ideal. If movies reflected what H. L. Mencken called "the moron majority," they also exemplified a glamorous life-style most Americans longed to emulate. Director George Cukor recalled meeting a woman from Akron, Ohio, who told him that the movies were her greatest solace during the Great Depression. For thirty cents she could enter a great marble palace and lose herself in romance and celluloid fantasies. "That gave her strength," Cukor said. "People were quite emotional about this. They loved the movie queens and wanted them to look marvelous." Off-screen, with the help of studio publicists, Hollywood became the Glamour Capital of the World, conspicuous in its display of clothes and mansions and exhibitionist in its conduct.

Most of the famous movie personalities were young, suddenly thrust into a pseudo-aristocratic world where they were pampered and rich and celebrated. "Hollywood's wealth is first-generation wealth," Leo Rosten explained, "possessed by people who have not inherited it, spent by people who have not been accustomed to handling it, earned as a reward for talent (or luck) rather than heritage." (Rosten, *Hollywood*) The movie colony produced a new population of nouveaux riches, not unlike those earlier in banking, railroads, or real estate, while the social mores of Beverly Hills and Bel Air paralleled those of Oyster Bay and Newport.

At the same time Hollywood also had a reputation as a godless, depraved city; ministers, educators, and clubwomen attacked it regularly as a synonym for debauchery and evil. Hollywood's night life was reputedly the most wicked, its sexual behavior the most promiscuous and unconventional; in Hollywood divorce seemed an inevitable satellite to marriage. Actually the divorce rate in Hollywood ran no higher than in New York or Chicago or Detroit, and screen actresses probably were no more responsible for the breakup of marriages than graduate students on university campuses. What differed was the publicity and the larger-than-life image projected by the fan magazines and press.

In many respects Hollywood remained a parochial town, its citizens nervous about their social standing. "It's a very frightened community," actress Carol Bruce noted. "One has to learn how to live with that." But many of Hollywood's denizens were destroyed by the excitement and uncertainty of a squirrel-cage existence. "Hollywood was hardly a nursery for intellectuals," actor David Niven wrote, "it was a hotbed of false values, it harbored an unattractive percentage of small-time crooks and con artists." (Niven, *Empty Horses*) In contrast to earlier American heroes, Hollywood stars were expected to be spendthrifts and to demonstrate profligate ways. Theirs was supposedly a life of sunshine, pleasure, and endless lovemaking, where the living was easy amid orange blossoms and sea breezes, and everyone stayed eternally young and devastatingly beautiful.

New Yorkers particularly scoffed at the image. "Hollywood is like a play with a bad cast," said Wilson Mizner, part-owner of the Brown Derby restaurant. (Wilkerson and Borie, *Hollywood Reporter*) Others complained that the town was an intellectual desert. "It was a world of showing off and tennis playing and cars," Broadway actress Celeste Holm said. "I couldn't stand it." Songwriter Harry Warren agreed: "I thought Hollywood was a horrible place. I hated it and kept trying to get back to New York. It was like being in Iowa." Still others found the town innocent and strangely old-fashioned, its restaurants and nightclubs prosaic. "I never knew Hollywood as being particularly exciting," Douglas Fairbanks, Jr., claimed. "There were only three good restaurants, really—Chasen's, Romanoff's, and Perino's." (Fairbanks int., Wagner) Most of the entertaining took place in private homes. "Hollywood is a peculiar town," observed director Vincente Minnelli. "[In New York] you can decide at seven o'clock where you want to have dinner, what theater you want to see, and you can usually get in. But in Los Angeles there's nothing much to do except go to a few nightclubs and entertain in people's homes." (Minnelli int., Schickel) Easterners were inclined to rush back to Manhattan as often as possible to see shows and find stimulation. "By the late 1930s," James Cagney wrote, "my pattern of living was fairly well set. Learn the words, do the scenes, and then when the picture was finished, without any delay, back East to Martha's Vineyard." (*Cagney by Cagney*) Character actor Arnold Moss thought Hollywood's values were "somewhat distorted" and continued to live in New York, commuting to California only when making a picture. Most of Moss's friends assumed he objected to his children's growing up in Los Angeles. "I think it's fine for children," the actor informed them. "I'm not so sure it's all that good for adults."

While there were hundreds of gifted people living in Hollywood, neither the intellectual life nor the cultural life could compare with that of New York, London, or Paris. "I don't think the motion picture industry really allows for it," observed director Martin Ritt. "It only allows for success." Film talent tended to think and talk only of movies. Everybody appeared to know what everyone else was working on. "Privacy was something that did not exist," said Arnold Moss. "You lived in a fishbowl. You couldn't do anything without the world of Hollywood knowing it. If you were working, you were a great fella. If you were between jobs, you weren't so great." It was easy to become self-absorbed. "There's nothing worse than success in this town," producer Ross Hunter remarked, "especially if you're young."

In the early days Mack Sennett was Hollywood's recognized social leader, entertaining in his large Mexican-style house. "He'd give a dinner party," producer Walter Wanger remembered, "and if you didn't take the young lady on your right upstairs between the soup and the entree, you were considered a homosexual." (Wanger int., Rosenberg and Silverstein) But Wanger found such behavior harmless. Later in the silent era, Mary Pickford and Douglas Fairbanks headed a more organized social system, and their estate in Beverly Hills—Pickfair—became "the Buckingham Palace of Hollywood." Dinners at Pickfair were formal, with dozens of butlers in pantaloons waiting to serve guests. In the hills beyond Pickfair was Rudolph Valentino's Falcon's Lair, an exotic mausoleum with the walls painted black and the windows hung with ebony draperies. After the sound revolution David and Irene Selznick emerged as social leaders, but none ever surpassed William Randolph Hearst and Marion Davies.

Newspaper tycoon Hearst figured as the film colony's most lavish host, while his long-standing blonde mistress, movie actress Marion Davies, was perhaps Hollywood's most beloved hostess. During the 1930s members of the industry knew they had arrived when they were invited to spend a weekend at San Simeon, Hearst's hilltop castle two hundred miles north of the movie capital. A private train left Glendale station on Friday evenings, carrying Hearst's forty or fifty guests for the weekend, and arrived at San Luis Obispo around midnight. A fleet of limousines met the party and took them up to the castle, where Hearst and Davies greeted their weekend guests amid regal splendor. Since Hearst looked upon San Simeon as "the Ranch," life there was casual. Visitors played tennis or croquet, rode horseback, or swam in the glorious Neptune Pool. Still, the surroundings and guest list could be intimidating. Journalist Adela Rogers St. Johns remembered a dinner one evening at the Ranch, when she sat on Hearst's left with Gloria Swanson on his right and Marion Davies across the table, with Samuel Goldwyn on *her* right. "It was really incredible," character actor Leon Ames said. "I spent a weekend there early in my career. You'd see all of these big stars, and you were afraid to say hello to them. When you're young and you come into this business practically off the farm, and suddenly you're shoved in with the top of the heap, mixing with these famous people, it's hard to adjust. Here I was in this great baronial castle, at a table that held maybe twenty-five people, and I just sat there in awe." His studio had assigned Ames a date to take up to the castle—Lupe Velez, the "Mexican Spitfire. "I practically didn't know her," the actor said later. "She didn't pay any attention to me, and I didn't pay any attention to her. She was the twistiest, most sensuous-looking thing I've ever seen. But I don't think I said two words to her while we were up there."

Since San Simeon is located on the coast midway between Los Angeles and San Francisco, Hearst built another castle for Davies more convenient to her film work, a white palace on the ocean sands at Santa Monica, consisting of over a hundred rooms and fifty-five bathrooms. At her beach house, Davies gave an annual costume party that was the social event of the season. Everyone tried to wangle an invitation, to see and be seen. Davies's studio dressing room, a fourteen-room bungalow, was also the scene of much socializing, the reception quarters for European royalty, famous authors, and renowned heads of state. When Davies left Metro-Goldwyn-Mayer for Warner Bros., following a dispute with L. B. Mayer over roles, her bungalow was towed in sections the ten miles from Culver City to Burbank, where Jack Warner welcomed her, fully aware of the publicity value of the Hearst press that came with her.

In Hollywood, as elsewhere, the rich hastened to flaunt their wealth by building huge, opulent homes. Beverly Hills and Bel Air became studded with English manors, Italian villas, and Spanish haciendas, all with de rigueur swimming pools and tennis courts. Lawns were manicured, flower beds ablaze with tropical blooms, and terrace views were magnificent. Social occasions were elegant—and permeated by the snobbery of success. "The guest lists of the highly publicized big parties reeked of it," David Niven wrote. "The successful and the established were invited; the struggling and the passe were not." Most of the conversation revolved around "What are you doing now? What are you going to do next? How much are you getting? Is this executive going to stay on or is another one coming in?" (Niven, *Empty Horses*) Personal scandals also made for juicy gossip, but rarely did the talk rise above the business itself.

There were exceptions. In a few households dinner parties might include a scientist seated next to a politician, but those occasions usually took place when an international visitor was being entertained. Sometimes the host and hostess demonstrated impeccable taste and an instinct for selecting interesting guests. Director Lewis Milestone and his wife, for instance, combined charm and an enjoyable atmosphere with the best of the theater and motion picture worlds. Usually only people in the same income bracket socialized together. Studio executives normally didn't mix with cameramen and certainly not with grips or technicians. Whereas cast parties in the theater extended to the crew and everyone involved, outside the studio the Hollywood hierarchy was far more rigid. "It was not a healthy, cosmopolitan environment like New York," actress Dolly Haas observed, "where you meet people from different backgrounds. Only the people who are very successful mingle together, and everyone was sort of categorized. Consequently you kept meeting the same people."

Actor Cesar Romero, who escorted many of the glamorous stars, found Hollywood exciting during its Golden Era. "You'd go to these big affairs in white tie and tails," he recalled. "Everybody was dressed beautifully and seemed to take great pride in being what they were. It was an interesting town at one time." Pamela Mason remembered coming from England and loving Hollywood. "I've always been very social," she said, "and I thought it was all wonderful. There were parties every night, and everybody dressed up and it was very glamorous. Everywhere you went you actually saw people like Rita Hayworth." Her husband, James Mason, wasn't as impressed. "He wasn't in love with this whole business like I was. But I loved that side of it— tennis parties and swimming parties and all the babies getting together on Sundays. I thought it was a wonderful time."

Hollywood chroniclers liked to exaggerate the town's nightlife, which the press depicted as an endless, giddy whirl. During the 1930s and early 1940s the Cocoanut Grove in the Ambassador Hotel was *the* place to go. Located in the middle of swank Los Angeles, the Cocoanut Grove offered top quality entertainment and served as an internationally famous showcase for Hollywood's elite. "It was like being on a movie set," singer Jane Powell said; "it even had fake clouds blowing languidly across the ceiling. All the big acts played there, appearing amid the fake palm trees with fake monkeys dangling from them." (Powell, *Girl Next Door*) Tuesday nights at the Grove were popular, but any night dozens of stars could be seen there, rendezvousing with friends or flaunting their current romance.

In 1934 Billy Wilkerson opened the Trocadero on a site on the Sunset Strip previously occupied by La Boheme, and the club quickly became one of Hollywood's favorite night spots. Its view was a mass of glimmering lights, remarkable even by Los Angeles standards; soon many similar clubs appeared in the vicinity. Ciro's followed in 1940, where the old Club Seville had been located earlier, and the Mocambo opened a year later, both on the Sunset Strip and both instant hits, pictured repeatedly in the fan magazines. The Palomar Ballroom in Hollywood proper boasted an Arabian decor, while the massive Palladium dance floor accommodated seventy-five hundred dancers. Other famous nightspots included the Florentine Gardens on Hollywood Boulevard, the Polo Lounge in the sedate Beverly Hills Hotel, Don the Beachcomber's (where the Zombie, a famous drink, was invented), the grand ballroom of the Biltmore Hotel, and the Hollywood and Alexandria Hotels, both favorite meeting grounds for movie personnel. Anchored off the Santa Monica pier was the S.S. *Rex*, the king of the three-mile-limit crafts that offered gambling to the city's society crowd during the 1930s and remained a constant source of irritation for lawmen. Thirteen water taxis carried customers out to the *Rex*, a twelve-minute ride that was both comfortable and inexpensive.

Of the fashionable restaurants Dave Chasen's, which opened in 1936, served the best steaks; Perino's, famous for excellent food, stood opposite the Ambassador Hotel; and Romanoff's, located on Rodeo Drive just north of Wilshire Boulevard, featured the town's premier imposter, Prince Michael Romanoff, supposedly the cousin of the late czar. Eventually there would be three Brown Derby locations: two on Wilshire Boulevard and one on Vine Street, near Hollywood Boulevard, the latter famous for Eddie Vitch's caricatures of the stars. Musso-Franks on Hollywood Boulevard was an old-fashioned restaurant popular with writers and the older movie circle, while many of the younger set congregated at Schwab's Drugstore on Sunset Boulevard at the corner of North Crescent Heights, a two-minute walk from the Garden of Allah, a favorite hotel.

Seating at restaurants and clubs usually depended on one's current popularity, although status within the industry and income could be influential. Actor Lew Ayres automatically received choice seats at restaurants and nightclubs during his early stardom, but he suddenly found himself relegated to the third and fourth rows when his film popularity declined. "Hollywood has always been a success-oriented town," Jesse Lasky, Jr., observed, "and there's little sympathy for failure." (Lasky int., Wagner) When singer Carol Bruce arrived in Hollywood, following her success in Irving Berlin's *Louisiana Purchase* on Broadway, she was hailed as the newest discovery and squired all over town—Ciro's, Chasen's, the Mocambo. "I didn't know then that they kept score," said Bruce, "but it was a very option-conscious town. It could be vicious. At the end of filming, when my first picture turned out to be a bomb, suddenly I was aware of a new climate around me. No one was overtly unkind to me, but I was suddenly put in the back of the bus. I'd walk into a restaurant or nightclub and not be greeted with the same effusiveness by the maitre d' or captain of the club the way I had been."

Hollywood social life had to fit work schedules. "I couldn't carry on a social life except on weekends when I was working," director Vincente Minnelli said. "You lived more or less like a hermit for that particular time while you were making a picture." (Minnelli int., Schickel) Joan Crawford, a consummate professional, rarely accepted social engagements during the shooting of a picture. "We were working six days a week," singer Gordon MacRae recalled of a film he made with Crawford. "Joan would never go out when she was working, except maybe on Saturday nights."

Most of the parties took place between Friday evening and Monday morning. "The rest of the time the entire industry labored like ditchdiggers," Adela Rogers St. Johns said. (St. Johns, *Love, Laughter*) David O. Selznick and his second wife, screen actress Jennifer Jones, gave extended parties on Sundays, starting with tennis or croquet in the morning, followed by lunch and a film. Then guests went home, dressed, and came back for cocktails, dinner, and another film.

Studio moguls stayed current on each other's pictures, and private screenings were a favorite form of Hollywood amusement. New movies were circulated around the town's mansions before their release, sometimes with devastating results. "I've seen pictures destroyed at somebody's home," film executive Howard W. Koch said, "if it got bad word of mouth. The Bel Air circuit could be terrible." Performers who received negative reviews or were out of work or in debt might suffer similar fates. "You could be wiped out very quickly professionally," screenwriter Jesse Lasky, Jr., said, "scrubbed from the slate, a figure to be avoided." (Lasky int., Wagner)

Actors under contract to a studio did not necessarily mix socially with each other or with the executives. Dancer Dan Dailey was initially under contract to MGM but never associated with Louis B. Mayer outside the studio. After Dailey signed with 20th Century-Fox he got to know Darryl Zanuck's daughters, since they were near his age; Dailey water-skied with the girls and spent many weekends in the Zanucks' home. Lynn Bari, also under contract to Fox, knew Virginia Zanuck better than she knew Darryl, since they frequently played cards together, but Bari saw far more of Zanuck's lieutenant Lew Schreiber and his wife. "Every Saturday night I had dinner with them," the actress said, "and I was dragging. We worked six days a week, and I'd always go to sleep on the couch after dinner. All the people would be around talking or playing cards, but Lew and Joan Schreiber were always terribly nice to me."

Joan Fontaine was under contract to David Selznick, yet never knew fellow Selznick player Dorothy McGuire during that time and only met Vivien Leigh and Ingrid Bergman briefly in Selznick's office. Gregory Peck made *Spellbound* (1945), during which he became acquainted with the Swedish actress and liked her very much. "That's one of the rewarding things about this job," said Peck, "that you often do maintain those friendships over a long period of years. Sometimes I might not see Ingrid for three or four years, but when I did, we'd laugh and fall into each other's arms and have lots to talk and joke about. There's a bond there that doesn't entirely fade away."

Unmarried Patricia Morison was often invited to formal dinner parties as the partner of a single guest of honor. "The formalities were really strange," said Morison, "and the intrigue that went on was interesting to watch. But I didn't get involved." Others claimed they hated the phoniness of the Hollywood scene so much that they avoided it. "I wasn't much of a social person," Warner Bros. star Virginia Mayo declared, "and I think in a way that might have been a hindrance. I just didn't fit in with those parties. A lot of stars loved it, but I didn't." Jack Warner, who was a gracious host, gave some lavish affairs that Mayo did attend. "He had a birthday party on my thirtieth birthday at his mansion," she recalled. "He sat on my right at this gorgeous table, with a birthday cake and everything, and Clark Gable was on my left. That was some birthday party!"

There were others who maintained that social ineptitude prevented their careers from advancing. Actress Laraine Day came from a Mormon background and never moved comfortably in the Hollywood party whirl. One New Year's Eve she was invited to Tyrone Power's house and showed up elegantly dressed in a tight suit and with her hair piled on top of her head. "I looked sensational," Day recalled. "I made my entrance, shook hands with Tyrone Power and his wife, took one step onto a slippery floor, and fell flat. I slid right across the room, landing at L. B. Mayer's feet. Mayer turned to Mervyn LeRoy and said, 'Who is the girl that just fell down?' I'd been working for him for something like three years. That was the only contact I had with Mayer while I was at Metro." Her MGM contract ended shortly afterwards.

Irene Dunne said that while the film business was her job, her personal life was another matter entirely. Although she lived in Hollywood, she seldom considered herself part of its culture and cultivated friends outside the industry. "I was never movie-struck," Dunne claimed. "When I left filmmaking, I never missed it, because I had so many other things I wanted to do." Screenwriter Robert Buckner felt much the same, taking to heart advice producer Robert Lord had given him soon after Buckner went to Warner Bros.

"Stay away from the throne," Lord cautioned. "Don't get mixed up in the Beverly Hills circuit." When Buckner retired, he and his wife lived in Mexico, far from the pressures he had dealt with successfully for decades.

Margaret Tallichet, who married director William Wyler, was not convinced that the social life she and her husband enjoyed was typical of Hollywood. "We went to the big affairs," Mrs. Wyler conceded, "which were interesting and fine, but we had our own circle of friends. Many of them were very happily married couples and had been for as long as Willy and I were." Actress Barbara Rush said, "I could never understand where the orgy was. Nobody invited me. I kept looking for it, but nobody ever so much as made a pass at me." Most of Julius Epstein's friends were fellow writers. "I have been here since October, 1933," the screenwriter declared in the 1980s, "and I've never been to a party with cocaine. As far as my personal life goes, I'm living in a little town in the Midwest."

Actress Jayne Meadows taught Sunday school the entire time she was under contract to Metro. Donna Reed, Marie Windsor, Dorothy Malone, and many others lived at the Studio Club early in their careers, where they shared each other's excitement but were well chaperoned. "The memories of it just give me goose bumps," Marie Windsor said, full of nostalgia. H. L. Mencken, editor of the *American Mercury*, visited Hollywood in 1927 and wrote that the wildest nightlife he saw there was at Aimee Semple McPherson's tabernacle. "I saw no wildness among the movie folk," Mencken said. "They seemed to me, in the main, to be very serious and even somber people. And no wonder, for they are worked like Pullman porters or magazine editors. When they finish their day's labors, they are far too tired for any recreation requiring stamina. Immorality? Oh, my God! Hollywood seemed to me to be one of the most respectable towns in America. Even Baltimore can't beat it." (Brownlow, *Parade's Gone By*)

Television director Jay Sandrich, son of Hollywood film director Mark Sandrich, went to school and associated with friends whose parents were also successful in the motion picture business. "It was not a very realistic community to grow up in as far as life experiences go," the younger Sandrich admitted, "but my parents did their best to inculcate in me the fact that we were not living the normal, typical life. I've always been thankful that my folks forced me to face the fact that this was an unusual lifestyle and that we were not special. It was just that my father was very lucky at the time, to be doing as well as he was. I was thirteen when he died, but I grew up in the business. It had no aura of glamour to me."

Star couples in real life captured the public's fancy, and their studios tried to capitalize on such unions. Clark Gable and Carole Lombard ultimately married, Jean Harlow and William Powell became engaged, and Gary Cooper and Lupe Velez made headlines with their stormy romance, as had Greta Garbo and John Gilbert earlier. Balancing two demanding careers and marriage, however, proved difficult. Barbara Rush and actor Jeffrey Hunter finally divorced, as did countless others. "All I wanted to do was work," explained Rush, "and all Jeffrey wanted to do was work, too. We were separated all the time. I would be on location in England, and he was off on location in the Mediterranean or what have you. It was impossible."

Margaret Tallichet had been an actress before her marriage to William Wyler, but gave up her career shortly afterwards. "Willy did not particularly relish the idea of having an actress for a wife," Mrs. Wyler said later. "He'd had one actress for a wife, Margaret Sullavan, and he hadn't enjoyed it that much. He'd found two careers very trying to have in one family." When former child star Bonita Granville married businessman Jack Wrather, she put her career on hold with few regrets, devoting herself to her husband and children. "Jack was a very morally strict man," said Granville, "and he really and truly couldn't face the fact that I would be having love scenes with other men. It sounds absurd, but it bothered him and he admitted it. Jack felt a woman's place was in the home."

Actually there was never one Hollywood, but several—socially as well as geographically. Like the Wylers, many of the industry's psychologically well-adjusted workers surrounded themselves with a group of friends and savored those relationships over a lifetime. During the mid-1930s and early 1940s James Cagney, Pat O'Brien, Spencer Tracy, Ralph Bellamy, Frank McHugh, Lynne Overman, and Frank Morgan, all of whom knew each other from their theater days in New York, met for dinner every Thursday evening, sometimes at a restaurant but often in one of their homes. Soon they came to be known as "the Irish Mafia."

Gene Kelly was the center of another group, which gathered at his home on Sundays for volleyball and brunch. Frank Sinatra, Judy Garland, Stanley Donen, Saul Chaplin, Betty Comden, Adolph Green, Nancy Walker, Jules Munshin, and Keenan Wynn were all part of that group. "I watched them," Nancy Walker said later. "I thought they were crazy—all that running and screaming. I'd just say, 'Listen, give me a big glass of iced tea and put me in the shade, and I'll watch. I'm not interested in playing volleyball.'" A more sedate group met for Sunday brunch at Louis B. Mayer's house, where the guest list might include a Roman Catholic cardinal, a visiting statesman, or even in the 1940s the president of the United States. A younger crowd from MGM gathered on Sunday afternoons at Roddy McDowall's—Elizabeth Taylor, Jane Powell, Ricardo Montalban, Ann Blyth, and a dozen or so more.

None of them smoked or drank at that time—they were in their teens—but they swam, played badminton, danced to records, ate dinner, and went home around nine or ten o'clock. "It was all good, clean fun," said actor Marshall Thompson, "and we all got along beautifully. We formed kind of a clique."

Still another group from MGM got together regularly on Saturday nights and Sundays for marathon poker sessions. Actress Binnie Barnes maintained that she was often employed at the studio because she was good at the game and therefore Metro executives wanted to have her around. "I was the only woman allowed to play," said Barnes. "If I was working Monday, they'd let me off early on Sunday. If I wasn't, I'd have to play through Monday morning." The British actress also anted up at director Edmund Goulding's house, at a table that usually included Betty Grable and her husband, orchestra leader Harry James. "Betty was the greatest fun," Barnes recalled. "She was a very happy-go-lucky girl, and our poker games were a riot. We'd stop and have dinner and then start again. Nobody won or lost much money, but we had great fun, talking about everybody and about where we were working."

The Holmby Hills neighborhood formed another Hollywood enclave, consisting of Bing Crosby, Joan Bennett and Walter Wanger, Humphrey Bogart and Lauren Bacall, director Alan Crossland, and others. The Crosby boys used to play with Joan Bennett's girls, while Bogart and Bacall lived just down the street. Elsewhere small groups of the movie colony entertained in one another's homes, played tennis or attended football games together, or visited back and forth as neighbors. Actress Ella Raines recalled spaghetti dinners at Van Johnson's house, where Judy Garland sang until eleven or eleven-thirty at night, and Tippi Hedren remembered gourmet meals, with delicacies and wines flown in from all over the world, prepared by Alma Hitchcock, wife of director Alfred Hitchcock.

Young Robert Stack knew Clark Gable, Fredric March, Gary Cooper, Fred MacMurray, and Robert Taylor through their mutual interest in shooting sports; Darryl Zanuck headed a team that played polo and rode horseback up and down hills that in those days were open country in the San Fernando Valley. Other film celebrities formed bowling partnerships, played croquet, or got together after work for drinks and a casual dinner.

Rock Hudson remembered an evening at Elizabeth Taylor's house at the beginning of their work on *Giant* (1956), when the stars were just getting acquainted. "We had a terrific time," Hudson said, "so much so that we all got smashed. Suddenly it was four o'clock in the morning, and two hours later I had to get up and drive to the studio. I was still half-drunk and hung over, as was Elizabeth, and we had to shoot the wedding scene, where Elizabeth is matron of honor and I come in and stand behind her. In between shots we were running out and throwing up. All the women on the set—seamstresses, wardrobe women, and hairdressers—were sobbing and saying, 'Oh, what a moving scene.' We were so sick we could hardly move. But that's what made that scene work."

Hollywood's British colony remained largely insular, with Ronald Colman the unofficial king and Basil Rathbone's wife the self-appointed queen. Others included Sir Cedric Hardwicke, Herbert Marshall, C. Aubrey Smith, Ray Milland, Nigel Bruce, David Niven, Madeleine Carroll, Boris Karloff, and Ida Lupino. "Ronald Colman was a private person," Douglas Fairbanks, Jr., said, "and kept very much to himself. He didn't mingle a lot, but had very steady, close, good friends. His private life was like that of an Englishman in the colonies, in the sense that he was more English abroad than he would have been at home. His group rarely went out to public places, but stayed pretty much at home and circulated among themselves."

The Santa Monica residence of Polish-born screenwriter Salka Viertel became the Sunday afternoon gathering place for European exiles during the Nazi era. Many world-celebrated artists, writers, musicians, and intellectuals poured into Los Angeles during the late 1930s, not all of whom found success in Hollywood. "But they brought intelligence and flavor to this desert town," actress Dolly Haas observed. They also helped sharpen a political awareness that would ignite controversy in the decades ahead.

The Garden of Allah, a residence hotel on Sunset Boulevard, was another oasis for Hollywood intellectuals and foreign dignitaries. Originally the Spanish-style mansion of actress Alla Nazimova, the Garden of Allah consisted of a main house and twenty-five villas built around a huge pool; it opened as a hotel in 1927. More charming than elegant, the hotel offered privacy and an escape from the turmoil of the old studios. Over the years Robert Benchley, Marc Connelly, F. Scott Fitzgerald, Lillian Hellman, S. J. Perelman, Dorothy Parker, George Kaufman, Nathanael West, and John O'Hara stayed at the Garden of Allah, so that the hotel became known as the "Algonquin Round Table gone west." Guests lolled around its pool, while writers sat outside their quarters on wooden slat chairs typing, seemingly undisturbed by the frivolity about them. Errol Flynn and Humphrey Bogart took up residence there between marriages. "Yet intellectuals and celebrities from all over the world were to find it a convenient haven and a fascinating home," columnist Sheilah Graham commented. (Graham, *Garden of Allah*)

The Chateau Marmont, just down the street from the Allah, remained a favorite headquarters of New Yorkers and theater people; it was finally dubbed "Sardi's with beds." A turreted, seven-story castle with a breathtaking view, the Chateau Marmont was built in 1929 and boasted individually decorated rooms, an Old World atmosphere, and seclusion from the glitter and hype of studio publicists. Orson Welles and John Houseman lived at the Chateau Marmont during much of the preparation for *Citizen Kane* (1941); Robert Benchley, a regular at the Garden of Allah, was a frequent visitor, even though Benchley's fear of traffic along Sunset Boulevard was so great that his taxi rides from one side of the street to the other became legendary.

Newcomers to the industry faced adjustments in living publicly and ran the risk of believing what their publicists said about them. "It was a novelty," Gregory Peck said, "to suddenly be recognized more and more as I went about, to feel myself losing my anonymity. But I think because of my stage training and my years in New York, I had a respect for the craft of acting and a fascination with it, so that the intent to do work of high quality was always of more interest to me than being a movie star. I can honestly say I never went off the rails on becoming a celebrity. It never interested me nearly as much as trying to improve as a player." Others claimed fame had its price, eroding their personal lives and eclipsing the fortunes they assumed would be theirs. Hollywood could be a lonely place, even for stars at the peak of their brightness. "The great are so mighty that people are afraid to talk to them," columnist Sheilah Graham noted. "As for the women stars, you could call almost any one of them late on a Saturday afternoon, and they would be doing nothing that night." (Graham, *Garden of Allah*) While depression, alcoholism, and suicide haunted Hollywood's stars, the fan magazines continued to picture life among Hollywood's chosen as a lavish and delirious spree.

Shortly after World War II, however, the public's view of Hollywood began to change; suddenly the old excitement seemed to be missing. Nightclub attendance started to fall off, while Sunset Strip, the scene of nightly activity during the war, changed drastically. As motion picture production declined after 1946, the studios showed less interest in keeping their stars in the limelight, arranging fewer dates for them to be photographed at the Cocoanut Grove, the Mocambo, or Ciro's. With the coming of television, stars found it easier to stay home. When they entertained, it was more likely with a group of friends in intimate surroundings.

Much of the publicity that made headlines during the late 1940s and 1950s was scandalous, destructive to the image studio publicists had labored earlier to create. In 1948 actor Robert Mitchum was arrested in a Laurel Canyon home for smoking marijuana; he was handcuffed and taken in a police car to the Los Angeles county jail. Although handsome Wallace Reid had been involved in a narcotics charge in the 1920s and Errol Flynn and others had regularly smoked joints for years, Mitchum's arrest posed a serious threat to his career, in part because of the media coverage the incident received. With the sharp drop in box office returns, escalating production costs, and a shrinkage of the European market, Hollywood studios were not eager to incur moralistic attacks that could limit distribution in heartland America. For a time it seemed RKO might invoke its "morals clause" to cancel Mitchum's contract, but the actor's sleepy, sexy attitude remained popular at the box office.

Far more devastating was the announcement in December, 1949, that in three months actress Ingrid Bergman was expecting a baby fathered by Italian director Roberto Rossellini. Bergman was still the wife of Dr. Peter Lindstrom; her pregnancy out of wedlock rocked Hollywood and the moral dignity of the entire nation. The millions of fans who adored her were shocked. Having accepted the conviction and purity Bergman injected into her roles, her public felt betrayed. In a period of McCarthyism and panic over Communist infiltration, the actress soon became symbolic of an undermining of traditional values. As news circulated that she planned to divorce her husband, forsake her daughter, and marry Rossellini, the director of the Motion Picture Association of America's production code, Joseph Breen, sent Bergman a warning: "It goes without saying that these reports are the cause of great consternation among large numbers of our people who have come to look upon you as the first lady of the screen, both individually and artistically. . . . Such stories will not only not react favorably to your picture, but may very well *destroy your career as a motion picture artist.* They may result in the American public becoming so thoroughly enraged that your pictures will be ignored, and your box-office value ruined." (Bergman and Burgess, *My Story*) Breen concluded by urging Bergman to issue a denial of the rumors at the earliest possible moment.

When the actress held firm, Hollywood executives threatened not only to ban *Stromboli* (1949), the film she was making in Italy with Rossellini, but *Joan of Arc* (1948) and *Under Capricorn* (1949) as well. On March 14, 1950, Edwin C. Johnson of Colorado stood on the floor of the United States Senate and delivered a vitriolic attack against the Swedish actress, arguing that she had perpetrated "an assault upon the institution of marriage." From her ashes Johnson called for a better Hollywood. "If out of the degradation associated with Stromboli," the senator concluded, "decency and common sense can be established in Hollywood, Ingrid Bergman will not have destroyed her career for naught." Guilt-ridden and feeling that she had ruined herself and harmed others as well, Bergman assumed her career was finished. "I thought I would have to give up acting," she said later, "to save the world from disaster I mean—for it seemed that I had corrupted everybody in the world." (Bergman and Burgess, *My Story*)

Reality and image clearly were at odds. For a decade Hollywood publicists had created an image of Bergman as a saint—honest, talented, wholesomely beautiful, in contrast with the usual celluloid glamour girl. The mystique publicists had created boomeranged when she showed herself to be human on the island of Stromboli and had the courage to face her behavior openly—traits that heretofore had rarely been part of the Hollywood package. For seven years Bergman was barred from American films, while clubwomen and religious groups branded her "Hollywood's apostle of degradation." Suddenly the entire film industry seemed tarnished and out of control.

But there was more to come. In 1951 20th Century-Fox production chief Darryl Zanuck fell in love with his protégé, Bella Darvi, her name derived from combining the initial letters of his first name with those of his wife Virginia's, a blatant exercise in egotism even by Hollywood standards. Zanuck was determined to make Darvi a star, although her talent as an actress was limited. In 1956 he followed her to Europe, after his marriage had broken up. Zanuck's career, once the envy of Hollywood, had crested; Darvi became a compulsive gambler and eventually committed suicide. Together they mirrored the seediness and destruction to which Hollywood had fallen prey.

The playful, joyous atmosphere seemed gone; in its place were ruins, decayed morals, and perversion. The new tabloids, *Confidential* and later the *National Enquirer*, told all—the sicker and more twisted the details, the better. With the decline of the old studios, celebrities' status was reduced. Soon the gods and goddesses of the old studios had become museum pieces, relics of a bygone age when life was simpler and unquestioning audiences trusted in heroes, dreams, and magic.

17
Internal Friction

A delegation of the movie industry's Committee for the First Amendment, at the Capitol in 1947 to protest the investigation of the House Un-American Activities Committee. Photo by Martha Holmes, Life *magazine © 1947, Time Warner Inc.*

D espite persistent criticism, Hollywood maintained its golden image until the end of World War II. Internal rivalries and fragmentation had seldom been divisive and sometimes even proved healthy to the overall business climate. What served one studio generally served the rest, so that the major film companies looked upon the successes of their rivals as stimulants to a collective box office rather than as damaging competition. Policies were worked out behind the scenes, aimed at preserving the status quo and expanding the industry's financial returns. After the war, however, divisions that had appeared during the preceding decade became ruptures that ultimately wrecked the solidarity on which Hollywood's big studio system had been built. External pressures, shifting social forces, new entertainment preferences, and changing economic patterns widened the breach, resulting in anger, civil strife, and permanent despondency, as the Hollywood studios were rocked by tremor after tremor, until the Golden Age of the feature film was over.

Unionization posed an initial assault on Hollywood's paternalistic kingdom, literally splitting the industry left and right. The American Federation of Labor had attempted to invade filmland as early as 1916, to organize the studio crafts. Two years later Hollywood experienced its first serious labor trouble when five hundred studio workers struck for a salary increase to meet the rising cost of living that followed World War I; because of the strike, several studios were forced to shut down for a time. In July, 1921, more unionized craftsmen walked off their jobs. For the most part, however, Hollywood remained a non-union town until the Great Depression of the early 1930s.

The advent of talking pictures paved the way for more powerful Hollywood unions, since the technical revolution brought in Broadway actors and playwrights acquainted with Actors Equity, a major force in New York since 1919, and the other theatrical unions. Although Hollywood producers resisted the loss of absolute power, the Screen Writers Guild, which became the prototype for the other talent guilds, was formed in 1933, with John Howard Lawson as its first president. Studio heads, fearful of a new era of militant labor, had tried to offset the trend toward organization by creating in 1927 the Academy of Motion Picture Arts and Sciences, a company-formed union, but by April 1933, some two hundred writers had forsaken the Academy for the Screen Writers Guild. Even after the passage of the Wagner Act in 1935, which guaranteed labor's right to collective bargaining, Hollywood producers were not willing to accept their writers' efforts at unionization without a fight. "They used to have spies when we had a meeting, waiting outside, writing down the names of those who went in," screenwriter William Ludwig said. "In the morning, when you'd come to work, you'd be called upstairs and they'd say, 'What do you want to get mixed up with those people for? You know we'll take care of you.'"

But Hollywood writers felt exploited and wanted higher wages, regularized hiring practices, standardized contracts, and control over their screen credits. In 1936 producers tried to break the Screen Writers Guild with another company union, the Screen Playwrights. Writers at Metro-Goldwyn-Mayer, for example, were summoned into a projection room and given an ultimatum by production head Irving Thalberg: either resign from the Guild and join Screen Playwrights or be fired. Columnist Sheilah Graham remembered sitting around the Garden of Allah and listening to John Howard Lawson, Donald Ogden Stewart, Dashiell Hammett, Lillian Hellman, and Frances and Albert Hackett talk about the Screen Writers Guild. "From the time I arrived in Hollywood in 1936," wrote Graham, "until it was settled late in 1938, all I heard at the Garden was the fight between the Left and the Right to gain control of all the writers at the studios." (Graham, *Garden of Allah*)

Because screenwriters were intelligent, usually well-educated, articulate, and in most cases politically aware, their struggle with management took on ideological implications. The battle over trade union recognition and the right to bargain collectively politicized the screenwriters more than any other studio workers. By 1936 the Communist Party began to play an important role in the thinking of the Hollywood Left; for some it became nearly synonymous with serious political involvement. Marxist study groups appeared in Los Angeles, attended weekly by young members of Hollywood's Left, who engaged in cultural discussions and enjoyed an intellectual camaraderie not present at most industry gatherings. The outbreak of the Spanish Civil War and support for the Loyalists provided a focus on which most Hollywood progressives, both Communist and non-Communist, could agree, at the same time alarming the front offices of major studios. At MGM, studio manager Eddie Mannix, studio attorney M. E. Greenwood, and Greenwood's assistant, Floyd Hendrickson, placed a number of employees under surveillance. So began what story editor Sam Marx called "a new facet of studio operations—spying and undercover investigations."

(Schwartz, *Hollywood Writers' Wars*) Throughout the 1930s the conservative *Hollywood Reporter* engaged in name-calling and Red-baiting; by 1936 the trade paper had begun to link the activities of the Screen Writers Guild directly with Communism.

Hollywood liberals, as the decade progressed, were actively anti-Fascist and anti-Nazi. For a fund-raising dinner at the Biltmore Hotel for the Anti-Nazi League around 1937, both Dorothy Parker and Marc Connelly bought tickets for two tables, each seating twelve guests. Jews in particular were staunchly anti-Fascist because of the anti-Semitism then rampant in Europe. It was a time of frenzied political activity, a period of excitement for Hollywood Left-wingers, many of whom eventually joined the Communist Party. After the Hitler-Stalin Pact of 1939, with the Soviet Communists joining forces with the German Nazis, it was suddenly not as appropriate for American Left-wingers to be members of the Communist Party, especially for Jews. Communist Party members found themselves suspect, received coolly by their liberal friends, who viewed their new pacificism with disfavor.

The schisms in Hollywood expanded as the CIO, considered in the late 1930s a Left-wing union, began making deeper inroads into the Hollywood studios. Conservatives became convinced that Communists were running the guilds. "The Hollywood Right-wing leaders wanted to stop the unions short," said Edward Dmytryk. "They did not want any kind of Communist control of the unions." Even moderates suspected that the political sympathies of many guild leaders bordered on radicalism, while bitterness among the screenwriters burst into civil war. Life-long friendships were destroyed, and lasting animosities were created. Those caught in the middle felt blasted from two extremes. "I used to tell them all," writer Mary McCall, Jr., said, "that when the revolution comes, you can put me up against a cellophane wall and shoot at me from both sides." (Schwartz, *Hollywood Writers' Wars*)

Communists had been active in the Screen Writers Guild, and several Hollywood actors had been Party members for a brief period. "I think we had something like seven or eight Communists in the Directors Guild," said Edward Dmytryk, himself a one-time Party member. "In the Directors Guild we used our influence mostly in trying to get liberal people into office. There weren't that many Communists in Hollywood. I think at the height there may have been 200 or 250." Screenwriter Ring Lardner, Jr.—like Dmytryk one of the Hollywood Ten called before the House Un-American Activities Committee on charges of being a Communist— said of his political involvement during the late 1930s: "I was already identified as kind of a radical, mainly because I had taken part in the reorganization of the Screen Writers Guild in 1937 and had been co-opted to the Executive Board of the Guild to represent the younger writers. For that and some other activities, like raising money for Spain, I was branded a radical."

In 1940 Texas congressman Martin Dies, chairman of the newly formed House Committee on Un-American Activities, arrived in Hollywood to continue the attack on filmland's Communist sympathizers that he had launched earlier in the press. Dies met with a committee of producers and talked privately with several actors, but ultimately he reported that none were Communists or Communist sympathizers. Dies merely suggested they be more cautious about giving money to political causes. A year later isolationist senator Burton K. Wheeler spearheaded an attack on Hollywood for alleged propaganda designed to bring the United States into the war in Europe, a charge he linked to the monopoly that existed within the motion picture industry, another growing issue.

Once the United States entered the war, the Right and Left in Hollywood attempted to cooperate for the national good, but when the war was over the uneasy coalition turned dissonant. In 1944, while the conflict in Europe and the Pacific still raged, the Motion Picture Alliance for the Preservation of American Ideals was formed, the brainchild of extreme Rightists dedicated to stamping out Communism. The Motion Picture Alliance was engineered by a prominent group of Hollywood reactionaries who opposed Franklin Roosevelt and selected as their organization's first president the conservative director Sam Wood. With Roosevelt's death in April, 1945 and the end of the war in September, Hollywood's political in-fighting returned with a ferocity that seemed more intense for having been abandoned for four years.

In the fall of 1945, with the no-strike pledge of the war no longer in effect, the Conference of Studio Unions called a strike of Hollywood's painters, carpenters, office workers, readers, and other members of the craft unions. The strike quickly turned angry, led by Herb Sorrell of the Painters Union; Sorrell was known to have strong Left-wing leanings and was widely regarded as a Communist. As the strike dragged on, Sorrell decided to concentrate on Warner Bros. for mass picketing, in an effort to break the united resistance of the studios. Picketing began on October 5, and within three days goon squads from the Los Angeles sheriff's office met the strikers in open combat. Fire hoses were turned on the picket line outside Warner Bros.' Burbank studios, and tear gas was eventually used. Violent strikers overturned cars, while metropolitan police armed with rifles charged their lines. Fifty picketers were injured, and many more were arrested. Character actress Rosemary De-Camp was making *Night Unto Night* at Warners at the time: "You went in with an armed guard in the car in the morning, about six o'clock," she said. "It was rather grim to look out and see those guys with rifles." Ronald Reagan, then president of the Screen Actors Guild, was in the picture with her, and DeCamp recalled that Reagan arrived on the set each morning having been up most of the night helping to negotiate.

Jack and Harry Warner remained strongly anti-union and were determined not to yield any ground, even though the strike cost their studio millions of dollars. Angry pickets later massed at the gates of MGM, while RKO became a beleaguered compound, with employees staying at the studio around the clock rather than contend with picket lines to go home. The strike represented the first and the only overt effort of Left-wing unions to seize control in Hollywood. Producers decided they would break Sorrell regardless of the cost. Either laborers went back to work, or they would be starved out. The result was that splits in the already divided talent guilds widened still further. "A lot of us supported the strike and marched in picket lines and incurred Jack Warner's disfavor," said Ring Lardner, Jr. "That along with other events of 1946—the Churchill speech about the Iron Curtain at Fulton, Missouri, the start of the so-called Truman Doctrine in regard to Greece and Turkey, the congressional election of that year, which was a Republican victory, and just a general turning to the Right and the beginning of the Cold War—gave indications of what was happening."

In 1947 Hollywood became a special target for a more aggressive House Committee on Un-American Activities, then chaired by J. Parnell Thomas of New Jersey. Convinced that the recent strikes had been Communist-inspired, Thomas was zealous in launching a purge that would win banner headlines and public exposure for his Committee. As Nora Sayre wrote, "Hollywood's overwhelming attraction for the Committee was its celebrities: the investigators had a fixation on the famous." (Sayre, *Running Time*) Hollywood, the glamour factory, was in the business of manufacturing dreams, and the industry had long made a practice of publicizing images of the stars' real lives worthy of the silver screen. Thomas and the House Committee on Un-American Activities saw a chance to bask in the glow of Hollywood stars and earn enough publicity themselves to become Cold War heroes with the folks back home. "I think the attack on Hollywood took place for two reasons," Edward Dmytryk said. "In the first place the Committee knew the movie industry would be highly visible and that it would attract the attention of the whole nation. They would get television coverage, radio coverage, newspaper coverage, every kind of coverage, because obviously there were going to be many famous people involved on both sides—both the friendly and the so-called unfriendly witnesses. The other reason was that they wanted to knock down the unions."

While the politicians claimed they were ferreting out Communist subver-

sion in filmmaking, their main target, according to film historians Larry Ceplair and Steven Englund, was the populist and liberal themes that had often
appeared in films made by artists and intellectuals and the Jewish businessmen who had dominated the industry from the outset. "Like a beacon in the
darkening political night of postwar America," Ceplair and Englund wrote,
"Hollywood attracted the moths of reaction again and again." (Ceplair and
Englund, *Inquisition in Hollywood*) Although many Communists of the
1930s had denounced their party affiliations by 1950, popular thinking in the
early years of the Cold War linked anyone who had ever been a Communist
to espionage, making any former Party member a potential spy for the Soviet Union.

The film community had undergone major changes since Martin Dies's
investigation seven years before. Producers had opposed Dies's charges in
1940, and those named had been given an opportunity to clear themselves in
private. The attitude of producers in 1947 was uncertain, and the accused
were called to defend themselves before a glare of lights and a bank of newsreel cameras. Hollywood's once solid resistance was crumbling. "The studio
heads were frightened," Gregory Peck observed. "They fell in and denied the
charges at the same time, so that the industry suffered this very dim, shame-
ful period."

In the autumn of 1947 the House Committee on Un-American Activities
subpoenaed two groups of witnesses to hearings in Washington. The initial
group, which became known as "friendly" witnesses, was eager to identify for
the Committee fellow workers assumed to be members of the Communist
Party and to discuss cases in which Communist propaganda had been insert-
ed into Hollywood films. The second group, termed "unfriendly" witnesses,
consisted of nineteen writers, producers, directors, and actors who were suspected of Communist sympathies and therefore became the defendants in
the hearings. The Committee had the support of the Motion Picture Alliance for the Preservation of American Ideals, which remained convinced
that the industry was dominated by Communists, radicals, and Left-wing
crackpots. Preliminary hearings were held in Los Angeles in the spring, producing a climate of anxiety; the subpoenas arrived in September of 1947.

Of the nineteen "unfriendly" witnesses subpoenaed by the Committee,
eleven were called to testify in Washington during October. One, German
playwright Bertolt Brecht, answered the Committee's questions, denied he
was a Communist, and within hours returned to Europe, reducing the group
to the famous Hollywood Ten. The Ten consisted of screenwriters John
Howard Lawson, Dalton Trumbo, Ring Lardner, Jr., Alvah Bessie, Lester
Cole, Samuel Ornitz, and Albert Maltz, producer Adrian Scott, and directors

Edward Dmytryk and Herbert Biberman. "I was in the room at RKO when Adrian Scott and Dmytryk got their subpoenas," director Joseph Losey recalled. "It was complete chaos." (Losey int., Ciment) Dmytryk and Scott had recently made *Crossfire* (1947) together, a film that took a strong stand against anti-Semitism. A few weeks before they were subpoenaed, federal agents had come to the studio and demanded to see the picture. Scott was being touted as a young Irving Thalberg, and Dmytryk was rapidly emerging as one of RKO's top directors. "On Tuesday he was Mr. RKO, crowded around with friends," Dmytryk's wife, actress Jean Porter, remembered. "On Thursday he had a pink subpoena to appear before the House Un-American Activities Committee. All of a sudden I was getting phone calls from my friends saying, 'You're not going to continue going with this man are you?' Some of his friends dropped away; some of my friends dropped away. He didn't believe for one minute that he would be fired, but both of our careers were affected."

All nineteen "unfriendly" witnesses engaged lawyers and held a meeting at director Lewis Milestone's house before leaving Los Angeles. "Milestone was not a Communist," Edward Dmytryk said, "and I'm sure that he didn't know that most of us were members of the Communist Party." A unanimity principle was agreed upon, which Dmytryk later termed "a classic Communist maneuver. We couldn't take any action unless it was unanimously agreed upon. In other words, all nineteen had to vote yes or no. It meant that things could be railroaded through more easily than if you had independent votes." Meeting after meeting followed, with the group eventually deciding that Dalton Trumbo and John Howard Lawson would serve as pilot cases. "We stipulated that the rest of us would stand on the results of these two trials," Dmytryk said. (Dmytryk, *It's a Hell of a Life*) The decision was also made that the Ten would refuse to answer questions about their Party membership, essentially challenging the Committee's legality. They agreed to base their stand on the First Amendment's guarantee of freedom of speech rather than the Fifth Amendment's protection against self-incrimination, in the hope of obtaining a favorable verdict later from the Supreme Court.

A Committee for the First Amendment was formed by Hollywood professionals to protest the procedures of the House Committee and to offer support to the beleaguered witnesses. A plane was chartered to take the protesters to Washington, and a mass meeting was held at the Shrine Auditorium in Los Angeles the night before they left. "That idea was born at Lucy's restaurant on Melrose, near Paramount," actress Marsha Hunt recalled. "One day at lunch, when Willy Wyler, John Huston, and Philip Dunne could hardly enjoy their meal together because they were so spitting mad at what was taking place, these old friends said, 'We have to do something. We have to fight back. They're headline hunters, so let's meet headlines with headlines.'" Thirty-three people boarded the flight to Washington—actors, writers, producers, and directors, all outraged over what they considered a violation of constitutional rights. "This was so appalling to us!" said Marsha Hunt, who was on the flight herself. "The worst thing to us movie people was the suggestion that movies were somehow being subverted with Communist propaganda. We were so angry that headline-hunting Congressmen would use Hollywood and motion pictures as a step to their own self-glorification. That's why the flight was taken." Also on board were Humphrey Bogart, Lauren Bacall, Danny Kaye, Gene Kelly, John Huston, Ira Gershwin, Jane Wyatt, John Garfield, June Havoc, Evelyn Keyes, Philip Dunne, and Sterling Hayden. Members of the Committee for the First Amendment appeared on two network radio broadcasts and called press conferences to point out how difficult it would be for writers to subvert a movie. "We all sat there in the hearing room of the House," Marsha Hunt remembered, "as concerned citizens and filmmakers."

The "friendly" witnesses, on the other hand, made it clear that they considered the House Committee's investigation of Hollywood long overdue. Jack Warner gave the Congressmen a list of his employees he thought might be Communists and accepted the Committee's mandate that the studios rid themselves of "termites." Lela Rogers, Ginger Rogers's mother, testified that her daughter had had to say the line "Share and share alike—that's democracy" in *Tender Comrade* (1943), a wartime film directed by Edward Dmytryk and written by Dalton Trumbo. Mrs. Rogers also declared that she had forbidden her daughter to play the title role in Dreiser's *Sister Carrie*, a novel she described as "open propaganda." (Cogley, *Report on Blacklisting*) Robert Taylor told of his reluctance to accept the leading role in *Song of Russia* (1944), while conservative novelist Ayn Rand objected to "a suspicious number of smiling children" in the picture, as well as questionable elements in *The North Star* and *Mission to Moscow* (both 1943), even though all three movies had been made at the request of Washington officials to strengthen the U.S.-Soviet alliance during the war. (Christensen, *Reel Politics*) Walt Disney described Communist attempts to subvert Mickey Mouse by seizing control of the Cartoonists Guild.

Others pointed out that since screenwriting was largely a collaborative effort, fragmenting responsibility, it would be difficult for Communist propaganda to slip into scripts. "You'd have to be some kind of great magician to be able to get any real Communist propaganda into a film," Edward Dmytryk later claimed, "because the people who financed our pictures were the bankers, and they were all good one hundred percent American capitalists. It's crazy to think that I could have put propaganda into a film that the common man would understand but that the capitalists wouldn't. Besides, Communistic manifestos are very dull drama." Director Martin Ritt was in complete agreement: "What is in those films politically that has anything to do with destroying our way of life? Not a phrase!" Dorothy Jones's detailed scrutiny in 1956 of three hundred films on which Hollywood Communists worked revealed that while some writers did attempt to incorporate ideas that paralleled those of Communism, such efforts were not consistently made even by the most ardent members of the Communist Party. (Cogley, *Report on Blacklisting*)

The House Committee on Un-American Activities, however, was less concerned with reality than with publicity, and the hearings dragged on. The "unfriendly" witnesses continued to hold meetings with their attorneys, writing out separate statements and preparing evasive answers to possible questions, planning their strategy cautiously. "We decided the best policy was not to answer the questions at all," Ring Lardner, Jr., said, "and to challenge the Committee's right to ask them. That might get the issue into the courts in a way that would hamper the Committee. We thought that answering yes, that we had been members of the Communist Party, had the problem that they would then ask us about other people. Getting indicted on a perjury charge is something much worse than getting indicted on a contempt charge." Rather than become informers, the Hollywood Ten decided to argue their right to freedom of speech under the First Amendment. "We were guinea pigs," said Edward Dmytryk, "a test group. Nobody after that ever pleaded the First Amendment; they all pleaded the Fifth Amendment, which meant that they couldn't be put in jail. We pleaded the First in the interest of humanity. There's nothing that guarantees your staying out of prison in the First Amendment. I took a chance. I knew I might go to jail; we all did."

The "unfriendly" nineteen were put up in Washington at the Shoreham Hotel in October, 1947. "That nightmarish week in Washington will remain with me the rest of my life," said writer Howard Koch. "We were in our own capital, yet no foreign city could have been more alien and hostile. All our hotel rooms were bugged. When we, the nineteen 'unfriendly' witnesses, wanted to talk with each other or with our attorneys, we had to either keep twirling a metal key to jam the circuit or go out of doors." (Koch, *As Time Goes By*)

Meanwhile Hollywood's unity was shattered, with the Motion Picture Alliance for the Preservation of American Ideals at one extreme and the Committee for the First Amendment at the other. With box office returns declining after the industry's peak year of 1946, studio executives panicked. "They were cowardly, there's no question about it," Gregory Peck said. "There was a reactionary group out here, chargers on their white horses—Adolphe Menjou, Ward Bond, John Wayne to a lesser degree. And the studio heads were frightened at the thought of boycotts of their films. Anything that threatened the box office threatened them; they were certainly not courageous. The whole thing was ridiculous because all you have to do is read a script to determine whether or not there's Communist propaganda in it." But each segment of the business—the studios, the agents, the various guilds—seemed to be protecting its own goals, fighting for its own interests, even if that meant violating basic canons of loyalty and decency. "Pluralism," Victor Navasky concluded, "at least in the land of the happy endings, didn't work." (Navasky, *Naming Names*)

The appearance of Hollywood personalities before the House Committee remained the biggest show in Washington for several weeks. Since the Committee was not bound by safeguards provided in normal courtroom procedures, its tactics often became abusive. Many legal minds in retrospect consider the Hollywood hearings to raise larger questions about civil liberties than any other events of the Cold War period. "We had the sympathy of everybody in the country," said Edward Dmytryk of the Hollywood Ten, "until some of our people started behaving as badly as the Committee was behaving. Then it was all over." Dmytryk overestimated his group's support. "When the Hollywood Ten took the stand," former Communist writer Roy Huggins said, "I was appalled by their lack of candor, because they all pretended to be Jeffersonian Democrats, and they weren't. It was all so terribly dishonest. It was simply not a reflection of their total political commitment."

The hearings generated such bad publicity for the Committee around the world that chairman J. Parnell Thomas cut them short on October 30, 1947. Eight of the "unfriendly" nineteen were not heard, although some were called back to Washington later. Thomas advised the picture industry to "set about immediately to clean its own house and not wait for public opinion to force it to do so." (Cogley, *Report on Blacklisting*) The Hollywood Ten settled in for a two-year wait, hoping that the Supreme Court would eventually decide in their favor. Since the confrontation with Hollywood Rightists, led by Cecil B. DeMille, Sam Wood, Hedda Hopper, Adolphe Menjou, Robert Montgomery, and John Wayne, had been building for some time, the Ten initially approached the fight almost with elation. But social ostracism and unemployment quickly made the implications of the conflict real. "Once we were subpoenaed, we were all broke within six months,"

Dmytryk said, "because we couldn't work." Even liberal producers suddenly withdrew support. "When we got back from Washington," Dmytryk said, "they confronted us with the ultimatum—either we purge ourselves or we get fired." In restaurants and social situations people suddenly looked the other way when defenders of the controversial Ten walked in. "We were just back from our flight to Washington," Marsha Hunt recalled of the Committee for the First Amendment, "when some of the people on the flight began regretting having gone. Hedda Hopper, who was a particularly arch conservative columnist, took it upon herself to badger all of us who had protested the hearings in Washington and to make *our* loyalty suspect. It was appalling! The very people who had defended the industry were now being made suspect in the industry. It was an unbelievable turnabout."

On November 24, 1947, Congress voted to cite the "unfriendly" Ten for contempt. Three days later, after hours of meetings at the Waldorf-Astoria Hotel in New York, fifty top Hollywood executives formally announced their position on the Ten: they vowed to fire any accused workers who would not freely answer questions put before them by the House Committee on Un-American Activities or who could not publicly clear themselves of charges of having been a member of the Communist Party. While Darryl Zanuck, Harry Cohn, Dore Schary, and others opposed the decree, pressure from the financial circles that controlled the industry proved too great for political principle to triumph over fundamental economic considerations. Even the most independent of studio moguls, Samuel Goldwyn, who had spoken against the Waldorf-Astoria declaration in Manhattan, was unwilling to do battle once the statement was signed. When director William Wyler urged him later to hire one of the Hollywood Ten, Goldwyn replied, "I'm sorry. It would be a dishonorable thing to do." (Navasky, *Naming Names*)

Ring Lardner, Jr., was working on a script at 20th Century-Fox when the Waldorf-Astoria resolution was issued. Darryl Zanuck shunted the duty of firing him off on his assistant, Lew Schreiber. "Schreiber informed me that my contract was abrogated and that I should leave the studio that day, which I did," recalled Lardner. "But first I talked to Phil Dunne and George Seaton, both of whom wanted to walk out with me. We finally agreed that it wasn't a good idea unless they could get a number of writers to join them. They tried and weren't able to get anybody else."

Soon even those who had been on the plane to Washington in support of the Hollywood Ten's rights came under direct fire. Warner Bros. applied pressure on Humphrey Bogart, who placed huge advertisements in the trade papers affirming that he was not a Communist and never had been. "I couldn't get a job," recalled actress Jane Wyatt, who had been on the Washington flight. "It was just after I'd played with Gary Cooper and had been in a Goldwyn picture and was getting parts all the time. Suddenly my agent called up and said, 'I don't know what's the matter. I've had two wonderful parts for you, but at the last moment the producers pulled out.' I didn't put the two together until later."

Hollywood continued to deny that there was a blacklist, and at first each studio did its own sleuthing. "These self-appointed arbiters of loyalty were hired," Marsha Hunt said, "for sniffing out membership in this or that extracurricular activity. Things were distorted, turned around, and the informing began." Contracts were broken at will, using the morals clause as justification. People were tried and convicted without the benefit of a legal system. Many found work on Broadway, others went to England, while writers sometimes found jobs anonymously for less money than they had been making. "I was hired by Burgess Meredith and Franchot Tone," Ring Lardner, Jr., said. "They had formed a company and were going to make a film of a John Steinbeck story, which they asked me to write the script on. I remember I had to go into a bank in Beverly Hills where Franchot Tone withdrew $10,000 in cash and gave it to me."

Virtually overnight the atmosphere in Hollywood became one of terror. Lives were wrecked, careers destroyed, marriages and families shattered as friend betrayed friend, sometimes after swearing devotion the night before. Disillusionment swept over the industry as it became clear that the studio heads were unwilling to protect either privacy or individual opinion. "The motion picture business was really running very, very scared," said blacklisted director Michael Gordon. "It was in the throes of a severe financial pinch, and the studio heads felt they were at the mercy of any kind of adverse publicity which could have a deleterious impact on their destiny. They were particularly susceptible to the kind of onslaught that occurred. All manner of very respectable names, totally untainted in terms of personal histories of political involvement, became vulnerable, since they were involved in protesting the mushroom cloud that ultimately did occur. There was an atmosphere of suspicion and mistrust and anxiety that pervaded this town as smog does today."

Liberal spokesmen for the Screen Directors Guild called the membership together to take a stand against the Waldorf-Astoria decree. At the meeting militant anti-Communists—C. B. DeMille, Sam Wood, George Marshall, Leo McCarey, and Victor Fleming, among others—came armed for combat. The liberals never had a chance. Early in the session someone moved that all votes be taken by secret ballot, an issue that was decided by a show of hands. As those in favor of secret ballots raised their arms, Michael

Curtiz rose to his feet, shouting, "Take their names! Take their names!" Immediately half the hands dropped. DeMille, who had broken with the unions over a compulsory payment of one dollar for a strike fund, hadn't been active in the Directors Guild for some time, but not only was he present on this occasion, he was given an honorary seat at the officers' table. In the midst of the uproar that followed, DeMille jumped up and screamed, "This is war!" "That was the most disgusting exhibition I've ever seen," Edward Dmytryk claimed years later. "All the in-betweens were scared to death."

Still the net of hysteria was not as widely spread as it would be later. In April 1950, the Supreme Court by a substantial majority decided not to review the cases of the Hollywood Ten. "The fighting was over," Dmytryk wrote. "There was the usual meaningless second appeal, which merely took up more time. Finally we got the word that we would be sentenced on June 29, 1950." (Dmytryk, *It's a Hell of a Life*) The Ten were sent to prison in pairs for contempt of Congress. Lester Cole and Ring Lardner, Jr., were imprisoned at Danbury, Connecticut. Albert Maltz and Edward Dmytryk were sent to the prison camp at Mill Point, West Virginia — chained together in handcuffs and leg irons. Each of the Ten was fined $1,000, the maximum under law, although sentences varied from a year to six months. "It was scary to have your husband in jail," said Jean Porter, Dmytryk's wife. "You stick by what you believe. It wasn't that Eddie was there because he was a Communist; he wasn't a Communist. He had left the Party long before."

In 1950 a pamphlet called *Red Channels* appeared, purporting to list those persons in the film, radio, and television fields whose activities justified, in the minds of the publication's editors, suspicion of their loyalty as American citizens. "They were careful in the foreword not to charge those people named with being Communists," Marsha Hunt observed, "because that was a libelous thing to do. But they made it very clear that these people probably should not be allowed to work." Hunt was listed because she had been on the plane to Washington as a member of the Committee for the First Amendment; Jean Porter was listed because she was Edward Dmytryk's wife. Hundreds more were named who had neither testified before any committee nor been a member of the Communist Party. Suddenly *Red Channels* was on every producer's desk. All executives had to do was flip through its pages for someone to become unemployable. "I don't know who may or may not have named me," screenwriter Walter Bernstein said, "but my name appeared in *Red Channels*, and I was blacklisted in Hollywood from 1950 to 1958. I stayed in New York, always working under the table."

Actress Kim Hunter, who won an Academy Award for *A Streetcar Named*

Desire in 1951, was blacklisted because of petitions she had signed. "I couldn't have been more apolitical," said Hunter, "but I did feel strongly about human rights, civil rights, injustices of all kinds. At that time it was dangerous to express such feelings openly." By then the craft unions kept lists of people they suspected of pro-Communist activities and dossiers on everyone from Communist Party members to innocents whom watchdogs labeled "dupes" or "fellow travelers." Not only producers but agents and sponsors became frightened at the prospect of employing anyone suspected of subversive activity. "I didn't work in films for over five years," Kim Hunter said. "Then television went out, network by network. The worst effect was on your sense of self-confidence. There were days and nights, nights in particular, that I was absolutely convinced of my worthlessness as an actress. That's not surprising, since without being able to work there can be no growth. You can only slide backward, and the awareness of that began to eat away at me unmercifully. Deep depression kept recurring."

Anne Revere, who had won an Oscar for *National Velvet* in 1945, was soon blacklisted, as was actress Gale Sondergaard, wife of Herbert Biberman, one of the imprisoned Hollywood Ten. Actor Sam Jaffe, who had never been a Communist, told his future wife that he loved her very much but added, "I can't marry you because my future is behind me." Jaffe didn't work in American films for seven years.

Producer Sam Katz read Marsha Hunt a statement his lawyers had prepared that would have cleared her to work in Hollywood again had she signed it. The statement said that the controversial flight to Washington had not only been a mistake, but had been masterminded by Communists behind the scenes and that she unwittingly had been a dupe. The actress remembered pacing her hotel room, weighing her decision. Finally she determined, "I will not do this. I guess I don't want a career that would be at the expense of what I believe and what I know to be true." Another Hollywood producer told her, "Marsha, I couldn't agree with you more, but this is an ugly time. This is a time for expedience, not integrity."

When Joseph Mankiewicz, then president of the Screen Directors Guild, took an extended trip to Europe in July 1950, Cecil B. DeMille convinced the guild's board of directors to pass a mandatory loyalty oath. The implicit threat was that anyone who refused to sign the oath would be blacklisted. DeMille also tried to oust Mankiewicz, an outspoken liberal but never a Communist, as the guild's president, since Mankiewicz had attempted to keep the organization's membership on a moderate course. William Wyler, John Huston, Rouben Mamoulian, and others opposed the move and defeated it. "That was the kind of internecine conflict that was going on throughout the entire industry," director Michael Gordon observed, "and of course reflecting itself in the guilds to some degree."

With the Communist witch-hunt now threatening other molders of public opinion—higher education, the publishing world, the ministry—the House Committee on Un-American Activities felt strong enough by 1951 to resume its investigation of the film industry, throwing Hollywood into an even deeper panic. On March 21, actor Larry Parks spent a day before the Committee, becoming the first Hollywood witness to admit that he had been a Communist and, reluctantly, also becoming the first to give names of other Party members. On the brink of superstardom with two successful screen portrayals of Al Jolson behind him, Parks prophetically assessed that his movie career had been ruined. When asked to name other Party members, Parks pleaded with Congressman John S. Wood, then chairman of the Committee: "You know who the people are. I don't think this is American justice, to make me choose whether to be in contempt of this Committee or crawl through the mud for no purpose!" (Bentley, *Are You Now*) When he left the hearings, Parks seemed to have no fight left. He informed Harry Cohn he was willing to cancel his Columbia contract, an offer the mogul accepted. Parks and his wife, singer Betty Garrett, took to the road with a variety act. "We weren't out of the business," said Garrett, "but suddenly we had to make our own jobs. We had been in demand and were making big pictures—the Jolson pictures and MGM musicals. We could have been the most popular kids around, but that was cut off." Parks made only three more movies before his death, playing a supporting role in each.

When Robert Taylor testified before the House Committee in 1947, he said he didn't know whether character actor Howard DaSilva was a Communist or not, but that at Screen Actors Guild meetings DaSilva "always has something to say at the wrong time." After that DaSilva had trouble finding employment, changing agents four times within the next four years in an effort to improve his situation. Film executives repeatedly informed his agents, "We can't hire him—he's too hot." (Cogley, *Report on Blacklisting*) When DaSilva himself appeared before the House Committee on Un-American Activities in 1951, he was belligerently uncooperative. At RKO the actor had recently made *Slaughter Trail* (1951), in which he played a cavalry captain, for director Irving Allen. Since DaSilva was the first uncooperative witness after the Hollywood Ten, pressure was exerted within the industry to set an example. Allen quickly announced that DaSilva's performance would be cut out of his picture and reenacted by Brian Donlevy, a change that cost the studio an estimated $100,000.

Director Elia Kazan, at the height of his career, testified before the House

Committee twice. The first time he answered all questions except those asking him to name people he knew to be Communists during the two years he himself had been a Party member. The second time he told the Committee that he had come to the conclusion he had done wrong to withhold names earlier and cited eight members of the Group Theater along with other Party functionaries. Kazan's cooperative attitude and the rumor that a big-money deal with 20th Century-Fox was contingent upon his revealing names established the director on the Left as the ultimate betrayer. Kazan later did much to defend himself over naming Party members, but he lost many close friends, including playwright Arthur Miller, with whom he had previously collaborated successfully. "Kazan is one of those for whom I feel contempt," Dalton Trumbo said, a view shared by others. (Navasky, *Naming Names*)

Suddenly workers of the same persuasion no longer were speaking to one another. By the time the Hollywood Ten were released from prison the political climate had worsened. "By then the Rosenberg trial had taken place," Ring Lardner, Jr., said, "the Korean War had broken out during the week between the time we were convicted and sentenced, and generally the whole situation in the country was more tense." Lardner's mother had had a stroke, so he and his wife and their three children lived in her house in Connecticut for two years.

On April 25, 1951, Edward Dmytryk reappeared before the House Committee and named twenty-six people he had known to be Communists — seventeen writers, six directors, and three others. He told the congressmen he had changed his attitude because of the problematic world situation. After his second appearance before the Committee, the director faced antagonism from both extremes. "When I lost friends," Dmytryk said later, "was when I testified against the Communist Party. You can't imagine how vicious and bitter they were."

Among the people Dmytryk named in his testimony was director Michael Gordon, whose contract with 20th Century-Fox was dropped in midterm. "They paid off in full," Gordon recalled. "I was just invited to get my personal belongings and vacate the premises, which I did. Zanuck urged me to cooperate with the Committee." While screenwriters could work under assumed names, that option wasn't open to directors. Gordon, who had three children, sold his home in Los Angeles and prepared to leave the city. After his family had turned their house over to its new owner, a friend took them in for a few days. At dinner the night before the director was to leave for his appearance before the House hearings, he suddenly asked, "What if they ask me where I'm living? What will I say?" Gordon registered at and slept in a motel that night so he could give the motel as his address and not implicate his host. "I think that properly describes the terror under which we lived," he later remarked.

"Hollywood is a community of survivors of necessity," said Edward Dmytryk, "and the weak fall by the wayside and the wolves get them. That's life. The ten of us who went to jail knew what we were doing; we knew what we were being punished for. We had asked for it. But there were hundreds and hundreds of people who had done nothing who were blacklisted. They were the ones who really suffered."

Most of the informers went back to work, but morale in the industry was ravaged. The informers had polluted social life and destroyed trust and Hollywood's sense of community. As a company town, Hollywood had defined itself in part by its social life; as its community life died, so did much of the Tinseltown mystique. Hollywood, the home of happy endings, had been divided, its balance destroyed. Throughout the Golden Era the old studios had maintained a coherence despite contradictory tendencies: commerce vs. culture, formula vs. originality, social commentary vs. entertainment. The House Committee on Un-American Activities purge "helped to shatter this fragile network of delicate balance," Victor S. Navasky concluded, "but Hollywood was an active contributor to its own predicament. The architects of repression created the conditions under which good people and organizations betrayed their friends, but that is really all they created. They opened the door to the informer, but they did not determine who would hold the door open, who would walk through it, and who would stand idly by watching the traffic." (Navasky, *Naming Names*) Hollywood itself had to share responsibility for its wounds and ultimate demise.

As the blacklist lengthened during the early 1950s, many progressive filmmakers were driven from the industry. Some left the country permanently, others suffered traumas so deep it took them years to recover; a few committed suicide. From 1950 to 1952 there were fewer movies with social content and fewer films with the honesty or intensity of *The Best Years of Our Lives* (1946), *Crossfire* (1947), or *Naked City* (1948). Hollywood had become cautious and flabby at the very time it needed its strength. Late in the decade the blacklists began to break up as the Communist witch-hunters in Washington fell into disgrace, but not before Hollywood had been humbled and dealt a death blow. In 1959 Dalton Trumbo admitted that he had been the elusive "Robert Rich" who had won an Academy Award for writing *The Brave One* three years before under a made-up name. In 1960 producer-director Otto Preminger gave Trumbo screen credit for his work on *Exodus*, and Kirk Douglas openly hired him to write the script for *Spartacus*. But by then the damage to Hollywood was irreparable.

If the hearings of the House Committee on Un-American Activities and the aftermath divided and weakened Hollywood, the federal government's divorcement of the studios from their theaters destroyed the industry's basic structure. Since the major studios controlled all three phases of the motion picture business—production, distribution, and exhibition—the Justice Department initiated an antitrust litigation as early as 1938. Three additional federal actions were filed in 1939 against the nation's three largest independent theater circuits. In the *Crescent* case nine exhibition companies, having stock affiliations with one another, were judged to have conspired with the eight leading distributors to monopolize exhibition and restrain trade in seventy-eight towns in Kentucky, Tennessee, Alabama, Mississippi, and Arkansas. The *Paramount* case, which had been filed in 1938, was amended in 1940 and quickly became the major case in the divorcement proceedings. In the supplemental complaint the five major film companies—Paramount, Loew's, Warner Bros. 20th Century-Fox, and RKO—were charged with combining and conspiring to restrain trade unreasonably and to monopolize the production, distribution, and exhibition of motion pictures. Three minor defendants—Universal, Columbia, and United Artists—were charged with combining with the major five.

Under the practice of block booking, all of these distributors except United Artists licensed the showing of their pictures in indivisible blocks before the films had been made. Under most agreements an independent exhibitor had to accept an entire yearly output of the distributor's features, the bad with the good, which involved twenty-five to sixty pictures a year. Many block-booking agreements also forced exhibitors to take large numbers of short subjects, even if their theaters lacked sufficient screen time to show them. On November 20, 1940, the government issued a consent decree in the *Paramount* case that was to continue for three years. During that period the courts agreed not to press for divestment of theaters controlled by the major studios.

Even though the movie industry rushed to the government's aid during World War II, the Justice Department reactivated the *Paramount* case in August 1944. At that time Interstate Theater Circuit in Texas, for example, in large measure controlled by Paramount, operated forty-three theaters and monopolized first-run showings in the five major cities of the state. Paramount mogul Adolph Zukor pointed out that the system of block booking had been devised to insure exhibitors of sufficient product to fill their theaters for a year. Besides, Zukor claimed, there was no law forbidding the manufacturer of any merchandise to have its own outlet. After all, many shoe companies had their own shoe stores as well as their factories. But exhibitors had come to realize that by breaking up block booking they would be free to select the films shown in their theaters. The burden to make good films would then rest on the producers. "The stumbling block was that the producers controlled theaters," said Eugene Zukor. The divorcement proceedings had been prompted by these captive exhibitors. "If they could split their theaters away from the producers," the younger Zukor explained, "then the producers would really be at the exhibitors' mercy. It worked, much to their disappointment, resulting in financial losses of gigantic size."

The *Paramount* case was tried in 1945, and a district court the next year held block booking, pooling agreements, and certain other discriminatory license terms to be illegal. Paramount had been declared a vertical trust; no longer could a production company control its own outlets. Illegal also were the fixing of admission prices and uniform patterns of runs and clearances (the waiting period before pictures could be shown in second-run theaters). An appeal was made to the Supreme Court, but the ruling of the lower court was upheld. In 1945 Hollywood's major production companies had interests in 3,137 of the country's 18,076 motion picture theaters, most of them first-run houses. By monopolizing the biggest marketing outlets, the major firms had succeeded in curtailing independent producers. In 1946 the near-monopoly of production, distribution, and exhibition was broken. The studios had five years to divest themselves of their theater chains, but they were allowed to retain control over production and distribution.

Paramount was the first to comply, by 1949 becoming simply a production-distribution organization. The company's profits plummeted from $20 million in 1949 to $6 million in 1950. Loew's, Inc. was the last of the major firms to complete divorcement, setting up an exhibition subsidiary in 1954 but not concluding the process for another five years. Columbia and Universal, among the minor distributors, had also engaged in block booking, although United Artists never had. With no assured market the studios soon began cutting production. Whereas Hollywood's five major companies had released 243 pictures in 1940, they released only 116 pictures in 1956, a decrease of 52 percent.

Exhibitors no longer could provide pictures with the extensive promotion that movies had received earlier, when millions of dollars were spent to acquaint the public with Hollywood's current product. "An exhibitor didn't have the resources," Eugene Zukor said, "or he used one-column ads instead of four-column and half-pages the way we did to publicize our pictures properly. Many elements were withdrawn, because when it's your own baby, you really go much farther than if someone else is the custodian."

The antitrust decisions and the breakup of block booking signaled a new era for independent producers, although they were still looked down on in the beginning. Right after the war George Stevens, William Wyler, and Frank Capra created Liberty Films as an independent production company run by directors. "Our idea," said Wyler, "was that directors form their own units, look for material, get it financed by a studio, but control the whole project." Liberty's first film was Capra's *It's a Wonderful Life* (1947), now a classic but not a financial success at the time. The company soon dissolved, its concept a few years ahead of its time.

Director Lewis Milestone had grown discouraged with the major studios once the war was over and joined Enterprise Productions. Enterprise successfully made *Body and Soul* (1947), starring John Garfield, before Milestone directed Ingrid Bergman and Charles Boyer in *Arch of Triumph* (1948). Bertolt Brecht had been hired to do some writing on the script, but in the end the Bergman movie was severely edited and lost $4 million. "Enterprise just never could come back from the *Arch of Triumph* disaster," said Norman Lloyd, who acted as Milestone's assistant, "but it was the greatest setup ever." Milestone tried one more picture with the company, *No Minor Vices* (also 1948), which was not successful artistically. "The mistake was the management's," Milestone reflected later. "I thought the motive should have been to be slightly different from the usual product that a major studio would put out. But they tried to do the same thing that every other studio was doing—make money, no matter how."

In 1946 actress Rosalind Russell and her husband, producer Frederick Brisson, formed Independent Artists with writer Dudley Nichols; the next year they made their first picture, *The Velvet Touch*, using the RKO lot. "We were the first artist-producer team to form an independent company," said Brisson, "and we were going up against the famous major independents like David O. Selznick and Sam Goldwyn. For an actress, a producer, and a writer to form a company was then unheard of." Actress Joan Bennett, producer Walter Wanger, and director Fritz Lang quickly followed suit, launching Diana Productions with *Secret Beyond the Door* and *Hollow Triumph* (both 1948), made at Universal-International and Eagle-Lion studios. William Dozier, who became a vice president at Universal, welcomed independent producers on the lot and himself formed Rampart Productions, which made films there with Dozier's wife, actress Joan Fontaine.

The most adventuresome of the independent producers was Stanley Kramer, whose Screen Plays Inc. turned out a string of modest-budget message pictures that included *Champion* and *Home of the Brave* (both 1949) and *The Men* (1950), Marlon Brando's first picture. Actor Burt Lancaster and his agent, Harold Hecht, began their own production company with *The Crimson Pirate* (1952), later winning awards for such realistic films as *Marty* (1955) and *The Bachelor Party* (1957), neither of which had Hollywood's typical glossy look. "I suppose we were probably influenced by the Italian neorealism after the war," remarked Delbert Mann, who directed both features. "Paddy Chayefsky and I had lots of discussions with cinematographer Joe LaShelle on the type of photography we wanted—less light and more shadows, which would achieve a more gritty, realistic effect."

Gradually the production monopoly of the big studios ended, and a new look began to creep into American movies. "At the time I thought it was very good," film editor Elmo Williams recalled, "because the independents were gaining some strength and the hold of the major studios was being broken. The thing I had against the majors was their monopoly. When they owned the theaters, if an independent made a film, no matter how good it was, he could never get a booking for it except in a little art house somewhere."

Stars also were growing restless, unhappy with the old contract system and starting to demand percentages of their pictures' profits. A new trend began in 1950 when James Stewart, earlier under contract to Metro, succeeded in obtaining a percentage deal from Universal to star in *Winchester '73*, one of the decade's A westerns. "By 1950 in almost every part of the industry you could see that the big studios were changing," Stewart said later. "My contract with MGM ran out during the war. When I got home from the war, they offered me a new contract, but my agents, Lew Wasserman and Leland Hayward, advised me not to sign and to go independent. I took their advice. I think they knew very well that things were changing." Stewart's arrangement with Universal paved a new way for studios to attract stars without keeping them under long-term contract. "Stars discovered that working for the major companies meant they couldn't hold on to most of their earnings," publicist Arthur Mayer observed. "Each star thereupon sought to establish his own company. For example, if you own a company, you can arrange your taxes quite differently. . . . So the stars all went out for themselves." (Mayer int., Rosenberg and Silverstein) After 1950 one- and two-picture arrangements became common, which insured stars more money and greater freedom but broke the continuity that assured long careers.

The generation of actors and actresses who could excite an audience merely by appearing on the screen was coming to a close. In the late 1940s the "Method" actors began their invasion of Hollywood, mostly New York stage actors trained at the Actors Studio in a more naturalistic style. Marlon Brando's casual mumbling and scratching as Stanley Kowalski in *A Streetcar*

Named Desire (1951) was typical, as was Montgomery Clift's brooding sensitivity in *Red River* (1948), unique for its day. Veteran screen actress Ann Doran remembered watching James Dean go through all manner of gyrations preparing himself for a scene in *Rebel Without a Cause* (1955). "Jim Backus and I simply got hysterical," Doran said. "We had never seen this 'Method' of doing things. Backus and I both came from the same school: learn your lines, come in and do it the way the director says, and shut up." While many of the new breed were better actors than the old guard, few demonstrated the force of personality or durability of Gable, Bogart, Dietrich, Crawford, Cooper, Tracy, Davis, Hepburn, and dozens of other screen idols from the Golden Era.

With the proliferation of independent producers the old Production Code, the bastion of film censorship since the early 1930s, soon encountered flagrant violation. As early as 1943, when the Production Code was still rigorously applied, renegade producer Howard Hughes had released *The Outlaw* without the Motion Picture Association's Seal of Approval. Today, *The Outlaw* seems an innocuous western, but in 1943 Jane Russell's heaving cleavage, the strong suggestions of carnal sex, overtones of homosexuality, and a happy ending for immoral people made approval by the Motion Picture Production Association impossible. The obstinate Hughes, after a delay of several months amid mounting publicity, finally released *The Outlaw* to hundreds of independent second-run theaters across the country, along with a suggestive advertising campaign.

A decade later director Otto Preminger, by then an independent producer, openly defied the Production Code by refusing to eliminate such words as "seduce," "virgin," and "pregnant" from the script of *The Moon Is Blue* (1953). Based on F. Hugh Herbert's successful romantic comedy, *The Moon Is Blue* had played on Broadway for over two years and went on tour across the country. Preminger saw no reason to change a word of the script, which he had selected as his first independent production; he secured the nervous support of United Artists, the company that released the film. When the censorship office rejected the script unless six lines were altered, Preminger held fast, and the picture was distributed without the seal of approval. Although *The Moon Is Blue* was denounced from pulpits as evil and Catholic priests in small towns stood outside theaters taking down names of parishioners who went in to see it, the picture was a huge success in the large cities. Two years later Preminger violated the Code again with *The Man with the Golden Arm*, based on Nelson Algren's novel. The censorship office was firm in its opposition to this graphic story of drug addiction, and from the first day of shooting it was clear that the seal of approval would be withheld.

This time, however, the Catholic Legion of Decency gave its sanction, which it had not for *The Moon Is Blue,* and *The Man with the Golden Arm* proved among the year's biggest successes. The Production Code was extensively revised and finally abolished in 1968.

During the decade after the war nearly every Hollywood studio experienced internal friction between East Coast executives, allied with the bankers, and the old studio heads who had built their empires from nothing and were reluctant to surrender control. Metro-Goldwyn-Mayer was a prime example. With box office returns falling, the studio had adopted a committee approach to production, with the authority of producers increasingly curbed as to what they could spend and the type of pictures they could make. Producers soon had to obtain approvals from various sources. In 1948 Metro stepped up color production and cut the number of its low-budget pictures to offset the threat of television. But it became clear that Louis B. Mayer and Nicholas Schenck, the president of Loew's, Inc., MGM's parent company, were in frequent disagreement over the selection and treatment of material for Metro pictures. In 1948 the conflict between Mayer and Schenck, who had the support of stockholders, had reached a breaking point. Mayer was forced to retire as chief executive in charge of production; Dore Schary, formerly head of production at RKO, was brought in to replace him.

Mayer was livid, determined to show the world that his filmmaking career was not over. Demagogue though he was, Mayer was a showman, willing to gamble. Schenck and the New York office were more conservative, especially about finances. Although Mayer had often been a tyrant, he had created a comfortable atmosphere on the MGM lot that would never occur again. Neither would his sense of elegance and style.

Dore Schary preferred a more realistic approach, specializing in message pictures so long as the statement didn't become too political. He chose stories like *Asphalt Jungle* (1950) and *Blackboard Jungle* (1955), reflective of the new Italian realism in cinema—almost the antithesis of what L. B. Mayer had stood for. Schary's ideal actor and actress seemed to be James Whitmore and Nancy Davis, later the wife of Ronald Reagan, although neither Whitmore, fine actor though he was, nor Davis represented glamour. Gone were the vehicles tailor-made for Metro stars, and soon the beautiful faces, beautiful sets, beautiful jewelry, and beautiful clothes vanished, too. While Dore Schary was respected on the MGM lot, many of the old guard remained loyal to Mayer. "I knew when Schary came into MGM that that was the end of the studio as I knew it," drama coach Lillian Burns said. "That's why I wanted out." New producers were recruited, while old ones retired. Schary never enjoyed Mayer's freedom; restrictions were placed on him from the beginning. Although Schary was a warm, soft-spoken man, accessible to workers, many on the lot resented him for replacing Papa Mayer, even though he had earlier been Mayer's protégé. Suddenly frivolity seemed to disappear from the studio; there was nothing but work and fear over dwindling profits.

"Dore Schary was a charming person and couldn't have been sweeter," said Metro publicist Esmé Chandlee. "But he was a writer, not an organizer. He should never have been head of a studio. I think he found he was in over his head. My own feeling is that he was unhappy the whole time he was at MGM." His unhappiness was magnified by declining business and the fact that there were constant fights and power struggles with the New York office. "Dore Schary was sort of like a rabbinical student who feels badly about having become a mountebank," said costume designer Lucinda Ballard, wife of Metro's New York publicity director Howard Dietz. "He was so moralistic and always wanting to do something about God or the pilgrims, which people didn't want to see. It was one of the things that wrecked him in the end."

A stockholders' rebellion took place in 1955, and Schary was ousted within a year. Joseph R. Vogel took over, then Sol Siegel, and the atmosphere around Metro began to change again. "That's when the company really went to hell in a sled," veteran producer Pandro Berman declared. "It continued to go down, down, down even faster. I think you can say that the blame no longer attached itself to the studio; the blame attached itself to the New York office. The New York office then became a group, and that group, like most groups, functioned poorly. It needed a person to make decisions. I would call the president of the company in New York on some important matter and not get a phone call back for a week. It became intolerable; that's why I left MGM. It lost its ability to function, and everything began to fall apart."

At 20th Century-Fox, a schism had developed by the mid-1950s between Darryl Zanuck, head of production, and Spyros Skouras, the company president based in New York. The two had never been friends. Since Skouras technically could veto Zanuck's production decisions, there had been an undercurrent of friction. "Spyros was power-hungry," Zanuck's friend and producer Otto Lang said. "He wanted Darryl relegated to a secondary position. So there was a big intracompany fight going on between the Spyros Skouras contingent and the Darryl Zanuck contingent." Zanuck remained a dedicated, knowledgeable filmmaker; Skouras was a blustery Greek, a tough, conservative businessman. "Skouras was kind of a joke," 20th Century-Fox screenwriter Philip Dunne said, "an amateur as far as filmmaking was concerned. He understood theaters and he understood manipulations, but his interference with the studio was death."

Zanuck resigned as Fox's production head in 1956 to become an independent producer, spending most of his time in Europe. Producer Buddy Adler was appointed 20th Century-Fox's new production chief. "It became almost impossible to make a good picture after that," Philip Dunne said. "All of a sudden the studio was a place where people were frightened. Things were done for the wrong reasons. They were saying, 'What will the censors say to that?' or 'How will this play in Peoria?' They were not thinking of the quality of the picture at all, simply how much it cost." Most workers agreed that Adler was a bright, pleasant man, but never a force, certainly no Zanuck. The atmosphere on the lot seemed to be one of just getting by. "Buddy was very political," screenwriter Edmund North said. "He knew all the levers of power, and he was a master at handling that part of the business. That's how he got to be head of the studio. But when it came to production, Zanuck was head and shoulders above Adler; there was just no comparison." A pall hung over the lot once Zanuck left. "Everything was safe, sanitized, and second-rate," wrote Philip Dunne, "as the studio entered its long decline." (Dunne, *Take Two*)

RKO's discord was different, since the idiosyncratic tycoon Howard Hughes, who hovered over the studio like an invisible dictator, had bought the company in 1948. Always mysterious, Hughes now and then issued a proclamation, frequently fired people, made decisions old-timers in Hollywood found incomprehensible, yet remained reclusive. "Everybody talked about him," actress Mala Powers said, "yet nobody actually saw him." Once Hughes ordered the sets for his musical *Two Tickets to Broadway* (1951) trucked from RKO two miles over to the Samuel Goldwyn lot, where he had offices. There the sets were reassembled, inspected, and hauled back to RKO for shooting once Hughes had approved them. To acquire the services of French ballerina Jeanmaire, the multimillionaire engaged her entire ballet troupe, then never used the dancer in a single RKO picture. More and more the studio's releases consisted of features shot by independent producers and outright purchases of films made off the lot. By 1951 the studio was making only a few movies, and most of its talent was leaving, having had enough of Howard Hughes. Three years later RKO—once one of Hollywood's major studios—was in chaos, having lost nearly $40 million. Director Joseph Losey was convinced that Hughes had bought the studio as a tax liability. "He wanted to run it into the ground so he could take a huge tax loss," Losey argued. (Losey int., Ciment) Whatever the explanation, RKO never regained momentum.

Y. Frank Freeman, the production head at Paramount, was growing old and by the late 1950s was no longer in good health. David Selznick sold his studio in 1949, and Sam Goldwyn produced his last picture, *Porgy and Bess,* a decade later. Of the Hollywood pioneers Jack Warner stayed at the helm longest, finally selling out to Seven Arts Productions, a Canadian-based corporation, in 1967. Toward the end even Warner showed signs of fatigue. "Jack was getting older, and his domestic life wasn't too happy," screenwriter Robert Buckner observed. "He became disgruntled and not easy to work with. That's when many of us left. One after another we saw the handwriting on the wall at the studio. Warner wasn't buying the material we wanted, he wasn't listening to suggestions, and the studio began to go downhill. It was like working on a sinking ship."

By the time Seven Arts bought Warner Bros., David Selznick, Harry Cohn, and Louis B. Mayer were all dead. The industry was in a depressed state, operating on boom-or-bust policies, so that 20th Century-Fox had been forced to sell its backlot to a land development company to remain solvent after the debacle with its multimillion-dollar *Cleopatra* (1963) production. L. B. Mayer had failed to establish himself as an independent producer and turned to horseracing; he died in October 1957, a heartbroken man. His last words were to Howard Strickling, his faithful publicity head. "Don't let them bother you," Mayer told Strickling with bitterness. "Nothing matters . . . nothing matters." (Wayne, *Robert Taylor*) His passing proved emblematic.

18

The Demise

Actress Gloria Swanson amid the demolition of the Roxy Theater in New York. Courtesy of the Academy of Motion Picture Arts and Sciences.

In 1946 ninety million people went to the movies each week. In neighborhood theaters, where double features sometimes changed three times a week, patrons could see as many as six different pictures every seven days. Housewives often cut corners on grocery bills and children saved their allowances for at least one weekly visit to their favorite movie palace. It was part of the American way of life to go to the movies on Saturday night. Most people went to see stars and knew from a picture's cast which studio had produced it. Many of the old screen stories, the love stories especially, now seem silly, yet so powerful were the star personalities that they rose above mediocre roles and forgettable pictures. Throughout the first half of the twentieth century movies gripped the American psyche, remaining the nation's predominant form of entertainment until the advent of television.

For nearly half a century movies had molded popular thinking and helped shape the American character. "I've talked to so many people whose lives were influenced by films," Warner Bros. scriptwriter Robert Buckner recalled, "who tried to imitate the people that they saw on the screen—small town America particularly, the drugstore cowboy kind of American. Boys would imitate Cagney or Flynn, and girls would imitate Bette Davis or Barbara Stanwyck. Their lives were affected more than many of them realized by the characters they saw in motion pictures." Over the decades Hollywood's impact had grown world-wide. Foreigners came to see Americans essentially as the movies portrayed them. Soon other countries began imitating the material culture that the screen magnified. If certain dress styles or kitchen appliances, yard machinery, or typewriters were shown in a picture, orders soon began pouring in from Rumania, Bolivia, Tasmania, all over the world.

Much of Hollywood's portrayal of life was unrealistic, for the old studios remained dedicated to glamour. "When Metro did a shot with Joan Crawford dragging a fifteen- or twenty-thousand-dollar mink coat along the ground," MGM screenwriter William Ludwig said, "every woman in the audience thought, 'Boy, would I like to do that! That's the way to live!' Glamour's what the movies sold, that was the business." Producers, living in a fairyland themselves, often lost touch with the common man, mistaking fantasy for reality.

Ludwig remembered writing a script with David Hertz at Metro in the 1940s that was intended to portray an average, middle-class couple. The writers had a conference in the producer's office one day, at which the film's art director, cameraman, set designer, and others were present. The producer had begun telling the group the plot when Hertz spoke up and reminded him that the family in the film was supposed to be middle-class. "Oh, yeah," the producer said, "there's one important thing you people must remember. This is a typical, average, middle-class family. The guy, if he's lucky, makes fifty or sixty thousand dollars a year." By 1940s standards fifty or sixty thousand dollars was a fortune.

"I found filmmakers in the United States made things appear clean and polished and in place," said Latin American actress Linda Cristal. "In Mexico and in Europe people in films were more human; they scratched, you could almost sense that they smelled. Here they made prototypes; there they portrayed individuals. Here things had to look nice; there things were as they were, and you loved them for their imperfections." When Cristal arrived in Hollywood during the late 1950s, the old studio system was changing, although she was still in awe of its glamorous image. "I saw Hollywood as much bigger than it really was at that time," the actress said. "I saw it as enormous, without limits. But working there was difficult, because instead of coming to a river that was flowing, I had come to a river that was shrinking."

In the postwar period, with movie audiences becoming more sophisticated, Hollywood's approach began to look naive and old-fashioned. Much of the mystery, adventure, and romance the big studios had provided was too simplistic for a more complex, less idealistic world to accept. Young people whose lives had been disrupted by the Depression and World War II were eager to buy homes and start a family. Many of them moved to the suburbs, where mortgages and daily drives into the urban center for work left them with less time and money for entertainment. Inflation and a new amusements tax also cut into weekly movie attendance, with the result that box office receipts after 1947 dropped sharply. To make up for the lost audience, theater owners raised prices, making movies less affordable.

As early as 1948 television, providing free entertainment at home, began showing signs of potential competition. Two years later Americans were buying television sets at the rate of over seven million a year, while weekly movie attendance plummeted to fifty million, about half its postwar peak. With the completion of the coaxial cable in 1951, all the television shows emanating from the East were available across the country at the same time. Forests of television antennas sprang up in major cities and even small towns, signaling the breakdown of the nation's movie-going habit.

Hollywood studios recognized the villain but chose to ignore it, convinced that the fad would pass. The studio moguls were confident they had a corner on big talent and that their stars and directors shared their contempt for a craze that couldn't last. When Eugene McDonald of Zenith offered Warner Bros. a substantial amount of stock in his corporation for ten of the studio's movies, Jack Warner unceremoniously threw him off the lot. Actress Sally Forrest recalled seeing a mass of people on the steps of the Thalberg Building at MGM and Dore Schary's telling them that television was ridiculous. Darryl Zanuck and Harry Cohn felt the same way. "I remember sitting at a dinner table one night," said actor Cesar Romero, "and hearing Zanuck insist, 'Oh, television doesn't mean a damn thing. It's just a passing fancy.'"

Studio executives were so opposed to the new medium and so mired in their own stagnation that they repeatedly refused to buy into the television business when they had an opportunity. Gradually, as the competition became more evident, the moguls refused to allow their talent to appear on network shows. Not only were old films at first not sold to television, but even buying television advertising for movies was viewed with suspicion. The word "television" itself was not to be mentioned in feature films, except derisively. When Marilyn Monroe, playing a would-be actress in Joseph Mankiewicz's *All About Eve* (1950), inquires of her critic-mentor if they have auditions for television, the critic (acidly played in the picture by George Sanders) replies, "Auditions, my dear, are *all* they have on television." Sanders's comment summed up Hollywood's thinking on the upstart medium.

Early in 1950 the motion picture industry adopted a new slogan, "Movies Are Better Than Ever," displaying it everywhere, even on popcorn boxes. Contract players were sent on the road as Hollywood emissaries to tell America how wonderful movies still were, while studio moguls persisted in living like ostriches, blind to what was happening around them. Meanwhile box-office returns diminished at an even more alarming rate, as three thousand theaters closed between 1950 and 1953. One wit suggested that exhibitors show their pictures in the street, thereby driving people into the movie houses, but Hollywood executives weren't amused.

Hardest hit were the theaters that played "oater" westerns and B pictures, since low-budget programming was exactly the kind of fare early television offered. Monogram, on filmland's Poverty Row, gave way to Allied Artists in 1953 in an effort to get out of the smaller feature business and produce major films. Republic sold most of its backlog of pictures to television, then closed in 1958. RKO had abandoned motion picture production the year before; the studio was sold to Lucille Ball and Desi Arnaz in 1958 for filming the "I

Love Lucy" series and other shows owned by the couple's video company, Desilu. When Hedda Hopper noticed in August that the famous RKO symbol, with its jagged streaks of lightning, had been replaced on the studio's water tower by the Desilu trademark, the angry columnist was said to have muttered, "That bastard, television." (Lasky, *RKO*)

To compete, the remaining movie studios decided to offer what television could not—bigger screens. In the fall of 1952 *This Is Cinerama* opened in New York, temporarily dazzling both the public and the critics. The Cinerama process required three projection booths and a gigantic curved screen, which nearly duplicated the 160-degree range of peripheral vision. *This Is Cinerama* gave audiences a thrilling roller coaster ride, an inside look at La Scala Opera House, and a tour of America's scenic wonders, proving an immediate sensation wherever the picture played. Tickets for limited showings sold four months in advance. The problem for exhibitors was cost. Restructuring the few theaters large enough to accommodate the process varied from $40,000 to $75,000, but the profits made the conversion worthwhile. By 1957 the initial Cinerama feature, operating in only twenty cities in the United States and eight cities in foreign countries, had grossed more than $25 million. "The whole industry was thrown into turmoil when Cinerama came out," studio manager Raymond Klune declared. "It burst upon the scene with a great deal of success. Almost concurrently, television was playing havoc with us." (Klune int., AFI)

The same year that Cinerama appeared, radio dramatist Arch Oboler invested $10,000 of his own money and raised enough more outside the industry to film *Bwana Devil* in 3-D, a picture he wrote and directed. The three-dimensional technique had already been rejected as unworkable by major studios, but Oboler opened his completed 3-D feature without a distributor at two Paramount theaters in Los Angeles, earning $154,000 within the first week. United Artists quickly volunteered to distribute the movie, and despite poor reviews and the fact that audiences had to wear uncomfortable Polaroid glasses to experience the effect, *Bwana Devil* was shown across the country with huge success. Filmed on a tiny budget, the brown hills north of Los Angeles substituting for Africa, the picture had required an enormous camera that took two pictures at once. Although it was a bad film, *Life* magazine hailed the 3-D process as "the most frenzied boom since the birth of sound." (Dowdy, *"Movies Are Better"*) "*Bwana Devil* was a freak picture that made money," admitted Robert Stack, who starred in the film, "and the process zoomed to popularity overnight, then fell on its face."

The 3-D craze lasted scarcely more than six months. Warner Bros.' *House of Wax* (1953) was the first three-dimensional film from a major studio, and an American public, fascinated by technical gimmicks, rushed to see it. Jack Warner predicted that the troublesome Polaroid glasses would be no obstacle. "We are convinced," Warner said, "that the public will wear such viewers as effortlessly as they wear wristwatches or carry fountain pens." (Dowdy, "*Movies Are Better*") William Thomas at Paramount scrapped his production of *Sangaree* (1953) after twelve days of shooting and started over with a 3-D camera, adding $400,000 to his budget. "We did each scene twice," recalled Arlene Dahl, the picture's leading lady, "once for 3-D and once for the regular Technicolor camera. So it took twice as long." By late 1953 audiences were tired of having objects hurled at them from the screen, so that 3-D became restricted to a few science fiction and horror films, such as *Creature from the Black Lagoon* and *Gog* (both 1954), and then was abandoned.

Early in 1953 Darryl Zanuck boldly announced that 20th Century-Fox was converting its entire production output to CinemaScope, a widescreen process aimed at capitalizing on the impact of Cinerama. Less expensive than Cinerama, the technique Fox pioneered utilized a reduction lens that squeezed a wider image onto standard thirty-five-millimeter film; then a compensating lens on a projector spread the image out again. On the screen the picture stretched about twice as wide as those shot with a traditional lens and only a third smaller than Cinerama, eliminating the distracting lines down the screen that accompanied the three-projector process.

The very dimensions of CinemaScope encouraged spectacle on the screen, and 20th Century-Fox launched its new process with *The Robe* (1953), a Biblical epic based on Lloyd C. Douglas's bestseller that proved an instant blockbuster. Within a short time the picture was second only to *Gone With the Wind* in box office returns. Unlike the 3-D scramble, CinemaScope was no mediocre magic show, but a well-planned bid for public favor, complete with a symphonic introduction in stereophonic sound. Twentieth Century-Fox based its future on the process and devised a series of pictures with sufficient variety to entice audiences back into the theaters.

Set decorators at the studio loved the extra dimension; for them, as department head Lyle Wheeler said, "the bigger the spread, the better it was." The lens also worked well for location filming, especially for shooting long, horizontal landscapes. But most directors and cameramen hated CinemaScope, finding the proportions wrong. "I think it's the worst shape ever invented," director Rouben Mamoulian said. "There's no depth to the focus," claimed George Cukor, "and they threw away everything we'd learned before and said, 'Oh, you've got to act on one plane.' Later we examined it and found that wasn't necessary at all." Director George Stevens commented,

"CinemaScope is fine, if you want a system that shows a boa constrictor to better advantage than a man." (MacCann, *Hollywood in Transition*) Cinematographer Leon Shamroy, who filmed *The Robe*, found the process infuriating. "Those early Bausch and Lomb lenses were hell," Shamroy declared. "You couldn't even do close-ups, because they'd distort so horribly. But though it wrecked the art of film for a decade, wide screen saved the picture business." (Shamroy int., Higham)

Despite the problems with CinemaScope, every studio in the business accepted the process except Paramount. Paramount developed its own version called VistaVision, advertised as "Motion Picture High Fidelity." The advantage of VistaVision was that the foreground and background of a picture were both in focus at the same time, producing a sharper image. "We called it the Chinese Camera," said screenwriter Melville Shavelson, "because the film ran sideways through the camera."

The period from 1954 to 1956 saw a "CinemaScope rebound" in the industry. The multimillion dollar *Around the World in Eighty Days* (1956), shot in the Todd A-O wide screen process, was a great success. Later the all-star *How the West Was Won* (1962) utilized Cinerama to tell its story, which created special problems on the set. "Instead of a set having a corner, they had to bend the corner," remembered actress Carolyn Jones, who was in the Cinerama picture. "If you were standing twenty feet from the camera, you might as well have been a naked lady, because nobody would know whether you had clothes on or not. It was so distorted and peculiar. There were very few close-ups and very few of the things we understood as being part of film-making. The camera was the star."

By 1958 the downward trend in movie attendance had resumed, forcing studio heads to cultivate segments of the public whose tastes were not satisfied by the repetitious entertainment they could now watch for free on home sets. Filmmakers either had to invent more imaginative story content or join the enemy, television. Studio heads decided to do both.

Initially Hollywood executives had formed the Motion Picture Producers Association and agreed not to sell any of their pictures to the networks. Only independent producers, such as Harry Sherman, who made the *Hopalong Cassidy* series, allowed their movies to be shown on television. By 1955 the Producers Association had reversed its position, and the old films of RKO and Republic soon began running on late-night channels. Early in 1958 all the major studios opened their vaults to television, unloading their pre-1948 features as a short-term boost to operating revenue. Paramount negotiated the most profitable transaction when it leased 750 of its movies to the Music Corporation of America for $50 million. But Paramount executive Barney

Balaban recognized that the transaction marked the end of an era. "When a family starts to sell its silverware," Balaban said, "it's the end of the family. And that's what is happening here." (Ludwig int.)

By that time all the film studios still in existence had moved into television production themselves. Columbia was the first to enter the field, setting up a subsidiary called Screen Gems in 1951 that provided half-hour shows for video programming on three-day production schedules. Walt Disney premiered his "Disneyland" series in October 1954, and the show became an immediate hit. Within a year Warner Bros., 20th Century-Fox, and Metro Goldwyn-Mayer all had their own programs on the air, using them to promote the studio and its forthcoming releases. As the studios gradually embraced the television industry, Hollywood was recolonized, since more and more network activity shifted from New York to Los Angeles. Universal, although still making feature pictures, began renting most of its space for television production, and others followed. While television programming was still looked upon as a stepchild and major stars refused to have anything to do with it, by the end of the decade many actors were employed in both media.

Having tried budget cuts, shorter shooting schedules, and fewer movies from original scripts, studio heads by 1958 decided to emphasize the special picture, making fewer but better features. If the programmer had disappeared with the initial panic, the medium-budget picture also ceased to exist once the "CinemaScope rebound" had run its course. With ticket sales down 12 percent in 1958 over the year before, executives decided to move into the realm of forbidden subjects, offering adult material to mature audiences. After the breakdown of the Production Code and with European films increasing in popularity throughout urban America, mature themes seemed a possible answer to Hollywood's problems. In such adult pictures as *Cat on a Hot Tin Roof*, *The Long Hot Summer* (both 1958), and *Suddenly Last Summer* (1959), the story rather than technology became the focus as Hollywood aimed at a more sophisticated audience. The old studios were changing. In 1960 Bob Hope quipped, "Our big pictures this year have had some intriguing themes—sex, perversion, adultery, and cannibalism. We'll get those kids away from their TV sets yet." (Goodman, *Fifty-Year Decline*)

Patrons were becoming more selective in their movie-going, but it was still the lavish film with costly production values that resulted in the longest runs and the highest grosses, even at inflated ticket prices. *Ben-Hur* (1959), *Spartacus* (1960), *El Cid* (1961), and *The Greatest Story Ever Told* (1965) succeeded at the box office, but the stakes were higher, with no room for failure. Budgets of $6 to $8 million became common. "Hollywood got very arty and very extravagant," observed actress Joan Fontaine, "and they began taking a long, long time on special pictures. The director began putting his name above the stars on the marquee, and the more he could spend on the movie, the more respect he had and the more he was thought of as an intellectual and artistic genius." While an occasional film recouped its high costs and went on to make a profit, most did not.

By the mid-1950s it was estimated that 60 percent of the movie audience was between twelve and twenty-four years of age. Nearly a quarter of all revenue came from drive-in theaters. To appeal to the youth market Hollywood offered more sex and violence, as well as a steady stream of science fiction and horror films. *Untamed Youth, Hot Rod Ramble, I Was a Teenage Werewolf* (all 1957), and *High School Confidential* (1958) fared well on the drive-in circuits.

Hollywood continued to lose its royal position in the late 1950s. When *Oh Men! Oh Women!* was shot on the 20th Century-Fox lot in 1957, there were only two or three movies in production at the once busy studio. The same was true at Warner Bros. Two years later on one of the gigantic sound stages at MGM, four men sang a Pepsi Cola jingle: mighty Metro-Goldwyn-Mayer had been reduced to renting out its space by the day to shoot commercials. No longer was there a steady flow of product, not nearly enough to support a profitable studio. "That's what it takes to maintain a studio," dancer Dan Dailey observed, "that steady income from picture after picture after picture. So the big studios declined, and the stars left, and the grass grew in the streets, which was very sad."

Suddenly the old studios had lost their magic. The lots were nearly empty, and studio executives appeared confused, trying to decide where they should turn next. The old star glitter was vanishing, as one by one the contract players were let go when their terms expired. "All concerned were somewhat nostalgic about the demise," said publicist Walter Seltzer, "because something we knew and cherished was gone and would never come back."

Overnight it seemed that everyone was expendable except the big names. Supporting players with high salaries were among the first to go. "I was frankly scared," confessed actor Marshall Thompson, when his Metro contract came to an end, "because I didn't know whether I'd be working anyplace else or not." Cesar Romero reacted much the same when he was terminated by 20th Century-Fox. "It was a very funny feeling," said Romero, "after all those years of being under contract to a studio. You sort of felt all of a sudden like an orphan. You felt as if you didn't have a home. It was an uncomfortable feeling; the future seemed very dark, having always had the protection of a studio."

Stars who had been pampered and spoiled found they were no longer protected. Dressing rooms and many of the extras had declined by Golden Age standards. At Metro the situation began to change shortly after Dore Schary took control. "The limousine that used to pick me up and drive me from one stage to the other was not there," Arlene Dahl remembered. "Bicycles were there instead." Maria Montez, a mainstay at Universal during the early 1940s, was forced on Douglas Fairbanks, Jr., for a bit part in *The Exile* (1947) to finish her contract. Lana Turner, a major attraction at MGM, held on almost a decade longer, ending her tenure with the studio in 1956. "My last few years at Metro were like working amid the ruins," Turner said. "Familiar faces disappeared. The wardrobe and the prop departments began to thin out, and publicity people I'd known for so long were dismissed. . . . It was all doom and gloom." (Turner, *Lana*) Greer Garson and Clark Gable had been dropped by the studio two years before, as part of a retrenchment that affected even janitors and secretaries. Publicist Esmé Chandlee was with Gable the day he left the MGM lot, after twenty-three years at the studio. Since Chandlee had handled Gable's fan mail, she went down to his set, arriving just as the star was coming out of his dressing room. "Do you want to ride to the gate with me?" he asked her. Chandlee got in his car with him and rode to the front gate. "Well, this is it," Gable said. "You'll be coming back," Chandlee replied. "No," Gable repeated, "this is it." And he drove off, without fanfare or ceremony.

Costs in filmmaking continued to soar, as the unions became more powerful and as agents pushed the price of freelance stars higher and higher. The original screen musical quickly vanished, in part because of the enormous expense involved. "Warner Bros. dropped my option in 1954," dancer Gene Nelson said. "I had a seven-year contract, but they were phasing out musicals at that point. During the three or four years I was under contract to the studio, musicals had gotten increasingly expensive, because prices kept going up. A musical always costs quite a bit more, since you have to figure in rehearsal time, preproduction scoring, and all the arrangements, whether the music used is original or not."

By the decade's end studio orchestras had been abolished, and music departments were being dissolved. "We were all let go," composer Henry Mancini said of the music department at Universal, "but we were called back on an independent basis. It happened quickly; it didn't happen over a period of time." By 1958 contract writers were rare in the studios. Screenwriter Burt Kennedy recalled being the only person left in the writers building at Warner Bros. "I was the last one," said Kennedy; "everybody else had gone. I was with Warners for a year, and then they dropped my option." When playwright Horton Foote went to Warners to work on a script around 1960, he found the writers building completely empty. "I had to punch a clock," Foote recalled. "Faulkner had been there earlier and Christopher Isherwood. So there were a lot of ghosts around."

Producers and directors also began drifting away as the number of feature films dwindled. Many who worked fast enough moved into television. Budget director Fred DeCordova, later producer of "The Tonight Show," made the transition smoothly. "I was actually functioning in television pretty much the way I had in motion pictures," DeCordova claimed, "quickly and organized. I was somewhat talented rather than outstandingly talented and able to get along with my stars and featured players. I found the hurry-up system of television not a great deal different from my hurry-up system in budget pictures."

As the number of productions shot on location increased, the need for craftsmen within the studios decreased. Clothes for contemporary movies were mostly bought in stores. By 1967 the makeup and hairdressing departments at Paramount had closed, at least so far as feature films were concerned. MGM had started paring down its publicity department a decade earlier. "I was a victim of the cutbacks around 1958," Esmé Chandlee said. "Three of us were let go the same day. But production had gone to nothing; you could roll a bowling ball down the main street of MGM." Ann Straus left the following year, another cutback. Howard Strickling, the department head, stayed until 1970, the year Metro's publicity office in New York closed.

In 1969 Warner Bros. decided to stop distributing short subjects, including cartoons, and halted production immediately. Special effects departments had also closed, as the effects men turned to freelance arrangements. Pauline Kessinger charted the decline of the old studios by what happened to the Paramount commissary. First the flowers, linen napkins, and tablecloths went, and Formica tops came in. Then the menu was cut down and the number of waitresses reduced. Finally in 1970 the commissary closed, and Kessinger, its manager, was forced to retire. By then the family atmosphere at all the studios was gone. "Suddenly there were different directors and different production people, and nothing was the same," said veteran stuntman Gil Perkins. Catherine McLeod remembered working on a television show at the old Republic lot, where she earlier had been under contract to studio boss Herbert Yates. When Yates returned to the lot and tried to visit the actress, he wasn't permitted on her set.

More and more empty sound stages were rented to independent producers—some for television, a few for feature films. The immense treasury of scenery and props that the old studios had once kept in storage became as useless in the new age, critic Hollis Alpert pointed out, "as the great nineteenth century palaces at Newport, with morning rooms and billiard rooms, and a dozen upstairs servants, and were far too imperial for a distressed industry serving a contracting market." (Alpert, *Dreams and the Dreamers*) Actress Jayne Meadows recalled going back to MGM to film an episode of "Medical Center" for television and finding dark sheets over everything in the makeup department. "It was like a morgue," said Meadows, "and I remembered the first day I walked into the hair department there and saw this row of wig stands with hair and the various stars' names on them. In my ears I could still hear Esther Williams and June Allyson and Deborah Kerr and Gloria DeHaven laughing and talking early in the morning before they went off to their sets. And I thought, 'This is a dream. I was never in the movies; I was never here.' It was like an earthquake had hit and all those stars had vanished." Actress Ann Rutherford in later years also had occasion to return to Metro. "The wardrobe department was an empty shell," she said, "that echoed and reverberated with those naked iron racks and clattered with forlorn wire coat hangers." (Rutherford int., Wagner) Even as early as 1956, MGM reminded Carolyn Jones "of an ancient dowager that was divesting herself of her jewels before going to bed at night, and she was about halfway through."

By 1958, 65 percent of Hollywood's movies were made by independent producers, as stars moved from one to another, taking advantage of deals negotiated by agents, lawyers, bankers, and promoters. Rather than a studio publicity office, each actor began hiring his or her own publicist. "There came a time," publicity veteran Gail Gifford said, "when if we wanted to do anything with a star, we had to call Atlanta and talk to his lawyer." Stars with aggressive agents had become powerhouses, dictating packages to studio heads with little compromise. "Hollywood today is a series of making deals, rather than making pictures," long-time observer Ken Murray said. Many who disliked the notion of raising money and negotiating terms dropped out of the business altogether, as veteran director King Vidor did after finishing *Solomon and Sheba* in 1959. "I just didn't feel like being a promoter," Vidor explained. "With the big studios I didn't ever have to do any of that. I didn't have to put packages together."

In 1967 twenty-two-year-old George Lucas, later director of the *Star Wars* trilogy, walked onto the Warner Bros. lot to begin an apprenticeship with director Francis Ford Coppola during the making of *Finian's Rainbow*. The day Lucas arrived happened to be the same day Jack Warner cleaned out his office and left the studio for good. Warner had sold out to Seven Arts, a television packaging firm, and his studio would soon be renamed the Burbank Studio. When Lucas appeared on the lot, Warners looked like a ghost town, since *Finian's Rainbow* was the only feature then before the cameras. When the novice filmmaker visited the animation department, he found a lone executive sitting behind an empty desk waiting for the phone to ring. The big studio system was dead.

The once powerful motion picture companies had all been taken over by conglomerates and had become little more than real estate operations and distributing organizations for independent producers. The men who ran them were industrialists who looked upon their studios not as glamour factories but as industrial compounds. When James T. Aubrey, Jr., became president of MGM in 1969, he realized he needed money fast. Aubrey quickly raised cash by selling off some of Metro's assets to an auctioneer—warehouses of costumes, furniture, props, cameras, carriages, vintage cars, and even the landlocked showboat *Cotton Blossom*. The auctions and private sales that followed reaped a profit of $12 million but reduced the studio, once the proudest of the majors, to a shell. "The Lion no longer roared," publicist Ann Straus said. "This once magical place from the Golden Era had been diminished to something less than imitation gold." By 1983 Metro had ceased production on feature films entirely, and Lorimar, initially a television firm, took over the studio. Metro-Goldwyn-Mayer, as a business organization, diversified its investments and earned substantial profits from its hotels and casinos in Las Vegas and Reno. Lot Three of the old studio had become Raintree County Condominiums, and Lot Four had been converted into a shopping mall.

Paramount Pictures was sold in 1967 to Gulf and Western, a huge conglomerate in which the studio represented no more than 5 percent. Paramount became a major supplier for television, eventually taking over Desilu, the old RKO studio, and the adjacent lots merged. "I think all of the companies who were absorbed by conglomerates changed," veteran producer Hal Wallis observed, "because they became pawns and a very small part of a conglomerate operation. They just figured in the balance sheet, but it became impersonal."

After the *Cleopatra* disaster 20th Century-Fox couldn't meet its payroll and had to resort to selling off its backlot, which became Century City. Darryl Zanuck, who had been resurrected and made president of Fox in an effort to salvage the situation, found his rescue mission difficult. "I was a victim of *Cleopatra*," Zanuck claimed. "The goddamned asp was biting me!" (Gussow, *Don't Say Yes*) Old-timers stood on the sidelines, saddened by the passing of an era. "I boycotted Century City for a whole year," said former Fox contract player Vanessa Brown, "not setting foot in the place until long after the studio had sold the land for it."

The Transamerica Corporation acquired United Artists in 1967; Columbia was purchased by Coca-Cola in 1981, diversifying into cable and pay television, and later was bought by Sony. In the age of conglomerates Universal finally surpassed the former Big Five studios, making millions of dollars both in television production and feature films. Decca Records took control of Universal in 1952; then when Decca merged with the Music Corporation of America in 1959, former agent Lew Wasserman became the dominant figure in the studio. As part of the MCA conglomerate Universal began to expand at a time when other firms were being forced out of business. In 1964 the studio resumed its tours, expanding them into a highly profitable segment of the company's operations. Players who had worked at the studio earlier, however, suddenly found the atmosphere cold and impersonal, referring to the administration building as the Black Tower and the executives who worked there as the Suits. "It ceased being a creative studio and became a business," producer Ross Hunter said. "The old guard was slowly let go. I frankly had to buy my contract out." Earlier the old studios had been accused of being assembly lines; by the 1980s their successors had become computerized. "Now everything goes through the computers," Edward Dmytryk said in 1979. "The whole downstairs of the main building at Universal is a computer center. Today the business is run by agents, who know nothing about making films. They're basically promoters and salesmen."

For Hollywood workers as well as fans, there is nostalgia for the big studio era; all agree it was a Golden Age that will never be seen again. "I still get a kick every time I drive down Warner Boulevard to play golf at the Lakeside Country Club," said former Warner Bros. player Gordon MacRae. "Some of the old buildings at the studio still have Warner Bros. on them, so there are nice happy pangs of remembrance." "When I arrived there in the late 1930s, Warner Bros. was a wonderland," actor Eddie Albert remarked. "But that was true of all of Hollywood. I was in awe of all the stars—Flynn, Cagney, Bette Davis, and the rest. Warners was a huge machine. I must have made twenty pictures in two or three years. We were belly-aching all the time, but the way the business is now, we had it pretty good then." Arlene Dahl, who spent four years under contract to Metro, said, "I was happy to have been there from 1947 to 1951 to see the Golden Era of Hollywood in its full bloom. I came in on the tail end, but I'm happy I was there at all."

Hollywood, veteran film workers maintain, isn't the same place it once was. "I couldn't get rid of the feeling that any minute I'd look out and see tumbleweeds come rolling past," director Joseph Mankiewicz declared.

(Freedland, *Warner Brothers*) "There's not much stardust left anymore," lamented Dan Dailey. "That's because the old stars were kept separated from the public, and the public only learned what the studio wanted people to know about them. Anything else was hushed up; now it's all wide open." Director George Sidney felt the magic was gone: "Today everyone knows that that darn shark over at Universal isn't real," said Sidney, referring to the movie *Jaws* (1975). "In our time we would have never told that."

"Now, studios are nothing but the Ramada Inn," said veteran director Billy Wilder. "You rent space, you shoot, and out you go." (Wilder int., Columbus) Actor Gregory Peck came to dislike the way movies were made by the 1980s, finding conglomerate ownership of the studios constrictive and inhibiting. "Today's movies are test-marketed ad nauseam," said Peck. "Of course, the inspiration goes, the daring goes, the controversial ideas fall by the wayside. It has become more of a heartless business. . . . A different kind of executive has taken over, a kind of crap-shooter. He stays in his job just as long as he keeps throwing sevens. If he craps out too many times, the studio gets in a new executive." Peck claimed he feels no nostalgia for the old days and accepts the changes that have occurred. But he added, "I used to *like* those old boys—Zanuck, Jack Warner, Goldwyn, Cohn—more than I like the executives today, because they had a visceral instinct for what was good entertainment, and they got personally involved. They had zest and red blood in them, and they ran those studios with their guts."

Although many of the new breed in Hollywood felt that anybody over twenty-five was outmoded, some were intrigued with the lingering mystique of the big studios, pumping old-timers for stories about the great legends of the silver screen. "I would like to bring back glamour," said Ross Hunter, whose idol was Sam Goldwyn, "bring back the fantasy world, to live a dream again. I would like to do movies about the beautiful people. They may not be great works of art, they may not give you a lesson in literature, but they will entertain you, which I strongly believe is what the motion picture business is all about. We all need an escape from our humdrum lives."

Much was lost with the passing of the big studio system, but the New Hollywood has also brought gains. With the new era actors more consistently were *actors* rather than personalities, moving comfortably between stage and screen. Screenplays developed characters of greater depth, no longer relying so strongly on enticing stories. Most of all, the screen became more honest, less timid about speaking out on contemporary issues, at times becoming profound. Yet as John Wayne's son Michael pointed out, "Movie stars carry films for producers. What my father did was guarantee a certain bottom. The old stars guaranteed that producers could make a picture and know that the studio wouldn't lose money. Now actors are indulging themselves in their acting rather than being stars."

The stability of the old studio system, with its consistent chain of command ruled over by great moguls for decades, is clearly missing from the conglomerate approach. Within a two-week period during the summer of 1984 the top management of three major Hollywood studios experienced a complete turnover in personnel. During the three years following the *Heaven's Gate* debacle in 1980, perhaps the biggest financial disaster in motion picture history, the management of every major company in the industry except one changed. "Not one production head in Hollywood today is where he was three and a half years earlier," Steven Bach wrote in 1985. (Bach, *Final Cut*)

Film for film the independent picturemakers have probably shown more concern for movies as a serious art form than the old moguls did, but their inability to produce a steady flow of successful movies has created problems for studio administrators. Also there's no overall planning; one major mistake can sink an entire company. While there are fewer pictures, substance and social commentary have definitely increased. Audiences have grown more sophisticated; many of the old conventions are no longer necessary to keep the public from becoming lost or bored, giving the filmmaker a wider swath for creativity. Sequels and the teenage market remain important to the industry, while home video rentals have become a major consideration. Still, excellent movies are being made every year, interpreting times past and present for a changing world. Without question filmmaking has continued to improve technologically, and human relationships can now be probed in their full range of complexity. What's missing is glamour and escapism and the cult of beautiful people.

Special effects have become far more important during recent decades, in many cases emerging as the real star of the picture. Seasoned directors and lasting stars are perhaps the greatest loss since the demise of the old studios, for there are no coordinated publicity offices to build talent effectively and consistently over the course of a lengthy career. In a throw-away culture the public can become saturated quickly, tiring of faces that appear too often, but also forgetting those who drop out at an inopportune moment. "It's a very strenuous, stressful, ruthless, bewildering, competitive kind of business to be in," Eddie Albert said. "I myself have been lucky. I have a medium amount of talent and got in early before it got really rough like it is now. When I entered the business, people didn't come and go as quickly, nor did they rise up to the top as quickly. I would be totally bewildered if I came out to Hollywood now. It's a bitch, it really is a bitch."

By 1970 mass audiences probably revered rock stars more than movie stars, while the Hollywood legends had either disappeared or taken to the straw hat circuits, except for the handful of megastars who managed to combine feature films with stage work and television. Current players are able to assume a greater variety of roles, enabling them to build individualized characterizations on the screen rather than relying on the stereotypes that persisted throughout the Golden Era. With the premium no longer on physical perfection, actors are freer to probe the depths of roles without fear of disappointing or alienating fans. In that respect the American film industry has matured.

Yet the age of the tough-minded, colorful pioneers in Hollywood is a chapter in American history that continues to fascinate successive generations, much as the exploits of the frontiersmen of the Old West do. The old studio heads shaped the movie industry and guided it to world fame, both gauging public taste and creating it. In the process they reflected ideals that had embedded themselves in American culture during the previous century—material success, social mobility, and individual recognition—amid more contemporary notions of romance, personal fulfillment, lasting happiness, and euphoric sex.

"Hollywood, after all, was only a picture of America run through the projector at triple speed," observed writer Budd Schulberg, who grew up in the industry. "If the Hollywood party was excessive, it was only because Hollywood had always been an excessive, speeded-up, larger-than-life reflection of the American Way." (Schulberg, *Moving Pictures*)

Americans, and the world, wanted to believe in Tinseltown. Hollywood's projection of romance and glamour offered a retreat from the difficult and sometimes painful process of growing up, both for individuals and society as a whole. The big studios fed its public's innocence, pampered and indulged audiences, encouraged them to remain the gullible idealists the majority preferred to be. For most adults, childhood—no matter how difficult—appears in retrospect a wondrous, unbounded time of youthful dreams and unfulfilled hopes that alternately haunts and warms and comforts. The pioneer filmmakers in Hollywood's glamour factories photographed the dream that was young America, packaged it with sophistication, added a sound track, and molded the attitudes and aspirations of at least two generations, selling that dream around the globe.

*Sound stages down the main street of Warner Bros. in Burbank.
On the left is the studio commissary. Courtesy of the Academy
of Motion Picture Arts and Sciences.*

Sources

Interviews in the SMU Oral History Collection

Unless otherwise stated these oral histories were conducted by the author.

Abbott, George. Miami, Florida, November 16, 1979
Abbott, John. Los Angeles, August 12, 1988
Abel, Walter. New York, January 5, 1979
Adams, Julie. Los Angeles, July 19, 1984
Alberghetti, Anna Maria. Dallas, February 26, 1981
Albert, Eddie. Los Angeles, July 18, 1991
Albright, Lola. Los Angeles, May 31, 1985
Alda, Robert. Dallas, January 3, 1978
Allen, Lewis. Los Angeles, July 17, 1985
Ameche, Don. Dallas, March 1, 1977
Ames, Leon. Corona Del Mar, California, August 18, 1983
Andrews, Maxene. Los Angeles, July 21, 1984
Astaire, Fred. Los Angeles, July 31, 1976
Autry, Gene. Los Angeles, July 24, 1984
Ayres, Lew. Los Angeles, August 4, 1981
Baiano, S. J. "Solly." Los Angeles, August 12, 1986
Baker, Herbert. Los Angeles, May 17, 1977
Ball, Lucille. Los Angeles, August 21, 1980
Ballard, Lucinda. New York, January 12, 1985, March 27, 29, 1986,
    October 11, 1986, March 19, 1987
Bari, Lynn. Los Angeles, August 18, 1986
Barnes, Binnie. Los Angeles, August 9, 1983
Bartlett, Hall. Los Angeles, July 18, 1985
Bellamy, Ralph. Los Angeles, May 18, 1977
Bennett, Charles. Los Angeles, August 14, 1980
Bennett, Joan. Scarsdale, New York, January 11, 1985
Berman, Pandro S. Los Angeles, August 21, 1978
Bernstein, Walter. New York, March 28, 1986
Blane, Ralph. Broken Arrow, Oklahoma, March 12, 1979
Boetticher, Budd. Ramona, California, August 13, 1988
Bogart, Paul. Los Angeles, July 20, 1984
Bolger, Ray. Los Angeles, August 5, 1976
Boone, Pat. Los Angeles, July 26, 1990
Bowdon, Dorris. Los Angeles, August 11, 1981

Bracken, Eddie. Dallas, July 31, 1986

Brisson, Frederick. Los Angeles, August 11, 1981, July 22, 1982

Broidy, Steven. Los Angeles, August 18, 1988

Brown, Vanessa. Los Angeles, August 8, 1983

Bruce, Carol. Los Angeles, July 20, 1979

Brynner, Yul. Los Angeles, August 1, 1975

Buckner, Robert. San Miguel de Allende, Mexico, July 23, 1988

Burns, Lillian. Los Angeles, August 17, 1986

Buttons, Red. Los Angeles, January 4, 1977

Buzzell, Edward. Los Angeles, July 17, 1982

Cahn, Sammy. Los Angeles, July 25, 1975

Carey, Harry, Jr. Los Angeles, July 23, 1984

Chakiris, George. Dallas, September 15, 1982

Chandlee, Esmé. Los Angeles, May 23, 1985

Chaplin, Saul. Los Angeles, July 29, 1976, interview conducted by Ann Burk

Chayefsky, Paddy. New York, October 17, 1979

Cherry, James O. Dallas, March 9, 1974, interview conducted by Dennis Hillman

Clayton, Jan. Los Angeles, August 19, 1980

Clooney, Rosemary. Dallas, January 25, 1975

Cohn, Joseph J. Los Angeles, July 25, 1991

Collins, Richard J. Los Angeles, July 25, 1990

Condon, Richard. Dallas, May 7, 1991

Conried, Hans. Dallas, November 30, 1973, interview conducted by Jeffrey Holmes

Cooper, Jackie. Los Angeles, July 17, 1979

Cristal, Linda. Los Angeles, July 17, 1991

Cukor, George. Los Angeles, May 27, 1977

Cummings, Robert. Dallas, July 22, 1974

Curtis, Donald. Dallas, November 10, 1988

Da Costa, Morton. West Redding, Connecticut, March 26, 1986

Dahl, Arlene. Dallas, September 24, 1975

Dailey, Dan. Dallas, July 13, 25, 1974, interview conducted by Sally Cullum

Davis, Ann B. Dallas, April 2, 1975

Day, Doris. Carmel, California, October 18, 1983

Day, Laraine. Los Angeles, July 17, 1979

Dean, Eddie. Woodland, Hills, California, August 20, 1988

Deason, A. D. Dallas, March 21, 1974, interview conducted by Dennis Hillman

DeCamp, Rosemary. Los Angeles, July 13, 1982

DeCordova, Frederick. Los Angeles, July 16, 1982

DeFore, Don. Los Angeles, August 21, 1986

DeToth, Andre. Los Angeles, July 15, 1987

Deutsch, Armand. Los Angeles, July 28, 1982
DiPaolo, Dante. Dallas, March 15, 1979
Dmytryk, Edward. Austin, Texas, December 2, 1979
Donen, Stanley. Los Angeles, August 9, 1983
Donnell, Jeff. Los Angeles, August 11, 1983
Doran, Ann. Los Angeles, August 10, 15, 1983
Douglas, Gordon. Los Angeles, July 23, 1987
Drake, Alfred. New York, March 24, 1986
Duff, Howard. Montecito, California, August 18, 1988
Dunne, Irene. Los Angeles, July 23, 1982
Dunne, Philip. Los Angeles, January 8, 1983
Eckart, Jean. Dallas, May 14, 1986
Eckart, William. Dallas, May 12, 1986
Epstein, Julius J. Los Angeles, August 27, 1986
Essex, Harry. Los Angeles, July 25, 1989
Evans, Dale. Fort Worth, Texas, April 26, 1982
Ewell, Tom. Dallas, January 24, 1974
Fabray, Nanette. Los Angeles, August 5, 12, 1975
Fairbanks, Douglas, Jr. Dallas, October 7, 12, 14, 1982
Foch, Nina. Los Angeles, August 16, 1983
Fontaine, Joan. Dallas, April 12, 1979
Foote, Bruce. Dallas, Fall, 1975, interview conducted by Ann Burk
Foote, Horton. Dallas, November 10, 1987
Ford, Glenn. Dallas, June 2, 1990
Forrest, Sally. Los Angeles, May 22, 1985
Fowler, Gene, Jr. Los Angeles, July 20, 1985
Fowler, Marjorie. Los Angeles, August 14, 1986
Frankovich, Mike. Los Angeles, July 16, 1985
Freeman, Kathleen. Los Angeles, July 24, 1984
Garde, Betty. Los Angeles, August 13, 1986
Garland, Beverly. Los Angeles, August 10, 1988
Garrett, Betty. Los Angeles, August 23, 1978
Geer, Will. Los Angeles, July 29, 1975
Gifford, Gail. Los Angeles, August 20, 1986
Gish, Lillian. New York, January 4, 1979
Gordon, Gale. Dallas, June 16, 1975
Gordon, Michael. Los Angeles, August 7, 12, 1981, July 26, 1982
Granville, Bonita. Los Angeles, August 4, 1976
Gray, Coleen. Los Angeles, January 14, 1983
Greaves, William. New York, October 9, 1985
Green, John. Los Angeles, July 17, 19, 21, 1975, interview conducted by Ann
   Burk
Haas, Dolly. New York, October 6, 1986

Hale, Barbara. Los Angeles, July 19, 1984

Harman, Estelle. Los Angeles, July 16, 1987

Harris, Julie. Dallas, February 10, 1989

Harris, Lynn. Dallas, March 7, 1974, interview conducted by Dennis Hillman

Hasso, Signe, Los Angeles, August 16, 1980

Hathaway, Henry. Los Angeles, January 6, 1983

Havoc, June. Stamford, Connecticut, April 14, 1990

Hayes, Helen. Nyack, New York, October 19, 1979

Head, Edith. Los Angeles, July 29, 1975, interview conducted by Ann Burk

Healy, Mary. New Rochelle, New York, October 2, 1982

Hedren, Tippi. Acton, California, July 24, 1982

Henderson, Florence. Los Angeles, August 18, 1983

Hendricks, Bill. Los Angeles, July 26, 1979

Henreid, Paul. Los Angeles, May 27, 1985

Heston, Charlton. Los Angeles, July 18, 21, 1989, July 20, 1990

Hill, Arthur. Los Angeles, January 5, 1977

Holliman, Earl. Los Angeles, July 24, 1989

Holm, Celeste. New York, March 30, 1988

Houseman, John. Los Angeles, July 19, 1979

Hudson, Rock. Los Angeles, August 24, 1983

Huggins, Roy. Los Angeles, August 16, 1986

Hunt, Marsha. Los Angeles, August 12, 1983

Hunter, Kim. New York, January 2, 1979

Hunter, Ross. Los Angeles, July 17, 1984

Hussey, Ruth. Carlsbad, California, July 30, 1984

Hyer, Martha. Los Angeles, July 30, 1982

Ireland, John. Montecito, California, July 27, 1990

Jackson, Felix. Los Angeles, July 16, 1984

Jacobson, Arthur. Los Angeles, July 19, 1989

Jaffe, Sam. Los Angeles, August 14, 1978

Jeffreys, Anne. Los Angeles, August 15, 1983

Jones, Allan. New York, January 9, 1985

Jones, Carolyn. Los Angeles, August 3, 1976

Jones, Shirley. Los Angeles, July 27, 1984

Jordon, James. Los Angeles, August 14, 1978

Jurow, Martin. Dallas, October 1, 1985, January 21, 1986, April 7, 1986

Kanter, Hal. Los Angeles, August 16, 1988

Keefer, Don. Los Angeles, May 21, 1985

Kelly, Gene. Dallas, June 20, 21, 25, 1974

Kennedy, Burt. Los Angeles, July 22, 1987

Kerns, Hubie. Los Angeles, July 25, 1989

Kessinger, Pauline. Los Angeles, July 29, 1976

Kirsten, Dorothy. Los Angeles, August 19, 1986

Koch, Howard W. Los Angeles, August 10, 1981

Koster, Henry. Los Angeles, August 13, 1980

Kramer, Stanley. Los Angeles, August 19, 1988

Lamas, Fernando. Los Angeles, August 6, 14, 1981

Lang, Otto. Los Angeles, August 7, 1981

Lardner, Ring, Jr. New York, January 11, 1985

Laurie, Piper. Los Angeles, July 21, 1989

Lederer, Francis. Los Angeles, August 10, 1988

Lee, Anna. Los Angeles, August 13, 1981

Lee, Pinky. Fort Worth, Texas, August 28, 1975

Leigh, Janet. Los Angeles, July 25, 1984

LeMassena, William. New York, January 11, 12, 1979

Leonard, Sheldon. Los Angeles, July 24, 1989

LeRoy, Mervyn. Los Angeles, May 16, 1977

Leslie, Joan. Los Angeles, August 13, 1981

Lindfors, Viveca. New York, January 8, 1979

Lloyd, Norman. Los Angeles, July 23, 24, 1979

Loos, Mary Anita. Los Angeles, July 26, 1990

Lorring, Joan. New York, January 7, 11, 1985

Loughton, Phyllis. Los Angeles, July 23, 1979

Lubin, Arthur. Los Angeles, July 15, 1985

Ludwig, William. Los Angeles, July 17, 1989

Lund, John. Los Angeles, August 17, 1988

Lydon, James. Los Angeles, July 26, 1989

Lynn, Betty. Los Angeles, July 26, 1984

McGuire, Dorothy. Los Angeles, August 29, 1986

McLeod, Catherine. Los Angeles, August 8, 1983

MacRae, Gordon. Dallas, October 25, 1979

Maharis, George. Dallas, April 3, 1975

Mamoulian, Rouben. Los Angeles, August 19, 1980, August 15, 1981

Mancini, Henry. Los Angeles, July 18, 1979

Mann, Daniel. Los Angeles, July 19, 1989

Mann, Delbert. Los Angeles, July 21, 1984

Mara, Adele. Los Angeles, August 26, 1986

Martin, Mary. Dallas and San Miguel de Allende, Mexico, June 11, 1984,
  July 18, 19, 1988

Mason, Pamela. Los Angeles, July 20, 1982

Mayo, Virginia. Dallas, November 30, 1973

Meade, Julia. New York, January 10, 1985

Meadows, Jayne. Dallas, November 11, 1975

Merrill, Dina. New York, January 8, 1985

Merriman, Nan. Los Angeles, August 6, 1976, interview conducted by Ann
  Burk

Milestone, Lewis. Los Angeles, July 23, 1979

Miller, Winston. Los Angeles, November 6, 1991

Minnelli, Vincente. Los Angeles, August 14, 1980

Mirisch, Walter. Los Angeles, August 19, 1986

Montgomery, George. Los Angeles, July 26, 1982

Moore, Terry. Los Angeles, August 27, 1986

Moran, Peggy. Camarillo, California, July 21, 1982

Morison, Patricia. Los Angeles, August 25, 1983

Moss, Arnold. New York, January 9, 1985

Murray, Don. Santa Barbara, California, July 20, 1989

Murray, Ken. Los Angeles, August 10, 1981

Nathan, Robert. Los Angeles, August 12, 1981

Natwick, Mildred. New York, January 4, 1979

Neame, Ronald. Los Angeles, July 18, 1985

Nelson, Barry. Dallas, October 16, 1985

Nelson, Gene. Los Angeles, August 15, 1980

Newman, Joseph. Los Angeles, July 23, 1984

Newmar, Julie. Dallas, November 3, 1977

Ney, Richard. Los Angeles, August 10, 1983

Nolan, Lloyd. Los Angeles, July 28, 1976

North, Edmund. Los Angeles, August 15, 1986

O'Brian, Hugh. Los Angeles, August 17, 1988

O'Brien, Pat. Dallas, February 4, 1975

Oswald, Gerd. Los Angeles, July 20, 1979

Paige, Janis. Los Angeles, August 6, 11, 1981

Pan, Hermes. Los Angeles, January 12, 1983

Parrish, Robert. Bridgehampton, New York, October 4, 1991

Peck, Gregory. Dallas and Los Angeles, November 3, 1974, May 27, 1977,
    August 15, 21, 1978, July 28, 1979, August 22, 1980

Peppard, George. Los Angeles, July 23, 1991

Perkins, Gil. Los Angeles, August 21, 1986

Pevney, Joseph. Carlsbad, California, July 23, 1989

Pirosh, Robert. Los Angeles, August 14, 1986

Porter, Jean. Dallas and Austin, Texas, April 18, 1980, April 4, 1981

Powell, Jane. Dallas, October 24, 1979

Powers, Mala. Los Angeles, July 26, 1984

Preminger, Otto. New York, January 5, 1982

Prinz, LeRoy. Los Angeles, July 16, 1975

Raines, Ella. Los Angeles, August 17, 1983

Raitt, John. Dallas, October 22, 1976

Raksin, David. Los Angeles, August 2, 1976, interview conducted by Ann Burk

Randall, Tony. New York, January 10, 1985

Rapper, Irving. Los Angeles, August 13, 1980

Ray, Johnnie. Dallas, December 16, 1976
Raymond, Gene. Los Angeles, August 26, 1986
Reid, Elliott. Los Angeles, August 22, 1986, July 20, 1987
Reynolds, William. Los Angeles, July 26, 1989
Riddle, Nelson. Los Angeles, August 18, 1983
Ritt, Martin. Los Angeles, July 13, 14, 1987
Robertson, Cliff. New York, December 30, 1987
Robertson, Dale. Yukon, Oklahoma, April 22, 1983
Robin, Leo. Los Angeles, July 24, 1979
Rogers, Charles "Buddy." Los Angeles, August 19, 1988
Romero, Cesar. Dallas, February 26, 1979
Rush, Barbara. Dallas, October 3, 1985
Ruttenberg, Joseph. Los Angeles, August 18, 24, 1978
Sandrich, Jay. Los Angeles, July 20, 1985
Saul, Oscar. Los Angeles, August 15, 17, 1988
Saxon, John. Los Angeles, May 28, 1985
Schafer, Natalie. Los Angeles, August 20, 1988
Scott, Lizabeth. Los Angeles, July 27, 1984, May 30, 1985
Scott, Martha. Los Angeles, August 16, 1988
Segal, Vivienne. Los Angeles, September 23, 1981
Seltzer, Walter. Los Angeles, August 25, 1986
Shavelson, Melville. Los Angeles, August 8, 1981
Sherman, Vincent. Los Angeles, August 5, 1981
Sidney, George. Los Angeles, August 22, 1980
Silver, Joseph. Los Angeles, July 14, 1982
Siodmak, Curt. Los Angeles, August 15, 1981
Smith, Dean. Dallas, July 6, 1979
Springsteen, R. G. Coronado, California, August 24, 1986
Stack, Robert. Los Angeles, August 12, 1975
Starrett, Charles. Laguna, Beach, California, August 13, 1983
Stevens, Risë. New York, January 8, 1982
Stewart, James. Dallas, February 24, 1989
Stewart, Peggy. Los Angeles, July 19, 1985
Stickney, Dorothy. New York, January 9, 1979
Stinson, Bill. Los Angeles, August 28, 1986
Stirling, Linda. Los Angeles, October 9, 1992
Stone, Andrew. Los Angeles, July 15, 1985
Stone, Milburn. Rancho Santa Fe, California, August 11, 1976
Straus, Ann. Los Angeles, May 29, 1985
Styne, Jule. New York, April 2, 1985
Swope, Herbert Bayard, Jr. Los Angeles, August 29, 1986
Thompson, Marshall. Los Angeles, August 22, 1980
Torchia, Emily. Los Angeles, July 16, 1984

Tormé, Mel. Dallas, February 9, 1976

Totter, Audrey. Los Angeles, August 12, 1983

Trescony, Al. Los Angeles, August 20, 1986

Tuttle, William. Los Angeles, August 12, 1976, interview conducted by Ann Burk

Vallee, Rudy. Dallas, October 4, 1975

Van Heusen, James. Rancho Mirage, California, July 18, 1987

Venuta, Benay. New York, January 5, 1979

Vidor, King. Los Angeles, August 4, 1975

Vorkapich, Slavko. Los Angeles, August 11, 1975

Walker, Nancy. Los Angeles, August 18, 1978

Wallis, Hal B. Los Angeles, July 20, 1982

Walston, Ray. Los Angeles, August 11, 1988

Walters, Charles. Los Angeles, August 21, 1980

Warren, Charles Marquis. Los Angeles, August 14, 20, 1980

Warren, Harry. Dallas and Los Angeles, August 23, 1977, October 24, 1977

Wayne, Michael. Los Angeles, July 22, 1991

Wayne, Patrick. Los Angeles, November 6, 1991

Wheeler, Lyle. Los Angeles, August 25, 1986

Whiting, Margaret. Dallas, March 15, 1979

Wilcoxon, Henry. Los Angeles, August 25, 1983

Wilde, Cornel. Los Angeles, August 12, 15, 20, 1980

Williams, Elmo. Brookings, Oregon, September 7, 8, 1989

Windsor, Marie. Los Angeles, August 18, 1983

Wise, Robert. Los Angeles, July 26, 1979

Withers, Jane. Los Angeles, July 17, 1985

Witten, Gladys. Los Angeles, January 3, 1983

Wood, Yvonne. Los Angeles, January 4, 1983

Worden, Hank. Los Angeles, August 14, 1981

Wyatt, Jane. Los Angeles, August 24, 1983

Wyler, Margaret Tallichet. Los Angeles, July 19, 1982

Wyler, William. Los Angeles, July 19, 1979

Wynn, Keenan. Los Angeles, July 18, 1984

Zukor, Adolph. Los Angeles, July 29, 1975

Zukor, Eugene. Los Angeles, July 31, 1975

Supplemental Interviews, Published and Unpublished

Ames, Preston. Mike Steen, *Hollywood Speaks: An Oral History* (New York: Putnam, 1974), 225–39

Ayres, Lew. Walter Wagner, *You Must Remember This* (New York: Putnam, 1975), 164–73

Ball, Lucille. Joseph McBride (ed.), *Filmmakers on Filmmaking*, I (Los Angeles: Tarcher, 1983), 83–93

Berkeley, Busby. Steen, *Hollywood Speaks*, 296–302

Berman, Pandro S. Steen, *Hollywood Speaks*, 167–83

Blanke, Henry. American Film Institute, Los Angeles, January 17, 24, February 5, 1969, interview conducted by Barry Steinberg

Bletcher, Billy. Bernard Rosenberg and Harry Silverstein, *The Real Tinsel* (London: Macmillan, 1970), 299–312

Booth, Margaret. Irene Kahn Atkins (interviewer), *Focus on Film*, XXV (Summer-Autumn, 1976), 51–57

Brahm, John. American Film Institute, Los Angeles, October 13, 1971, interview conducted by Joel Greenberg

Capra, Frank. Richard Schickel, *The Men Who Made the Movies* (New York: Atheneum, 1975), 57–92

Carle, Teet. American Film Institute, Los Angeles, February 17, March 5, 1969, interview conducted by Rae Lindquist

Cole, Jack. John Kobal, *People Will Talk* (New York: Knopf, 1985), 593–607

Cortez, Stanley. Charles Higham, *Hollywood Cameraman* (Bloomington: Indiana University, 1970), 98–119

Cukor, George. Gavin Lambert, *On Cukor* (New York: Capricorn, 1973)

Daniels, William. Higham, *Hollywood Cameraman*, 57–74

Davis, Bette. Joseph McBride (ed.), *Filmmakers on Filmmaking*, II (Los Angeles: Tarcher, 1983), 101–14

Diamond, I. A. L. McBride, *Filmmakers on Filmmaking*, I, 57–69

Engstead, John. Kobal, *People Will Talk*, 515–34

Fairbanks, Douglas, Jr. Wagner, *You Must Remember This*, 95–103

Field, Walter "Cap." Wagner, *You Must Remember This*, 21–29

Fields, Verna. McBride, *Filmmakers on Filmmaking*, I, 139–49

Fonda, Henry. Steen, *Hollywood Speaks*, 15–70

Ford, John. Peter Bogdanovich, *John Ford* (Berkeley: University of California, 1978)

Ford, John. Wagner, *You Must Remember This*, 55–65

Freed, Arthur. Kobal, *People Will Talk*, 635–53

Garmes, Lee. Higham, *Hollywood Cameraman*, 35–55

Gillespie, Arnold "Buddy." Steen, *Hollywood Speaks*, 285–96

Goldwyn, Frances. Wagner, *You Must Remember This*, 104–14

Green, John. Steen, *Hollywood Speaks*, 326–45

Harryhausen, Ray. Elliott Stein (interviewer), "The Thirteen Voyages of Ray Harryhausen," *Film Comment*, XIII (November-December, 1977), 24–28

Haskin, Byron. Joe Adamson (interviewer), *Byron Haskin* (Metuchen, N.J.: Scarecrow, 1984)

Hawks, Howard. Joseph McBride, *Hawks on Hawks* (Berkeley: University of California, 1982)

Hawks, Howard. Schickel, *The Men Who Made the Movies*, 95–128

Head, Edith. McBride, *Filmmakers on Filmmaking*, II, 163–75

Head, Edith. Steen, *Hollywood Speaks*, 247–58

Head, Edith. Wagner, *You Must Remember This*, 224–34

Henderson, Randell. Steen, *Hollywood Speaks*, 200–208

Heston, Charlton. McBride, *Filmmakers on Filmmaking*, II, 83–99

Hitchcock, Alfred. Schickel, *The Men Who Made the Movies*, 271–303

Horner, Harry. McBride, *Filmmakers on Filmmaking*, II, 149–61

Howe, James Wong. Higham, *Hollywood Cameraman*, 75–97

Howe, James Wong. McBride, *Filmmakers on Filmmaking*, I, 95–108

Howe, James Wong. Steen, *Hollywood Speaks*, 208–25

Kazan, Elia. Michel Ciment, *Kazan on Kazan* (New York: Viking, 1974)

Kimball, Ward. Wagner, *You Must Remember This*, 264–82

King, Henry, American Film Institute, Los Angeles, January 19, 1971, interview conducted by Thomas R. Stempel

Klune, Raymond. American Film Institute, Los Angeles, October 15, December 12, 1968, April 3, 1969, interview conducted by John Dorr

Koster, Henry. Irene Kahn Atkins (interviewer), *Henry Koster* (Metuchen, N.J.: Scarecrow, 1987)

Kramer, Stanley. Wagner, *You Must Remember This*, 283–96

Krams, Arthur. Steen, *Hollywood Speaks*, 240–47

Lang, Fritz. Peter Bogdanovich, *Fritz Lang in America* (New York: Praeger, 1969)

Lang, Fritz. Rosenberg and Silverstein, *The Real Tinsel*, 333–48

Lasky, Jesse, Jr. Wagner, *You Must Remember This*, 151–63

Lawrence, Catalina. Steen, *Hollywood Speaks*, 345–53

Leisen, Mitch. David Chierichetti, *Hollywood Director* (New York: Curtis, 1973)

Losey, Joseph. Michel Ciment, *Conversations with Losey* (London: Methuen, 1985)

Louis, Jean. Kobal, *People Will Talk*, 439–48

McCarey, Leo. American Film Institute, Los Angeles, November 25, 1968, interview conducted by Peter Bogdanovich

McLean, Barbara. American Film Institute, Los Angeles, December 22, 1971, interview conducted by Thomas R. Stempel

Mandell, Daniel. American Film Institute, Los Angeles, June 4, 1969, interview conducted by Barry Steinberg

Manley, Nellie. Steen, *Hollywood Speaks*, 274–85

Mayer, Arthur. Rosenberg and Silverstein, *The Real Tinsel*, 157–70

Miller, Arthur. Higham, *Hollywood Cameraman*, 134–54

Minnelli, Vincente. Schickel, *The Men Who Made the Movies*, 243–68

Moorehead, Agnes. Steen, *Hollywood Speaks*, 103–17

Pan, Hermes. Kobal, *People Will Talk*, 621–33

Pickford, Mary. Wagner, *You Must Remember This*, 11–20

Pratt, James. Steen, *Hollywood Speaks*, 183–91

Russell, Rosalind. Steen, *Hollywood Speaks*, 71–103

Rutherford, Ann. Wagner, *You Must Remember This*, 193–211

Schary, Dore. Rosenberg and Silverstein, *The Real Tinsel*, 125–53

Shamroy, Leon. Higham, *Hollywood Cameraman*, 18–34

Shaw, Wini. Rosenberg and Silverstein, *The Real Tinsel*, 265–73

Sherman, Vincent. Kobal, *People Will Talk*, 549–70

Smith, Fred Y. Steen, *Hollywood Speaks*, 302–14

Steiner, Max. Rosenberg and Silverstein, *The Real Tinsel*, 387–98

Stinson, Bill. Yale University School of Music, Newport Beach, California, October 12, 13, 1977, interview conducted by Irene Kahn Atkins

Vidor, King. Nancy Dowd and David Shepard (interviewers), *King Vidor* (Metuchen, N.J.: Scarecrow, 1988)

Vidor, King. Schickel, *The Men Who Made the Movies*, 131–60

Wallis, Hal B. McBride, *Filmmakers on Filmmaking*, I, 9–25

Walsh, Raoul. Schickel, *The Men Who Made the Movies*, 15–54

Wanger, Walter. Rosenberg and Silverstein, *The Real Tinsel*, 80–99

Wellman, William. Schickel, *The Men Who Made the Movies*, 191–240

Wellman, William. Steen, *Hollywood Speaks*, 155–67

Westmore, Perc. Steen, *Hollywood Speaks*, 258–73

Wilder, Billy. Chris Columbus, "Wilder Times," *American Film* (March, 1986) 22–28

Wilder, Billy. McBride, *Filmmakers on Filmmaking*, I, 57–69

Books and Articles

Adler, Bill, *Fred Astaire: A Wonderful Life* (New York: Carroll and Graf, 1987)

Affron, Charles, *Star Acting: Gish, Garbo, Davis* (New York: Dutton, 1977)

Aherne, Brian, *A Dreadful Man* (New York: Simon and Schuster, 1979)

Aherne, Brian, *A Proper Job: The Autobiography of an Actor's Actor* (Boston: Houghton Mifflin, 1969)

Allvine, Glendon, *The Greatest Fox of Them All* (New York: Lyle Stuart, 1969)

Allyson, June with Frances Spatz Leighton, *June Allyson* (New York: Putnam, 1982)

Alpert, Hollis, *Burton* (New York: Putnam, 1986)

Alpert, Hollis, *The Dreams and the Dreamers* (New York: Macmillan, 1962)

Andersen, Christopher P., *A Star, Is a Star, Is a Star! The Lives and Loves of Susan Hayward* (Garden City, N.Y.: Doubleday, 1980)

Arce, Hector, *Gary Cooper* (New York: Morrow, 1979)

Arce, Hector, *Groucho* (New York: Putnam, 1979)

Arden, Eve, *Three Phases of Eve* (New York: St. Martin's, 1985

Armour, Robert A., *Fritz Lang* (Boston: Twayne, 1977)

Astaire, Fred, *Steps in Time* (New York: Harper, 1959)

Astor, Mary, *A Life on Film* (New York: Delacorte, 1971)

Astor, Mary, *My Story: An Autobiography* (Garden City, N.Y.: Doubleday, 1959)

Aumont, Jean-Pierre, *Sun and Shadow: An Autobiography* (New York: Norton, 1977)

Bacall, Lauren, *By Myself* (New York: Knopf, 1979)

Bach, Steven, *Final Cut* (New York: Morrow, 1985)

Bach, Steven, *Marlene Dietrich: Life and Legend* (New York: Morrow, 1992)

Bainbridge, John, *Garbo* (New York: Holt, Rinehart and Winston, 1971)

Baker, Carroll, *Baby Doll: An Autobiography* (New York: Arbor, 1983)

Balio, Tino (ed.), *The American Film Industry* (Madison: University of Wisconsin, 1976)

Balio, Tino, *United Artists: The Company Built by the Stars* (Madison: University of Wisconsin, 1976)

Bare, Richard L., *The Film Director* (New York: Macmillan, 1971)

Barrow, Kenneth, *Helen Hayes: First Lady of the American Theater* (Garden City, N.Y.: Doubleday, 1985)

Bart, Peter, *Fade Out: The Calamitous Final Days of MGM* (New York: Morrow, 1990)

Basquette, Lina, *Lina: DeMille's Godless Girl* (Fairfax, Va.: Denlinger, 1990)

Behlmer, Rudy, *Inside Warner Bros. (1935–51)* (New York: Viking, 1985)

Behlmer, Rudy (ed.), *Memo from David O. Selznick* (New York: Viking, 1972)

Bellamy, Ralph, *When the Smoke Hit the Fan* (Garden City, N.Y.: Doubleday, 1979)

Benchley, Nathaniel, *Humphrey Bogart* (Boston: Little, Brown, 1975)

Bennett, Joan and Lois Kibbee, *The Bennett Playbill* (New York: Holt, Rinehart and Winston, 1970)

Bentley, Eric, *Are You Now or Have You Ever Been: The Investigation of Show Business by the Un-American Activities Committee, 1947–1958* (New York: Harper and Row, 1972)

Berg, A. Scott, *Goldwyn: A Biography* (New York: Knopf, 1989)

Bergan, Ronald, *The United Artists Story* (New York: Crown, 1986)

Bergman, Andrew, *We're in the Money: Depression America and Its Films* (New York: New York University, 1971)

Bergman, Ingrid and Alan Burgess, *Ingrid Bergman: My Story* (New York: Delacorte, 1980)

Bergreen, Laurence, *As Thousands Cheer: The Life of Irving Berlin* (New York: Viking, 1990)

Bertin, Celia, *Jean Renoir: A Life in Pictures* (Baltimore: Johns Hopkins, 1991)

Bickford, Charles, *Bulls, Balls, Bicycles, and Actors* (New York: Eriksson, 1965)

Black, Shirley Temple, *Child Star: An Autobiography* (New York: McGraw-Hill, 1988)

Bloom, Claire, *Limelight and After; The Education of an Actress* (New York: Harper and Row, 1982)

Bogarde, Dirk, *A Postillion Struck by Lightning: A Memoir* (New York: Holt, Rinehart and Winston, 1977)

Boller, Paul F., Jr. and Ronald L. Davis, *Hollywood Anecdotes* (New York: Morrow, 1987)

Bonanno, Margaret Wander, *Angela Lansbury: A Biography* (New York: St. Martin's, 1987)

Bosworth, Patricia, *Montgomery Clift* (New York: Harcourt Brace Jovanovich, 1978)

Bradford, Sarah, *Princess Grace* (New York: Stein and Day, 1984)

Brady, Frank, *Citizen Welles: A Biography of Orson Welles* (New York: Scribner, 1989)

Braun, Eric, *Deborah Kerr* (New York: St. Martin's, 1977)

Broman, Sven, *Conversations with Greta Garbo* (New York: Viking, 1992)

Brosnan, John, *Movie Magic* (New York: St. Martin's, 1974)

Brown, Peter Harry, *Kim Novak: Reluctant Goddess* (New York: St. Martin's, 1986)

Brownlow, Kevin, *Hollywood: The Pioneers* (New York: Knopf, 1979)

Brownlow, Kevin, *The Parade's Gone By* (New York: Knopf, 1968)

Brownlow, Kevin, *The War, the West, and the Wilderness* (New York: Knopf, 1979)

Brynner, Rock, *Yul: The Man Who Would Be King* (New York: Simon and Schuster, 1989)

Buckley, Gail Lumet, *The Hornes: An American Family* (New York: Knopf, 1986)

Burke, Billie, *With a Feather on My Nose* (New York: Appleton-Century-Crofts, 1949)

Cagney, James, *Cagney by Cagney* (Garden City, N.Y.: Doubleday, 1976)

Callow, Simon, *Charles Laughton: A Difficult Actor* (New York: Grove, 1988)

Calvet, Corinne, *Has Corinne Been a Good Girl?* (New York: St. Martin's, 1983)

Cannom, Robert C., *Van Dyke and the Mythical City Hollywood* (Culver City, Calif.: Murray and Gee, 1948)

Capra, Frank, *The Name Above the Title: An Autobiography* (New York: Macmillan, 1971)

Carey, Gary, *All the Stars in Heaven: Louis B. Mayer's MGM* (New York: Dutton, 1981)

Carey, Gary, *Anita Loos: A Biography* (New York: Knopf, 1988)

Carey, Gary, *Doug and Mary: A Biography of Douglas Fairbanks and Mary Pickford* (New York: Dutton, 1977)

Carey, Gary, *Judy Holliday: An Intimate Life Story* (New York: Seaview, 1982)

Carey, Gary, *Marlon Brando: The Only Contender* (New York: St. Martin's, 1985)

Carey, Macdonald, *The Days of My Life* (New York: St. Martin's, 1991)

Carpozi, George, Jr., *The Gary Cooper Story* (New Rochelle, N.Y.: Arlington, 1970)

Carringer, Robert L., *The Making of "Citizen Kane,"* (Berkeley: University of California, 1985)

Cary, Diana Serra, *The Hollywood Posse* (Boston: Houghton Mifflin, 1975)

Cary, Diana Serra, *Hollywood's Children* (Boston: Houghton Mifflin, 1979)

Caspary, Vera, *The Secrets of Grown-ups: An Autobiography* (New York: McGraw-Hill, 1979)

Casper, Joseph Andrew, *Stanley Donen* (Metuchen, N.J., Scarecrow, 1983)

Cassini, Oleg, *In My Own Fashion: An Autobiography* (New York: Simon and Schuster, 1987)

Ceplair, Larry and Steven Englund, *The Inquisition in Hollywood* (Garden City, N.Y.: Anchor, 1980)

Chaplin, Charles, *My Autobiography* (New York: Simon and Schuster, 1964)

Charyn, Jerome, *Movieland: Hollywood and the Great American Dream Culture* (New York: Putnam, 1989)

Chierichetti, David, *Hollywood Costume Design* (New York: Harmony, 1976)

Christensen, Terry, *Reel Politics* (New York: Blackwell, 1987)

Christian, Linda, *Linda: My Own Story* (New York: Crown, 1962)

Clooney, Rosemary with Raymond Strait, *This for Remembrance* (Chicago: Playboy, 1977)

Cody, Iron Eyes with Collin Perry, *Iron Eyes: My Life As a Hollywood Indian* (New York: Everest, 1982)

Cogley, John, *Report on Blacklisting: Movies* (New York: Fund for the Republic, 1956)

Cole, Lester, *Hollywood Red: The Autobiography of Lester Cole* (Palo Alto, Calif.: Ramparts, 1981)

Collier, Peter, *The Fondas: A Hollywood Dynasty* (New York: Putnam, 1991)

Collins, Joan, *Past Imperfect: An Autobiography* (New York: Simon and Schuster, 1984)

Colman, Juliet Benita, *Ronald Colman: A Very Private Person* (New York: Morrow, 1975)

Conant, Michael, *Antitrust in the Motion Picture Industry* (Berkeley: University of California, 1960)

Considine, Shaun, *Bette and Joan: The Divine Feud* (New York: Dutton, 1989)

Cook, Bruce, *Dalton Trumbo* (New York: Scribner, 1977)

Cooper, Jackie with Dick Kleiner, *Please Don't Shoot My Dog: The Autobiography of Jackie Cooper* (New York: Morrow, 1981)

Copland, Aaron, *The New Music, 1900–1960* (New York: Norton, 1968)

Coppedge, Walter, *Henry King's America* (Metuchen, N.J.: Scarecrow, 1986)

Cotten, Joseph, *Vanity Will Get You Somewhere* (San Francisco: Mercury, 1987)

Crichton, Kyle, *The Marx Brothers* (Garden City, N.Y.: Doubleday, 1950)

Croce, Arlene, *The Fred Astaire and Ginger Rogers Book* (New York: Galahad, 1972)

Cronyn, Hume, *A Terrible Liar: A Memoir* (New York: Morrow, 1991)

Crowther, Bosley, *Hollywood Rajah: The Life and Times of Louis B. Mayer* (New York: Holt, 1960)

Crowther, Bosley, *The Lion's Share: The Story of an Entertainment Empire* (New York: Dutton, 1957)

Curcio, Vincent, *Suicide Blonde: The Life of Gloria Grahame* (New York: Morrow, 1989)

Curtis, James, *Between Flops: A Biography of Preston Sturges* (New York: Harcourt Brace Jovanovich, 1982)

Curtis, James, *James Whale* (Metuchen, N.J.: Scarecrow, 1982)

Curtiss, Thomas Quinn, *Von Stroheim* (New York: Farrar, Straus and Giroux, 1971)

Dalton, David, *James Dean: The Mutant King* (San Francisco: Straight Arrow, 1974)

Davidson, Bill, *Spencer Tracy: Tragic Idol* (New York: Dutton, 1987)

Davies, Marion, *The Times We Had* (Indianapolis: Bobbs-Merrill, 1975)

Davis, Bette, *The Lonely Life: An Autobiography* (New York: Putnam, 1962)

Davis, Bette with Michael Herskowitz, *This 'N That* (New York: Putnam, 1987)

Davis, Ronald L., *Hollywood Beauty: Linda Darnell and the American Dream* (Norman: University of Oklahoma, 1991)

DeCarlo, Yvonne with Doug Warren, *Yvonne: An Autobiograpahy* (New York: St. Martin's, 1987)

DeCordova, Fred, *Johnny Came Lately* (New York: Simon and Schuster, 1988)

Deutsch, Armand, *Me and Bogie* (New York: Putnam, 1991)

Dick, Bernard F., *Billy Wilder* (Boston: Twayne, 1980)

Dick, Bernard F., *Joseph L. Mankiewicz* (Boston: Twayne, 1983)

Dietz, Howard, *Dancing in the Dark* (New York: Quadrangle, 1974)

Dmytryk, Edward, *It's a Hell of a Life But Not a Bad Living* (New York: Times Books, 1978)

Dmytryk, Edward, *On Film Editing* (Boston: Focal, 1986)

Dmytryk, Edward, *On Screen Directing* (Boston: Focal, 1984)

Douglas, Helen Gahagan, *A Full Life* (Garden City, N.Y.: Doubleday, 1982)

Douglas, Kirk, *The Ragman's Son: An Autobiography* (New York: Simon and Schuster, 1988)

Dowdy, Andrew, *"Movies Are Better Than Ever": Wide-Screen Memories of the Fifties* (New York: Morrow, 1973)

Drinkwater, John, *The Life and Adventures of Carl Laemmle* (London: William Heinemann, 1931)

Dunne, Philip, *Take Two: A Life in Movies and Politics* (New York: McGraw-Hill, 1980)

Eames, John Douglas, *The MGM Story* (New York: Crown, 1975)

Eames, John Douglas, *The Paramount Story* (New York: Crown, 1985)

Easton, Carol, *The Search for Sam Goldwyn* (New York: Morrow, 1976)

Edwards, Anne, *The DeMilles: An American Family* (New York: Abrams, 1988)

Edwards, Anne, *Early Reagan* (New York: Morrow, 1987)

Edwards, Anne, *Judy Garland: A Biography* (New York: Simon and Schuster, 1975)

Edwards, Anne, *A Remarkable Woman: A Biography of Katharine Hepburn* (New York: Morrow, 1985)

Edwards, Anne, *Shirley Temple: American Princess* (New York: Morrow, 1988)

Edwards, Anne, *Vivien Leigh: A Biography* (New York: Simon and Schuster, 1977)

Eells, George, *Final Gig* (New York: Harcourt Brace Jovanovich, 1991)

Eells, George, *Hedda and Louella* (New York: Putnam, 1972)

Eells, George, *Robert Mitchum* (New York: Watts, 1984)

Eells, George and Stanley Musgrove, *Mae West* (New York: Morrow, 1982)

Eisner, Lotte H., *Fritz Lang* (New York: Oxford, 1977)

Englund, Steven, *Grace of Monaco: An Interpretive Biography* (Garden City, N.Y.: Doubleday, 1984)

Epstein, Lawrence J., *Samuel Goldwyn* (Boston: Twayne, 1981)

Eyles, Allen, *James Stewart* (New York: Stein and Day, 1984)

Eyles, Allen, *Rex Harrison* (London: W. H. Allen, 1985)

Eyman, Scott, *Mary Pickford, America's Sweetheart* (New York: Donald Fine, 1990)

Fairbanks, Douglas, Jr., *The Salad Days: An Autobiography* (New York: Doubleday, 1988)

Faith, William Robert, *Bob Hope: A Life in Comedy* (New York: Putnam, 1982)

Farber, Stephen, *Hollywood Dynasties* (New York: Delilah, 1984)

Fell, John L., *A History of Films* (New York: Holt, Rinehart and Winston, 1979)

Fell, John L., Stephen Gong, Neil Harris, and others, *Before Hollywood: Turn-of-the Century American Film* (New York: Hudson Hills, 1987)

Fernett, Gene, *Poverty Row* (Satellite Beach, Fla.: Coral Reef, 1973)

Ferris, Paul, *Richard Burton* (New York: Coward, McCann and Geoghegan, 1981)

"Film Editors Forum," *Film Comment*, XIII (March-April, 1977), 24–29

Finch, Christopher, *Special Effects: Creating Movie Magic* (New York: Abbeville, 1984)

Finch, Christopher and Linda Rosenkrantz, *Gone Hollywood: The Movie Colony in the Golden Age* (Garden City, N.Y.: Doubleday, 1979)

Flamini, Roland, *Ava* (New York: Coward, McCann and Geoghegan, 1983)

Flynn, Errol, *My Wicked, Wicked Ways* (New York: Putnam, 1959)

Fonda, Henry with Howard Teichmann, *Fonda: My Life* (New York: New American Library, 1981)

Fontaine, Joan, *No Bed of Roses* (New York: Morrow, 1978)

Ford, Dan, *Pappy: The Life of John Ford* (Englewood Cliffs, N.J.: Prentice-Hall, 1979)

Fordin, Hugh, *The World of Entertainment: Hollywood's Greatest Musicals* (Garden City, N.Y.: Doubleday, 1975)

Fountain, Leatrice Gilbert, *Dark Star* (New York: St. Martin's, 1985)

Francisco, Charles, *Gentleman: The William Powell Story* (New York: St. Martin's, 1985)

Frank, Gerold, *Judy* (New York: Harper and Row, 1975)

Freedland, Michael, *The Goldwyn Touch* (London: Harrap, 1986)

Freedland, Michael, *Gregory Peck: A Biography* (New York: Morrow, 1980)

Freedland, Michael, *The Secret Life of Danny Kaye* (New York: St. Martin's, 1985)

Freedland, Michael, *The Warner Brothers* (New York: St. Martin's, 1983)

Friedrich, Otto, *City of Nets: A Portrait of Hollywood in the 1940s* (New York: Harper and Row, 1986)

Frischauer, Willi, *Behind the Scenes of Otto Preminger* (New York: Morrow, 1974)

Fulton, A. R., *Motion Pictures: The Development of an Art from Silent Films to the Age of Television* (Norman: University of Oklahoma, 1960)

Fussell, Betty Harper, *Mabel* (New Haven, Conn.: Ticknor and Fields, 1982)

Gabler, Neal, *An Empire of Their Own: How the Jews Invented Hollywood* (New York: Crown, 1988)

Gallagher, Tag, *John Ford: The Man and His Films* (Berkeley: University of California, 1986)

Gardner, Ava, *Ava: My Story* (New York: Bantam, 1990)

Gardner, Gerald, *The Censorship Papers: Movie Censorship Letters from the Hays Office, 1934 to 1968* (New York: Dodd, Mead, 1987)

Garnett, Tay with Fredda Dudley Balling, *Light Your Torches and Pull Up Your Tights* (New Rochelle, N.Y.: Arlington, 1973)

Geduld, Harry M., *The Birth of the Talkies: From Edison to Jolson* (Bloomington: Indiana University, 1975)

Geist, Kenneth L., *Pictures Will Talk: The Life and Films of Joseph L. Mankiewicz* (New York: Scribner, 1978)

Gil-Montero, Martha, *Brazilian Bombshell: The Biography of Carmen Miranda* (New York: Donald Fine, 1989)

Gingold, Hermione, *How to Grow Old Disgracefully: An Autobiography* (New York: St. Martin's, 1988)

Gish, Lillian with Ann Pinchot, *The Movies, Mr. Griffith, and Me* (Englewood Cliffs, N.J.: Prentice-Hall, 1969)

Golden, Eve, *Platinum Girl: The Life and Legends of Jean Harlow* (New York: Abbeville, 1991)

Goldman, Herbert G., *Fanny Brice: The Original Funny Girl* (New York: Oxford, 1992)

Goldman, Herbert G., *Jolson: The Legend Comes to Life* (New York: Oxford, 1988)

Goldstein, Malcolm, *George S. Kaufman: His Life, His Theater* (New York: Oxford, 1979)

Gomery, Douglas, *The Hollywood Studio System* (New York: St. Martin's, 1986)

Gomery, Douglas, *Shared Pleasures: A History of Movie Presentation in the United States* (Madison: University of Wisconsin, 1992)

Goodman, Ezra, *The Fifty-Year Decline and Fall of Hollywood* (New York: Simon and Schuster, 1961)

Goodman, Ezra, "How to Be a Hollywood Producer," *Harper's Magazine*, CXCVI (May, 1948), 413–23

Grady, Billy, *The Irish Peacock* (New Rochelle, N.Y.: Arlington, 1972)

Graham, Don, *No Name on the Bullet: A Biography of Audie Murphy* (New York: Viking, 1989)

Graham, Sheilah, *The Garden of Allah* (New York: Crown, 1970)

Granger, Stewart, *Sparks Fly Upward* (New York: Putnam, 1981)

Griffith, Linda Arvidson, *When the Movies Were Young* (New York: Blom, 1968)

Grobel, Lawrence, *The Hustons* (New York: Scribner, 1989)

Gronowicz, Antoni, *Garbo* (New York: Simon and Schuster, 1990)

Guiles, Fred Lawrence, *Marion Davies: A Biography* (New York: McGraw-Hill, 1972)

Guiles, Fred Lawrence, *Norma Jean* (New York: McGraw-Hill, 1969)

Guiles, Fred Lawrence, *Stan: The Life of Stan Laurel* (New York: Stein and Day, 1980)

Guiles, Fred Lawrence, *Tyrone Power: The Last Idol* (Garden City, N.Y.: Doubleday, 1979)

Gussow, Mel, *Don't Say Yes Until I Finish Talking: A Biography of Darryl F. Zanuck* (Garden City, N.J.: Doubleday, 1971)

Hamilton, Ian, *Writers in Hollywood, 1915–1951* (New York: Harper and Row, 1990)

Hampton, Benjamin B., *A History of the American Film Industry* (Magnolia, Mass.: Peter Smith, 1970)

Harris, Marlys J., *The Zanucks of Hollywood* (New York: Crown, 1989)

Harris, Warren G., *Cary Grant: A Touch of Elegance* (New York: Doubleday, 1987)

Harris, Warren G., *Gable and Lombard* (New York: Simon and Schuster, 1974)

Harris, Warren G., *Lucy and Desi* (New York: Simon and Schuster, 1991)

Harris, Warren G., *Natalie and R. J.: Hollywood's Star-Crossed Lovers* (New York: Doubleday, 1988)

Harris, Warren G., *The Other Marilyn* (New York: Arbor, 1985)

Harrison, Rex, *A Damned Serious Business: My Life in Comedy* (New York: Bantam, 1991)

Harrison, Rex, *Rex: An Autobiography* (New York: Morrow, 1975)

Harvey, James, *Romantic Comedy in Hollywood from Lubitsch to Sturges* (New York: Knopf, 1987)

Harvey, Stephen, *Fred Astaire* (New York: Pyramid, 1975)

Haskins, James with Kathleen Benson, *Lena: A Personal and Professional Biography of Lena Horne* (New York: Stein and Day, 1984)

Haver, Ronald, *David O. Selznick's Hollywood* (New York: Bonanza, 1985)

Hayes, Helen with Katherine Hatch, *My Life in Three Acts* (New York: Harcourt Brace Jovanovich, 1990)

Hayne, Donald (ed.), *The Autobiography of Cecil B. DeMille* (Englewood Cliffs, N.J.: Prentice-Hall, 1959)

Hayward, Brooke, *Haywire* (New York: Bantam, 1977)

Head, Edith and Paddy Calistro, *Edith Head's Hollywood* (New York: Dutton, 1983)

Heimann, Jim, *Out With the Stars* (New York: Abbeville, 1985)

Hemming, Roy, *The Melody Lingers On: The Great Songwriters and Their Movie Musicals* (New York: Newmarket, 1986)

Henderson, Robert M., *D. W. Griffith: His Life and Work* (New York: Oxford, 1972)

Henderson, Robert M., *D. W. Griffith: The Years at Biograph* (New York: Farrar, Straus and Giroux, 1970)

Henreid, Paul with Julius Fast, *Ladies Man: An Autobiography* (New York: St. Martin's, 1984)

Hepburn, Katharine, *The Making of "The African Queen,"* (New York: Knopf, 1987)

Hepburn, Katharine, *Me: Stories of My Life* (New York: Knopf, 1991)

Higham, Charles, *Audrey: The Life of Audrey Hepburn* (New York: Macmillan, 1984)

Higham, Charles, *Ava: A Life Story* (New York: Delacorte, 1974)

Higham, Charles, *Bette: The Life of Bette Davis* (New York: Macmillan, 1981)

Higham, Charles, *Brando: The Unauthorized Biography* (New York: New American Library, 1987)

Higham, Charles, *Cecil B. DeMille* (New York: Scribner, 1973)

Higham, Charles, *Charles Laughton: An Intimate Biography* (Garden City, N.Y.: Doubleday, 1976)

Higham, Charles, *Errol Flynn: The Untold Story* (Garden City, N.Y.: Doubleday, 1980)

Higham, Charles, *Hollywood at Sunset* (New York: Saturday Review, 1972)

Higham, Charles, *Kate: The Life of Katharine Hepburn* (New York: Norton, 1975)

Higham, Charles, *Lucy: The Real Life of Lucille Ball* (New York: St. Martin's, 1986)

Higham, Charles, *Marlene: The Life of Marlene Dietrich* (New York: Norton, 1977)

Higham, Charles, *Merchant of Dreams: Louis B. Mayer, M.G.M., and the Secret Hollywood* (New York: Donald Fine, 1993)

Higham, Charles, *Orson Welles: The Rise and Fall of an American Genius* (New York: St. Martin's, 1985)

Higham, Charles, *Sisters: The Story of Olivia de Havilland and Joan Fontaine* (New York: Coward-McCann, 1984)

Higham, Charles, *Warner Brothers* (New York: Scribner, 1975)

Higham, Charles and Roy Moseley, *Cary Grant: The Lonely Heart* (New York: Harcourt Brace Jovanovich, 1989)

Hirsch, Foster, *Joseph Losey* (Boston: Twayne, 1980)

Hirschhorn, Clive, *Gene Kelly: A Biography* (Chicago: Henry Regnery, 1974)

Hirschhorn, Clive, *The Universal Story* (New York: Crown, 1983)

Hirschhorn, Clive, *The Warner Bros. Story* (New York: Crown, 1979)

Holden, Anthony, *Laurence Olivier: A Biography* (New York: Atheneum, 1988)

Holtzman, Will, *Judy Holliday: A Biography* (New York: Putnam, 1982)

Hope, Bob with Melville Shavelson, *Don't Shoot, It's Only Me* (New York: Putnam, 1990)

Hopper, Hedda with James Brough, *The Whole Truth and Nothing But* (Garden City, N.Y.: Doubleday, 1963)

Hotchner, A. E., *Doris Day: Her Own Story* (New York: Bantam, 1976)

Houseman, John, *Front and Center* (New York: Simon and Schuster, 1979)

Hudson, Rock and Sara Davidson, *Rock Hudson: His Story* (New York: Morrow, 1986)

Hurst, Richard Maurice, *Republic Studios: Between Poverty Row and the Majors* (Metuchen, N.J.: Scarecrow, 1979)

Huston, John, *An Open Book* (New York: Knopf, 1980)

Hyams, Joe with Jay Hyams, *James Dean, Little Boy Lost* (New York: Warner, 1992)

Irwin, Will, *The House That Shadows Built* (Garden City, N.Y.: Doubleday, Doran, 1928)

Izod, John, *Hollywood and the Box Office, 1895–1986* (New York: Columbia University, 1988)

Jackson, Carlton, *Hattie: The Life of Hattie McDaniel* (Lanham, Md.: Madison, 1990)

Jacobs, Lewis, *The Rise of the American Film* (New York: Teachers College, 1968)

Jerome, Stuart, *Those Crazy Wonderful Years When We Ran Warner Bros.* (Secaucus, N.J.: Lyle Stuart, 1983)

Jewell, Richard B. with Vernon Harbin, *The RKO Story* (New York: Arlington, 1982)

Johnson, Dorris and Ellen Leventhal (eds.), *The Letters of Nunnally Johnson* (New York: Knopf, 1981)

Johnson, Nora, *Flashback: Nora Johnson on Nunnally Johnson* (Garden City, N.Y.: Doubleday, 1979)

Johnston, Alva, *The Great Goldwyn* (New York: Random House, 1937)

Jowett, Garth, *Film: The Democratic Art* (Boston: Little, Brown, 1976)

Kanin, Garson, *Hollywood* (New York: Viking, 1974)

Kazan, Elia, *Elia Kazan: A Life* (New York: Knopf, 1988)

Keith, Slim with Annette Tapert, *Slim: Memories of a Rich and Imperfect Life* (New York: Simon and Schuster, 1990)

Kelley, Kitty, *Elizabeth Taylor: The Last Star* (New York: Simon and Schuster, 1981)

Kelley, Kitty, *His Way: The Unauthorized Biography of Frank Sinatra* (New York: Bantam, 1986)

Keyes, Evelyn, *Scarlett O'Hara's Younger Sister* (Secaucus, N.J.: Lyle Stuart, 1977)

Kiernan, Thomas, *Sir Larry: The Life of Laurence Olivier* (New York: Times Books, 1981)

Kinden, Gorham (ed.), *The American Movie Industry* (Carbondale: University of Southern Illinois, 1982)

Knox, Donald, *The Magic Factory: How MGM Made "An American in Paris,"* (New York: Praeger, 1973)

Kobal, John, *Hollywood: The Years of Innocence* (New York: Abbeville, 1985)

Kobal, John, *Rita Hayworth: The Time, the Place and the Woman* (New York: Norton, 1977)

Kobler, John, *Damned in Paradise: The Life of John Barrymore* (New York: Atheneum, 1977)

Koch, Howard, *As Time Goes By: Memoirs of a Writer* (New York: Harcourt Brace Jovanovich, 1979)

Koppes, Clayton R. and Gregory D. Black, *Hollywood Goes to War* (New York: Free Press, 1987)

Kotsilibas-Davis, James and Myrna Loy, *Myrna Loy: Being and Becoming* (New York: Knopf, 1987)

LaGuardia, Robert, *Monty: A Biography of Montgomery Clift* (New York: Arbor, 1977)

LaGuardia, Robert and Gene Arceri, *Red: The Tempestuous Life of Susan Hayward* (New York: Macmillan, 1985)

Lahr, John, *Notes on a Cowardly Lion: The Biography of Bert Lahr* (New York: Knopf, 1969)

Lake, Veronica with Donald Bain, *Veronica: The Autobiography of Veronica Lake* (New York: Citadel, 1971)

Lamarr, Hedy, *Ecstasy and Me: My Life as a Woman* (New York: Bartholomew, 1966)

Lambert, Gavin, *Norma Shearer* (New York: Knopf, 1990)

Lamour, Dorothy with Dick McInnes, *My Side of the Road* (Englewood Cliffs, N.J.: Prentice-Hall, 1980)

Lanchester, Elsa, *Elsa Lanchester, Herself* (New York: St. Martin's, 1983)

Larkin, Rochelle, *Hail, Columbia* (New Rochelle, N.Y.: Arlington, 1975)

Lasky, Betty, *RKO: The Biggest Little Major of Them All* (Englewood Cliffs, N.J.: Prentice-Hall, 1984)

Lasky, Jesse L. with Don Weldon, *I Blow My Own Horn* (Garden City, N.Y.: Doubleday, 1957)

Lasky, Jesse L., Jr., *Whatever Happened to Hollywood?* (New York: Funk and Wagnalls, 1975)

Lawrence, Jerome, *Actor: The Life and Times of Paul Muni* (New York: Putnam, 1974)

Leamer, Laurence, *As Time Goes By: The Life of Ingrid Bergman* (New York: Harper and Row, 1986)

Leaming, Barbara, *Bette Davis: A Biography* (New York: Simon and Schuster, 1992)

Leaming, Barbara, *If This Was Happiness: A Biography of Rita Hayworth* (New York: Viking, 1989)

Leaming, Barbara, *Orson Welles* (New York: Viking, 1985)

Leff, Leonard J., *Hitchcock and Selznick* (New York: Weidenfeld and Nicolson, 1987)

Leigh, Janet, *There Really Was a Hollywood* (Garden City, N.J.: Doubleday, 1984)

Lenburg, Jeff, *Peekaboo: The Story of Veronica Lake* (New York: St. Martin's, 1983)

Leonard, Maurice, *Mae West: Empress of Sex* (Secaucus, N.J.: Birch Lane, 1992)

LeRoy, Mervyn with Dick Kleiner, *Mervyn LeRoy: Take One* (New York: Hawthorn, 1974)

Lindfors, Viveca, *Viveka . . . Viveca* (New York: Everest, 1981)

Linet, Beverly, *Ladd: The Life, the Legend, the Legacy of Alan Ladd* (New York: Arbor, 1979)

Linet, Beverly, *Star-Crossed: The Story of Robert Walker and Jennifer Jones* (New York: Putnam, 1986)

Linet, Beverly, *Susan Hayward: Portrait of a Survivor* (New York: Atheneum, 1980)

Lloyd, Ronald, *American Film Directors* (New York: Watts, 1976)

Logan, Joshua, *Movie Stars, Real People, and Me* (New York: Delacorte, 1978)

Loney, Glenn, *Unsung Genius: The Passion of Dancer-Choreographer Jack Cole* (New York: Watts, 1984)

Loos, Anita, *The Talmadge Girls* (New York: Viking, 1978)

MacAdams, William, *Ben Hecht: The Man Behind the Legend* (New York: Scribner, 1990)

McBride, Joseph, *Frank Capra: The Catastrophe of Success* (New York: Simon and Schuster, 1992)

McBride, Joseph and Michael Wilmington, *John Ford* (New York: DaCapo, 1975)

McCabe, John, *Babe: The Life of Oliver Hardy* (Secaucus, N.J.: Citadel, 1989)

McCabe, John, *Charlie Chaplin* (Garden City, N.Y.: Doubleday, 1978)

McCabe, John, *Mr. Laurel and Mr. Hardy* (New York: Grosset and Dunlap, 1966)

McCambridge, Mercedes, *The Quality of Mercy: An Autobiography* (New York: Times Books, 1981)

MacCann, Richard Dyer, *Hollywood in Transition* (Westport, Conn.: Greenwood, 1977)

McConathy, Dale and Diana Vreeland, *Hollywood Costume* (New York: Abrams, 1976)

McGerr, Celia, *Rene Clair* (Boston: Twayne, 1980)

McGilligan, Patrick, *Cagney: The Actor as Auteur* (San Diego: Barnes, 1982)

McGilligan, Patrick, *George Cukor: A Double Life* (New York: St. Martin's, 1991)

Macgowan, Kenneth, *Behind the Screen: The History and Techniques of the Motion Picture* (New York: Delacorte, 1965)

McGuire, Patricia Dubin, *Lullaby of Broadway: Life and Times of Al Dubin* (Secaucus, N.J.: Citadel, 1983)

McMurtry, Larry, *Film Flam: Essays on Hollywood* (New York: Simon and Schuster, 1987)

Madsen, Axel, *Billy Wilder* (Bloomington: Indiana University, 1969)

Madsen, Axel, *John Huston: A Biography* (Garden City, N.Y.: Doubleday, 1978)

Madsen, Axel, *William Wyler* (New York: Crowell, 1973)

Maeder, Edward, *Hollywood and History: Costume Design in Film* (New York: Thames and Hudson, 1987)

Maland, Charles J., *Frank Capra* (Boston: Twayne, 1980)

Mancini, Henry with Gene Lees, *Did They Mention the Music?* (Chicago: Contemporary Books, 1989)

Mandelbaum, Howard and Eric Myers, *Screen Deco* (New York: St. Martin's, 1985)

Mannering, Derek, *Mario Lanza: A Biography* (London: Robert Hale, 1991)

Marion, Frances, *Off With Their Heads!* (New York: Macmillan, 1972)

Markel, Helen, "Adrian Talks of Gowns—and of Goats," *New York Times Magazine* (May 27, 1945), 14, 25

Martin, Mary, *My Heart Belongs* (New York: Morrow, 1976)

Martin, Tony and Cyd Charisse, *The Two of Us* (New York: Mason/Charter, 1976)

Marx, Arthur, *Everybody Loves Somebody Sometime: The Story of Dean Martin and Jerry Lewis* (New York: Hawthorn, 1974)

Marx, Arthur, *Goldwyn: A Biography of the Man Behind the Myth* (New York: Norton, 1976)

Marx, Arthur, *The Nine Lives of Mickey Rooney* (New York: Stein and Day, 1986)

Marx, Arthur, *Red Skelton* (New York: Dutton, 1979)

Marx, Samuel, *A Gaudy Spree: The Literary Life of Hollywood in the 1930s* (New York: Watts, 1987)

Marx, Samuel, *Mayer and Thalberg: The Make-Believe Saints* (New York: Random House, 1975)

Mason, James, *Before I Forget: An Autobiography* (London: Sphere, 1982)

Massey, Raymond, *A Hundred Different Lives: An Autobiography* (Boston: Little, Brown, 1979)

Mast, Gerald, *Can't Help Singin': The American Musical on Stage and Screen* (Woodstock, N.Y.: Overlook, 1987)

Mast, Gerald, *Howard Hawks, Storyteller* (New York: Oxford, 1982)

May, Lary, *Screening Out the Past* (New York: Oxford, 1980)

Mayer, Arthur, *Merely Colossal* (New York: Simon and Schuster, 1953)

Meade, Marion, *Dorothy Parker: A Biography* (New York: Villard, 1988)

Merrill, Gary, *Bette, Rita, and the Rest of My Life* (Augusta, Me.: Tapley, 1988)

Meryman, Richard, *Mank: The Wit, World, and Life of Herman Mankiewicz* (New York: Morrow, 1978)

Meyer, William R., *Warner Brothers Directors* (New Rochelle, N.Y.: Arlington, 1978)

Milland, Ray, *Wide-Eyed in Babylon: An Autobiography* (New York: Morrow, 1974)

Miller, Ann with Norma Lee Browning, *Miller's High Life* (Garden City, N.Y.: Doubleday, 1972)

Miller, Don, *B Movies* (New York: Ballantine, 1988)

Millichap, Joseph R., *Lewis Milestone* (Boston: Twayne, 1981)

Mills, John, *Up in the Clouds, Gentlemen Please* (New Haven, Conn.: Ticknor and Fields, 1981)

Milne, Tom, *Rouben Mamoulian* (Bloomington: Indiana University, 1969)

Minnelli, Vincente with Hector Arce, *I Remember It Well* (Garden City, N.Y.: Doubleday, 1974)

Montalban, Ricardo with Bob Thomas, *Reflections: A Life in Two Worlds* (Garden City, N.Y.: Doubleday, 1980)

Moore, Colleen, *Silent Star* (Garden City, N.Y.: Doubleday, 1968)

Moore, Dick, *Twinkle, Twinkle, Little Star* (New York: Harper and Row, 1984)

Mordden, Ethan, *The Hollywood Musical* (New York: St. Martin's, 1981)

Mordden, Ethan, *The Hollywood Studios: House Style in the Golden Age of the Movies* (New York: Knopf, 1988)

Mordden, Ethan, *Movie Star: A Look at the Women Who Made Hollywood* (New York: St. Martin's, 1983)

Morella, Joe and Edward Z. Epstein, *Forever Lucy: The Life of Lucille Ball* (Secaucus, N.J.: Lyle Stuart, 1986)

Morella, Joe and Edward Z. Epstein, *Jane Wyman* (New York: Delacorte, 1985)

Morella, Joe and Edward Z. Epstein, *Lana: The Public and Private Lifes of Miss Turner* (New York: Citadel, 1971)

Morella, Joe and Edward Z. Epstein, *Loretta Young: An Extraordinary Life* (New York: Delacorte, 1986)

Morella, Joe and Edward Z. Epstein, *Paulette: The Adventurous Life of Paulette Goddard* (New York: St. Martin's, 1985)

Morley, Sheridan, *Gladys Cooper* (New York: McGraw-Hill, 1979)

Morley, Sheridan, *James Mason: Odd Man Out* (New York: Harper and Row, 1989)

Morley, Sheridan, *Tales from the Hollywood Raj: The British Film Colony On Screen and Off* (New York: Viking, 1983)

Mosley, Leonard, *Zanuck: The Rise and Fall of Hollywood's Last Tycoon* (Boston: Little, Brown, 1984)

Mueller, John, *Astaire Dancing* (New York: Knopf, 1985)

Munn, Michael, *Charlton Heston: A Biography* (New York: St. Martin's, 1986)

Munn, Michael, *Kid from the Bronx: A Biography of Tony Curtis* (London: W. H. Allen, 1984)

Munn, Michael, *Kirk Douglas* (New York: St. Martin's, 1985)

Murphy, William B., "Film Editing," *Films in Review*, VI (February, 1955), 70–72

Murray, Ken, *The Golden Days of San Simeon* (Garden City, N.Y.: Doubleday, 1971)

Naremore, James, *The Magic World of Orson Welles* (Dallas: Southern Methodist University, 1989)

Navasky, Victor S., *Naming Names* (New York: Viking, 1980)

Neal, Patricia with Richard DeNeut, *As I Am: An Autobiography* (New York: Simon and Schuster, 1988)

Negulesco, Jean, *Things I Did and Things I Think I Did* (New York: Linden, 1984)

Niven, David, *Bring On the Empty Horses* (New York: Putnam, 1975)

Niven, David, *The Moon's a Balloon* (New York: Putnam, 1972)

Nolan, William F., *John Huston, King Rebel* (Los Angeles: Sherbourne, 1965)

Norman, Barry, *The Story of Hollywood* (New York: New American Library, 1987)

O'Brien, Pat, *The Wind at My Back: The Life and Times of Pat O'Brien* (Garden City, N.Y.: Doubleday, 1964)

Olivier, Laurence, *Confessions of An Actor* (New York: Simon and Schuster, 1982)

Oppenheimer, Jerry and Jack Vitek, *Idol: Rock Hudson* (New York: Villard, 1986)

Oumano, Elena, *Paul Newman* (New York: St. Martin's, 1989)

Paris, Barry, *Louise Brooks* (New York: Knopf, 1989)

Parish, James Robert, *The Jeanette MacDonald Story* (New York: Mason/Charter, 1976)

Parish, James Robert and Steven Whitney, *Vincent Price Unmasked: A Biography* (New York: Drake, 1974)

Parker, John, *Five for Hollywood* (Secaucus, N.J.: Lyle Stuart, 1991)

Parrish, Robert, *Growing Up in Hollywood* (New York: Harcourt Brace Jovanovich, 1976)

Parrish, Robert, *Hollywood Doesn't Live Here Anymore* (Boston: Little, Brown, 1988)

Parsons, Louella O., *The Gay Illiterate* (Garden City, N.Y.: Doubleday, Doran, 1944)

Parsons, Louella O., *Tell It to Louella* (New York: Putnam, 1961)

Pasternak, Joe, *Easy the Hard Way: The Autobiography of Joe Pasternak* (London: W. H. Allen, 1956)

Pastos, Spero, *Pin-Up: The Tragedy of Betty Grable* (New York: Putnam, 1986)

Paul, William, *Ernst Lubitsch's American Comedy* (New York: Columbia University, 1983)

Peters, Margot, *The House of Barrymore* (New York: Knopf, 1990)

Phillips, Gene D., *George Cukor* (Boston: Twayne, 1982)

Pickford, Mary, *Sunshine and Shadow* (Garden City, N.Y.: Doubleday, 1955)

Poague, Leland A., *The Cinema of Ernst Lubitsch* (South Brunswick: Barnes, 1978)

Pollock, Dale, *Skywalking: The Life and Films of George Lucas* (New York: Harmony, 1983)

Powdermaker, Hortense, *Hollywood, the Dream Factory* (Boston: Little, Brown, 1950)

Powell, Jane, *The Girl Next Door . . . And How She Grew* (New York: Morrow, 1988)

Preminger, Otto, *Preminger: An Autobiography* (Garden City, N.Y.: Doubleday, 1977)

Quirk, Lawrence J., *Fasten Your Seat Belts: The Passionate Life of Bette Davis* (New York: Morrow, 1990)

Quirk, Lawrence J., *Norma: The Story of Norma Shearer* (New York: St. Martin's, 1988)

Ramsaye, Terry, *A Million and One Nights: A History of the Motion Picture* (New York: Simon and Schuster, 1926)

Randall, Tony and Michael Mindlin, *Which Reminds Me* (New York: Delacorte, 1989)

Reagan, Nancy with William Novak, *My Turn* (New York: Random House, 1989)

Reynolds, Debbie with David Patrick Columbia, *Debbie, My Life* (New York: Morrow, 1988)

Robbins, Jhan, *Everybody's Man: A Biography of Jimmy Stewart* (New York: Putnam, 1985)

Robbins, Jhan, *Inka Dinka Doo: The Life of Jimmy Durante* (New York: Paragon, 1991)

Robbins, Jhan, *Yul Brynner: The Inscrutable King* (New York: Dodd, Mead, 1987)

Roberson, Chuck with Bodie Thoene, *The Fall Guy* (North Vancouver, B.C.: Hancock, 1980)

Robinson, David, *Chaplin: His Life and Art* (New York: McGraw-Hill, 1985)

Robinson, Edward G. with Leonard Spigelgass, *All My Yesterdays* (New York: Signet, 1975)

Roddick, Nick, *A New Deal in Entertainment: Warner Brothers in the 1930s* (London: British Film Institute, 1983)

Rogers, Ginger, *Ginger: My Story* (New York: Harper Collins, 1991)

Rollyson, Carl E., Jr., *Marilyn Monroe: A Life of the Actress* (Ann Arbor, Mich.: UMI Research Press, 1986)

Rooney, Mickey, *I. E.: An Autobiography* (New York: Putnam, 1965)

Rooney, Mickey, *Life Is Too Short* (New York: Villard, 1991)

Rosenblum, Ralph and Robert Karen, *When the Shooting Stops . . . the Cutting Begins* (New York: Viking, 1979)

Ross, Murray, *Stars and Strikes: Unionization of Hollywood* (New York: Columbia University, 1941)

Rosten, Leo C., *Hollywood: The Movie Colony, the Movie Makers* (New York: Harcourt, Brace, 1941)

Rozsa, Miklos, *Double Life: The Autobiography of Miklos Rozsa* (New York: Hippocrene, 1982)

Russell, Harold with Dan Ferullo, *The Best Years of My Life* (Middlebury, Vt.: Eriksson, 1981)

Russell, Jane, *Jane Russell: My Path and My Detours* (New York: Watts, 1985)

Russell, Rosalind, *Life Is a Banquet* (New York: Random House, 1977)

St. Johns, Adela Rogers, *Love, Laughter and Tears; My Hollywood Story* (Garden City, N.Y.: Doubleday, 1978)

Sarlot, Raymond R. and Fred E. Basten, *Life at the Marmont* (Santa Monica, Calif.: Roundtable, 1987)

Saxton, Martha, *Jayne Mansfield and the American Fifties* (Boston: Houghton Mifflin, 1975)

Sayre, Nora, *Running Time: Films of the Cold War* (New York: Dial, 1982)

Schary, Dore, *Heyday: An Autobiography* (Boston: Little, Brown, 1979)

Schatz, Thomas, *The Genius of the System* (New York: Pantheon, 1988)

Schickel, Richard, *Brando: A Life in Our Times* (New York: Atheneum, 1991)

Schickel, Richard, *Intimate Strangers: The Culture of Celebrity* (Garden City, N.Y.: Doubleday, 1985)

Schickel, Richard, *The Disney Version: The Life, Times, Art, and Commerce of Walt Disney* (New York: Simon and Schuster, 1968)

Schickel, Richard, *D. W. Griffith: An American Life* (New York: Simon and Schuster, 1984)

Schulberg, Budd, *Moving Pictures: Memories of a Hollywood Prince* (New York: Stein and Day, 1981)

Schwartz, Nancy Lynn, *The Hollywood Writers' Wars* (New York: Knopf, 1982)

Selznick, Irene Mayer, *A Private View* (New York: Knopf, 1983)

Sennett, Mack, *King of Comedy* (Garden City, N.Y.: Doubleday, 1954)

Sennett, Ted, *Great Movie Directors* (New York: Abrams, 1986)

Sennett, Ted, *Hollywood Musicals* (New York: Abrams, 1981)

Sennett, Ted, *Warner Brothers Presents* (New Rochelle, N.Y.: Arlington, 1971)

Sharples, Win, Jr., "Prime Cut," *Film Comment*, XIII (March-April, 1977), 6–23

"She Cuts the Kisses," *American Magazine*, CXVVII (January, 1949), 99

Shepherd, Donald and Robert F. Slatzer, *Bing Crosby: The Hollow Man* (New York: St. Martin's, 1981)

Sheppard, Dick, *Elizabeth: The Life and Career of Elizabeth Taylor* (Garden City, N.Y.: Doubleday, 1974)

Sherk, Warren, *Agnes Moorehead: A Very Private Person* (Philadelophia: Dorrance, 1976)

Sherman, Eric, *Directing the Film: Film Directors on Their Art* (Boston: Little, Brown, 1976)

Silverman, Stephen M., *The Fox That Got Away: The Last Days of the Zanuck Dynasty at Twentieth Century-Fox* (Secaucus, N.J.: Lyle Stuart, 1988)

Silvers, Phil with Robert Saffron, *This Laugh Is On Me* (Englewood Cliffs, N.J.: Prentice-Hall, 1973)

Sinclair, Andrew, *John Ford* (New York: Dial, 1979)

Sinclair, Andrew, *Spiegel: The Man Behind the Pictures* (Boston: Little, Brown, 1987)

Sklar, Robert, *City Boys: Cagney, Bogart, and Garfield* (Princeton, N.J.: Princeton University, 1992)

Sklar, Robert, *Movie-Made America: A Social History of American Movies* (New York: Random House, 1975)

Slide, Anthony, *The Big V: A History of the Vitagraph Company* (Metuchen, N.J.: Scarecrow, 1976)

Smith, Ella, *Starring Miss Barbara Stanwyck* (New York: Crown, 1974)

Smith, H. Allen, *The Life and Legend of Gene Fowler* (New York: Morrow, 1977)

Smith, Steven C., *A Heart at Fire's Center: The Life and Music of Bernard Herrmann* (Berkeley: University of California, 1991)

Spada, James, *Grace: The Secret Lives of a Princess* (Garden City, N.Y.: Doubleday, 1987)

Spada, James, *Peter Lawford: The Man Who Kept the Secrets* (New York: Bantam, 1991)

Spehr, Paul C., *The Movies Begin: Making Movies in New Jersey* (Newark, N.J.: Newark Museum, 1977)

Spoto, Donald, *Blue Angel: The Life of Marlene Dietrich* (New York: Doubleday, 1992)

Spoto, Donald, *The Dark Side of Genius: The Life of Alfred Hitchcock* (Boston: Little Brown, 1983)

Spoto, Donald, *Laurence Olivier: A Biography* (New York: Harper Collins, 1992)

Spoto, Donald, *Madcap: The Life of Preston Sturges* (Boston: Little, Brown, 1990)

Spoto, Donald, *Marilyn Monroe* (New York: Harper Collins, 1993)

Spoto, Donald, *Stanley Kramer, Film Maker* (New York: Putnam, 1978)

Stack, Robert with Mark Evans, *Straight Shooting* (New York: Macmillan, 1980)

Stempel, Tom, *Screenwriter: The Life and Times of Nunnally Johnson* (San Diego: Barnes, 1980)

Stenn, David, *Clara Bow: Runnin' Wild* (New York: Doubleday, 1988)

Stern, Michael, *Douglas Sirk* (Boston: Twayne, 1979)

Stine, Whitney, *"I'd Love to Kiss You . . .": Conversations with Bette Davis* (New York: Pocket Books, 1990)

Stine, Whitney with Bette Davis, *Mother Goddam: The Story of the Career of Bette Davis* (New York: Hawthorn, 1974)

Strait, Raymond and Leif Henie, *Queen of Ice, Queen of Shadows: The Unsuspected Life of Sonja Henie* (New York: Stein and Day, 1985)

Strode, Woody and Sam Young, *Goal Dust* (Lanham, Md.: Madison, 1990)

Sturges, Preston and Sandy Sturges, *Preston Sturges: His Life in His Words* (New York: Simon and Schuster, 1990)

Swanson, Gloria, *Swanson On Swanson* (New York: Random House, 1980)

Swindell, Larry, *Body and Soul: The Story of John Garfield* (New York: Morrow, 1975)

Swindell, Larry, *Charles Boyer: The Reluctant Lover* (Garden City, N.Y.: Doubleday, 1983)

Swindell, Larry, *The Last Hero: A Biography of Gary Cooper* (Garden City, N.Y.: Doubleday, 1980)

Swindell, Larry, *Screwball: The Life of Carole Lombard* (New York: Morrow, 1975)

Swindell, Larry, *Spencer Tracy: A Biography* (New York: World, 1969)

Taylor, Elizabeth, *Elizabeth Taylor: An Informal Memoir* (New York: Harper and Row, 1965)

Taylor, John Russell, *Hitch: The Life and Times of Alfred Hitchcock* (New York: Pantheon, 1978)

Taylor, John Russell, *Strangers in Paradise: The Hollywood Emigres, 1933–1950* (New York: Holt, Rinehart and Winston, 1983)

Thomas, Bob, *Astaire: The Man, the Dancer* (New York: St. Martin's, 1984)

Thomas, Bob, *Bud and Lou: The Abbott and Costello Story* (Philadelphia: Lippincott, 1977)

Thomas, Bob, *Clown Prince of Hollywood: The Antic Life and Times of Jack L. Warner* (New York: McGraw-Hill, 1990)

Thomas, Bob, *Golden Boy: The Untold Story of William Holden* (New York: St. Martin's, 1983)

Thomas, Bob, *Joan Crawford: A Biography* (New York: Simon and Schuster, 1978)

Thomas, Bob, *King Cohn: The Life and Times of Harry Cohn* (New York: Putnam, 1967)

Thomas, Bob, *Marlon: Portrait of the Rebel as an Artist* (New York: Random House, 1973)

Thomas, Bob, *Selznick* (Garden City, N.Y.: Doubleday, 1970)

Thomas, Bob, *Thalberg: Life and Legend* (Garden City, N.Y.: Doubleday, 1969)

Thomas, Bob, *Walt Disney: An American Original* (New York: Simon and Schuster, 1976)

Thomas, Danny with Bill Davidson, *Make Room for Danny* (New York: Putnam, 1991)

Thomas, Tony, *Music for the Movies* (South Brunswick: Barnes, 1973)

Thomas, Tony and Aubrey Solomon, *The Films of 20th Century-Fox* (Secaucus, N.J.: Citadel, 1985)

Thomson, David, *Showman: The Life of David O. Selznick* (New York: Knopf, 1992)

Tierney, Gene with Mickey Herskowitz, *Self-Portrait* (New York: Wyden, 1979)

Tiomkin, Dimitri and Prosper Buranelli, *Please Don't Hate Me* (Garden City, N.Y.: Doubleday, 1959)

Tormé, Mel, *It Wasn't All Velvet* (New York: Viking, 1988)

Tornabene, Lyn, *Long Live the King: A Biography of Clark Gable* (New York: Putnam, 1976)

Tosches, Nick, *Dino: Living High in the Dirty Business of Dreams* (New York: Doubleday, 1992)

Truffaut, Francois, *Hitchcock* (New York: Simon and Schuster, 1984)

Turner, Lana, *Lana: The Lady, the Legend, the Truth* (New York: Dutton, 1982)

Tuska, Jon (ed.), *Close Up: The Contract Director* (Metuchen, N.J.: Scarecrow, 1976)

Underwood, Peter, *Horror Man: The Life of Boris Karloff* (London: Leslie Frewin, 1972)

Ustinov, Peter, *Dear Me* (Boston: Little, Brown, 1977)

VanDerBeets, Richard, *George Sanders: An Exhausted Life* (Lanham, Md.: Madison, 1990)

Vickers, Hugo, *Vivien Leigh* (Boston: Little, Brown, 1988)

Vidor, King, *On Film Making* (New York: McKay, 1972)

Vidor, King, *A Tree Is a Tree* (New York: Garland, 1977)

Viertel, Salka, *The Kindness of Strangers* (New York: Holt, Rinehart and Winston, 1969)

Vineberg, Steve, *Method Actors: Three Generations of an American Acting Style* (New York: Schirmer, 1991)

Von Sternberg, Josef, *Fun in a Chinese Laundry* (New York: Macmillan, 1965)

Wagenknecht, Edward, *The Movies in the Age of Innocence* (Norman: University of Oklahoma, 1962)

Walker, Alexander, *Elizabeth: The Life of Elizabeth Taylor* (New York: Grove Weidenfeld, 1990)

Walker, Alexander, *Stardom: The Hollywood Phenomenon* (New York: Stein and Day, 1970)

Walker, Alexander, *Vivien: The Life of Vivien Leigh* (New York: Weidenfeld and Nicolson, 1987)

Walker, Joseph and Juanita Walker, *The Light on Her Face* (Hollywood: ASC Press, 1984)

Wallis, Hal and Charles Higham, *Starmaker: The Autobiography of Hal Wallis* (New York: Macmillan, 1980)

Wallis, Martha Hyer, *Finding My Way: A Hollywood Memoir* (New York: Harper Collins, 1990)

Walsh, Raoul, *Each Man in His Time* (New York: Farrar, Straus and Giroux, 1974)

Wansell, Geoffrey, *Haunted Idol: The Story of the Real Cary Grant* (New York: Morrow, 1984)

Ward, Ken, *Mass Communications and the Modern World* (Chicago: Dorsey, 1989)

Warner, Jack with Dean Jennings, *My First Hundred Years in Hollywood* (New York: Random House, 1964)

Warren, Doug, *Betty Grable: The Reluctant Movie Queen* (New York: St. Martin's, 1981)

Warrick, Ruth with Don Preston, *The Confessions of Phoebe Tyler* (Englewood Cliffs, N.J.: Prentice-Hall, 1980)

Wayne, Jane Ellen, *Robert Taylor* (New York: St. Martin's, 1987)

Wayne, Jane Ellen, *Stanwyck* (New York: Arbor, 1985)

Webb, Michael (ed.), *Hollywood: Legend and Reality* (Boston: Little, Brown, 1986)

Weinberg, Herman G., *The Lubitsch Touch* (New York: Dover, 1977)

Weis, Elizabeth (ed.), *The Movie Star* (New York: Viking, 1981)

Wellman, William A., *A Short Time for Insanity: An Autobiography* (New York: Hawthorn, 1974)

Westmore, Frank and Muriel Davidson, *The Westmores of Hollywood* (Philadelphia: Lippincott, 1976)

Whiting, Charles, *Hero: The Life and Death of Audie Murphy* (Chelsea, Mich.: Scarborough, 1990)

Wilcoxon, Henry with Katherine Orrison, *Lionheart in Hollywood: The Autobiography of Henry Wilcoxon* (Metuchen, N.J.: Scarecrow, 1991)

Wiles, Buster with William Donati, *My Days with Errol Flynn: The Autobiography of Stuntman Buster Wiles* (Santa Monica, Calif.: Roundtable, 1988)

Wilk, Max, *The Wit and Wisdom of Hollywood* (New York: Atheneum, 1971)

Wilkerson, Tichi and Marcia Borie, *The Hollywood Reporter: The Golden Years* (New York: Howard-McCann, 1984)

Wilson, Earl, *Sinatra: An Unauthorized Biography* (New York: Macmillan, 1976)

Windeler, Robert, *Burt Lancaster* (New York: St. Martin's, 1984)

Windeler, Robert, *Sweetheart: The Story of Mary Pickford* (New York: Praeger, 1974)

Winters, Shelley, *Shelley* (New York: Morrow, 1980)

Winters, Shelley, *Shelley II: The Middle of My Century* (New York: Simon and Schuster, 1989)

Wood, Robin, *Howard Hawks* (London: British Film Institute, 1981)

Woodward, Ian, *Audrey Hepburn* (New York: St. Martin's, 1984)

Woulfe, Michael, "Costuming a Film," *Films in Review*, VI (August-September, 1955), 325–27

Wray, Fay, *On the Other Hand: A Life Story* (New York: St. Martin's, 1989)

Wynn, Keenan, *Ed Wynn's Son* (Garden City, N.Y.: Doubleday, 1959)

Wynn, Ned, *We Will Always Live in Beverly Hills: Growing Up Crazy in Hollywood* (New York: Morrow, 1990)

Yablonsky, Lewis, *George Raft* (New York: Signet, 1974)

Yallop, David A., *The Day the Laughter Stopped* (New York: St. Martin's, 1976)

Yule, Andrew, *Fast Fade* (New York: Delacorte, 1989)

Zierold, Norman J., *Garbo* (New York: Stein and Day, 1969)

Zierold, Norman J., *The Moguls* (New York: Coward-McCann, 1969)

Zinnemann, Fred, *A Life in the Movies* (New York: Scribner, 1992)

Zolotow, Maurice, *Billy Wilder in Hollywood* (New York: Putnam, 1977)

Zolotow, Maurice, *Shooting Star: A Biography of John Wayne* (New York: Simon and Schuster, 1974)

Zukor, Adolph with Dale Kramer, *The Public Is Never Wrong: The Autobiography of Adolph Zukor* (New York: Putnam, 1953)

Index

Eastwood, Clint, 85
Eddy, Nelson, 94, 114
Edens, Roger, 187, 188, 192
Edington, Harry, 53
Edison, Thomas, 2, 18, 24, 65
Editors, American Cinema, 281, 300
Eduoart, Farciot, 310
Einfeld, S. Charles, 142
Ekberg, Anita, 147
El Cid, 374
Elizabeth and Essex, 230
Elliott, Bill, 114, 263
Emerson, Faye, 94
Emmett (waiter), 319
Engel, Sam, 50
English, John, 78
Englund, Steven, 345
Engstead, John, 143, 155
Enright, Florence, 89
Enterprise Productions, 359
Epstein, Julius, 23, 24, 166, 167, 173,
    178–79, 332
Epstein, Philip, 23, 24, 166, 167, 173, 179
Evans, Dale, 12, 105, 114, 156, 196–97,
    252, 264, 265
Evans, Ray, 187
Ewell, Tom, 135
Executive Suite, 251
The Exile, 376
Exodus, 356
Eythe, William, 94

Fabray, Nanette, 144–45
Fairbanks, Douglas, Jr., 98, 110, 115, 156,
    175, 255, 268, 325, 335, 376
Fairbanks, Douglas, Sr., 10–11, 43, 98,
    104, 156, 326
Famous Players-Lasky, **1**, 31, 51, 228
A Farewell to Arms, 61, 226
The Farmer's Daughter, 129
The Farmer Takes a Wife, 255
Farnsworth, Arthur, **323**
Farrell, Edith, 162
Farrell, Glenda, 122
Farrell, Midge, 134
Father of the Bride, 169

Father's Little Dividend, 169
Faulkner, William, 172, 377
Fay, Frank, 177
Faye, Alice, 6, 114, 118, 194, 195
Fazan, Adrienne, 285, 292, 294, 297, 312
Feldman, Charles, 128
Felix, Seymour, 195
Fellows, Robert, 300
Ferrer, Jose, 110
Feuer, Cy, 197
Fields, Verna, 291
Fields, Virginia, 122
Fields, W. C., 44, 304
Fiesta, 201
Film Booking Offices (FBO), 51
Finian's Rainbow, 378–79
First National, 4, 228, 285, 298
The First Time, 152
Fitzgerald, Barry, 122
Fitzgerald, F. Scott, 38, 171, 314, 335
Fitzgerald, Roy, 90
Five Friends Plan (Five Finger Plan), 313
The Flame and the Arrow, 153
Flamingo Road, 49
Fleming, Rhonda, 103
Fleming, Victor, 61, 176, 351
Flesh and the Devil, 99
Flinn, John C., 151
Florey, Robert, 271
Flying Down to Rio, 169, 184, 310
Flynn, Errol, 46–47, 76, **97**, 109, 113,
    150–51, 166, 167, 174, 230, 269, 271,
    319, 335, 336–37, 368, 380
Foch, Nina, 10, 28, 30, 73, 115
Fogler, Gertrude, 87–88
Follow the Fleet, 185
Folsey, George, 82
Fonda, Henry, 101, 247, 255
Fontaine, Joan, 39, 102, 109, 111, 112,
    115–16, 156, 186, 230, 242, 250,
    330–31, 359, 375
Foote, Horton, 377
Footlight Parade, 183
Forbstein, Leo F., 198
Ford, Barbara, 267
Ford, Glenn, 10, 147

ducers, 35, 46, 56; finding/developing talent, 82, 85, 91; music, 196–97; and other studios, 110; television, 370, 373. *See also* Yates, Herbert J.

The Return of Frank James, 247

Revere, Anne, 353

Reynolds, Debbie, 84, 114, 145, 225

Reynolds, William, 284, 286, 290, 291, 295, 291

Rhapsody in Blue, 84, 108, 254

Rice, Della Owens, 88

Rich, Robert. *See* Trumbo, Dalton

Riding High, 257

Rio Grande, 267

Ripley, Arthur, 283

Riskin, Robert, 170

Ritt, Martin, 326, 348

Ritter, Thelma, 122, 129

River of No Return, 266

RKO, 5–6, 34, 223, 357, 364, 371; actors/actresses, 109, 112, 114, 124, 317, 337; commissary, 320; complaints against, 109, 112, 344; design, 212, 213, 222; directors, 72, 346; editors/editing, 281, 286, 288, 297–98; executives/producers, 34, 42, 44, 51–54, 359, 362; finding/developing talent, 85, 87; Hollywood Ten controversy, 346, 354; makeup/hairstyling, 224, 228; music/musicals, 184–86, 197, 198; and other studios, 5–6, 106, 111, 169, 357; special effects, 309–10; television, 370–71, 373

Roach, Hal, 68, 73

Road pictures (Crosby-Hope), 44, 87, 113, 170

The Robe, 77, 162, 260, 372, 373

Roberson, Chuck, 265, 267

Roberta, 173, 185

Roberts, Theodore, 45

Robertson, Dale, 115, 118, 175

Robin, Leo, 187

Robinson, Casey, 177

Robinson, Edward G., xi, 113

Robson, Mark, 282, 297, 301

Rodgers and Hart, 190

Rogers, Charles, 54

Rogers, Ginger, 42, 52, 85, **97–98**, 100, 114, 184–86, 193, 219, 222, 251, 310, 347

Rogers, Lela, 85, 87, 347

Rogers, Roy, 56, 78, 114, 197, 318

Rogers, Will, 6

Romance on the High Seas, 196

Roman Holiday, 276

Romanoff, Prince Michael, 329

Roman Scandals, 253

Romero, Cesar, 122, 133, 195, 328, 370, 375

Rooney, Mickey, 91, 93, **137**, 155, 186, 192

Roosevelt, Franklin, 343

Rose, Helen, 88, 215

Rose, Jack, 166

Roseanna McCoy, 243

Rose Marie, 188

Rosenberg, Aaron, 55

Rosenblum, Ralph, 280, 284

Rosenstein, Arthur, 88, 189

Rosenstein, Sophie, 85, 87

The Rose Tattoo, 45, 209, 237

Ross, Harold, 141

Rossellini, Roberto, 337

Rosten, Leo, 324

Roughin' It, 167

Rowland, Roy, 77

Royale, Selena, 122

The Royal Family of Broadway, 68, 283

Royal Wedding, 193

Rozsa, Miklos, 200

Rubenstein, Arthur, 254

Rush, Barbara, 90, 93, 102, 147, 240, 250, 266, 332, 333

Russell, Gail, 114, 292

Russell, Jane, 114, 318, 361

Russell, Rosalind, 113, 126, 237, 251, 359

Rutherford, Ann, 153, 217, 227, 378

Ruttenberg, Joseph, 82, 236–37, 311

Ryan, Peggy, 186

Ryan, Robert, 101

Ryman, Lucille, 80

About the Author

Photo by Hillsman Jackson

Ronald L. Davis is professor of history at Southern Methodist University where he heads the SMU Oral History Program and directs the DeGolyer Institute for American Studies. The author of eight previous books, including *Hollywood Anecdotes* (with Paul Boller), *Hollywood Beauty* (a biography of screen actress Linda Darnell), and the three-volume *History of Music in American Life*, he has conducted over four hundred interviews with film personalities. His most recent book has as its subject the director John Ford, and he is currently working on a biography of John Wayne.